# TRIUMPH STAG
## *Gold Portfolio*
## <u>1970-1977</u>

Compiled by
R.M.Clarke

ISBN 1 85520 3413

Brooklands
Books

BROOKLANDS BOOKS LTD.
P.O. BOX 146, COBHAM,
SURREY, KT11 1LG. UK

A-TT70GP

# Brooklands Books

## MOTORING

### BROOKLANDS ROAD TEST SERIES

Abarth Gold Portfolio 1950-1971
AC Ace & Aceca 1953-1983
Alfa Romeo Giulietta Gold Portfolio 1954-1965
Alfa Romeo Giulia Berlinas 1962-1976
Alfa Romeo Giulia Coupés 1963-1976
Alfa Romeo Giulia Coupés Gold P. 1963-1976
Alfa Romeo Spider 1966-1990
Alfa Romeo Spider Gold Portfolio 1966-1991
Alfa Romeo Alfasud 1972-1984
Alfa Romeo Alfetta Gold Portfolio 1972-1987
Alfa Romeo Alfetta GTV6 1980-1986
Allard Gold Portfolio 1937-1959
Alvis Gold Portfolio 1919-1967
AMX & Javelin Muscle Portfolio 1968-1974
Armstrong Siddeley Gold Portfolio 1945-1960
Aston Martin Gold Portfolio 1972-1985
Aston Martin Gold Portfolio 1985-1995
Audi Quattro Gold Portfolio 1980-1991
Austin A30 & A35 1951-1962
Austin Healey 100 & 100/6 Gold P. 1952-1959
Austin Healey 3000 Gold Portfolio 1959-1967
Austin Healey Sprite 1958-1971
Barracuda Muscle Portfolio 1964-1974
BMW Six Cylinder Coupés 1969-1975
BMW 1600 Collection No.1 1966-1981
BMW 2002 Gold Portfolio 1968-1976
BMW 316, 318, 320 (4 cyl.) Gold P. 1975-1990
BMW 320, 323, 325 (6 cyl.) Gold P. 1977-1990
BMW M Series Performance Portfolio 1976-1993
BMW 5 Series Gold Portfolio 1981-1987
Bricklin Gold Portfolio 1974-1975
Bristol Cars Gold Portfolio 1946-1992
Buick Automobiles 1947-1960
Buick Muscle Cars 1965-1970
Cadillac Allanté 1986-1993
Cadillac Automobiles 1949-1959
Cadillac Automobiles 1960-1969
Charger Muscle Portfolio 1966-1974
Chevrolet 1955-1957
Chevrolet Impala & SS 1958-1971
Chevrolet Corvair 1959-1969
Chevy II & Nova SS Muscle Portfolio 1962-1974
Chevy El Camino & SS 1959-1987
Chevelle & SS Muscle Portfolio 1964-1972
Chevrolet Muscle Cars 1966-1971
Chevy Blazer 1969-1981
Chevrolet Corvette Gold Portfolio 1953-1962
Chevrolet Corvette Sting Ray Gold P. 1963-1967
Chevrolet Corvette Gold Portfolio 1968-1977
High Performance Corvettes 1983-1989
Camaro Muscle Portfolio 1967-1973
Chevrolet Camaro Z28 & SS 1966-1973
Chevrolet Camaro & Z28 1973-1981
High Performance Camaros 1982-1988
Chrysler 300 Gold Portfolio 1955-1970
Chrysler Valiant 1960-1962
Citroen Traction Avant Gold Portfolio 1934-1957
Citroen 2CV Gold Portfolio 1948-1989
Citroen DS & ID 1955-1975
Citroen DS & ID Gold Portfolio 1955-1975
Citroen SM 1970-1975
Cobras & Replicas 1962-1983
Shelby Cobra Gold Portfolio 1962-1969
Cobras & Cobra Replicas Gold P. 1962-1989
Cunningham Automobiles 1951-1955
Daimler SP250 Sports & V-8 250 Saloon Gold P. 1959-1969
Datsun Roadsters 1962-1971
Datsun 240Z 1970-1973
Datsun 280Z & ZX 1975-1983
DeLorean Gold Portfolio 1977-1995
Dodge Muscle Cars 1967-1970
Dodge Viper on the Road
ERA Gold Portfolio 1934-1994
Excalibur Collection No.1 1952-1981
Facel Vega 1954-1964
Ferrari Dino 1965-1974
Ferrari Dino 308 1974-1979
Ferrari 328 • 348 • Mondial Gold Portfolio 1986-1994
Fiat 500 Gold Portfolio 1936-1972
Fiat 600 & 850 Gold Portfolio 1955-1972
Fiat Pininfarina 124 & 2000 Spider 1968-1985
Fiat-Bertone X1/9 1973-1988
Fiat Abarth Performance Portfolio 1972-1987
Ford Consul, Zephyr, Zodiac Mk.I & II 1950-1962
Ford Zephyr, Zodiac, Executive, Mk.III & Mk.IV 1962-1971
Ford Cortina 1600E & GT 1967-1970
High Performance Capris Gold Portfolio 1969-1987
Capri Muscle Portfolio 1974-1987
High Performance Fiestas 1979-1991
High Performance Escorts Mk.I 1968-1974
High Performance Escorts Mk.II 1975-1980
High Performance Escorts 1980-1985
High Performance Escorts 1985-1990
High Performance Sierras & Merkurs
Gold Portfolio 1983-1990
Ford Automobiles 1949-1959
Ford Fairlane 1955-1970
Ford Ranchero 1957-1959
Ford Thunderbird 1955-1957
Ford Thunderbird 1958-1963
Ford Thunderbird 1964-1976
Ford GT40 Gold Portfolio 1964-1987
Ford Bronco 1966-1977
Ford Bronco 1978-1988
Goggomobil Limited Edition
Holden 1948-1962
Honda CRX 1983-1987
International Scout Gold Portfolio 1961-1980
Isetta 1953-1964
Iso & Bizzarrini Gold Portfolio 1962-1974
Jaguar and SS Gold Portfolio 1931-1951
Jaguar XK120, 140, 150 Gold P. 1948-1960
Jaguar Mk.VII, VIII, IX, X, 420 Gold P. 1950-1970
Jaguar Mk.1 & Mk.2 Gold Portfolio 1959-1969
Jaguar E-Type Gold Portfolio 1961-1971
Jaguar E-Type V-12 1971-1975
Jaguar XJ12, XJ5.3, V12 Gold P. 1972-1990
Jaguar XJ6 Series I & II Gold P. 1968-1979
Jaguar XJ6 Series III 1979-1986
Jaguar XJ6 Gold Portfolio 1986-1994
Jaguar XJS Gold Portfolio 1975-1988
Jaguar XJS Gold Portfolio 1988-1995
Jeep CJ5 & CJ6 1960-1976
Jeep CJ5 & CJ7 1976-1986
Jensen Cars 1946-1967
Jensen Cars 1967-1979
Jensen Interceptor Gold Portfolio 1966-1986
Jensen Healey 1972-1976
Lagonda Gold Portfolio 1919-1964
Lamborghini Countach & Urraco 1974-1980
Lamborghini Countach & Jalpa 1980-1985
Lancia Aurelia & Flaminia Gold Portfolio 1950-1970
Lancia Fulvia Gold Portfolio 1963-1976
Lancia Beta Gold Portfolio 1972-1984
Lancia Delta Gold Portfolio 1979-1994
Lancia Stratos 1972-1985
Land Rover Series I 1948-1958
Land Rover Series II & IIa 1958-1971
Land Rover Series III 1971-1985
Land Rover 90 110 Defender Gold Portfolio 1983-1994
Land Rover Discovery 1989-1994
Land Rover Story Part One 1948-1971
Lincoln Gold Portfolio 1949-1960
Lincoln Continental 1961-1969
Lincoln Continental 1969-1976
Lotus Sports Racers Gold Portfolio 1953-1965
Lotus Seven Gold Portfolio 1957-1974
Lotus Caterham Seven Gold Portfolio 1974-1995
Lotus Elite 1957-1964
Lotus Elite & Eclat 1974-1982
Lotus Elan Gold Portfolio 1962-1974
Lotus Elan Collection No. 2 1963-1972
Lotus Elan & SE 1989-1992
Lotus Cortina Gold Portfolio 1963-1970
Lotus Europa Gold Portfolio 1966-1975
Lotus Elite & Eclat 1974-1982
Lotus Turbo Esprit 1980-1986
Marcos Cars 1960-1988
Maserati 1970-1975
Mazda RX-7 Gold Portfolio 1978-1991
Mercedes 190 & 300 SL 1954-1963
Mercedes 230/250/280SL 1963-1971
Mercedes G Wagen 1981-1994
Mercedes Benz SLs & SLCs Gold P. 1971-1989
Mercedes S & 600 1965-1972
Mercedes S Class 1972-1979
Mercedes SLs Performance Portfolio 1989-1994
Mercury Muscle Cars 1966-1971
Messerschmitt Gold Portfolio 1954-1964
MG Gold Portfolio 1929-1939
MG TA & TC Gold Portfolio 1936-1949
MG TD & TF Gold Portfolio 1949-1955
MGA & Twin Cam Gold Portfolio 1955-1962
MG Midget Gold Portfolio 1961-1979
MGB Roadsters 1962-1980
MGB MGC & V8 Gold Portfolio 1962-1980
MGB GT 1965-1980
Mini Gold Portfolio 1959-1969
Mini Gold Portfolio 1969-1980
High Performance Minis Gold Portfolio 1960-1973
Mini Cooper Gold Portfolio 1961-1971
Mini Moke Gold Portfolio 1964-1994
Mopar Muscle Cars 1964-1967
Morgan Three-Wheeler Gold Portfolio 1910-1952
Morgan Plus 4 & Four 4 Gold P. 1936-1967
Morgan Cars 1960-1970
Morgan Cars Gold Portfolio 1968-1989
Morris Minor Collection No. 1 1948-1980
Shelby Mustang Muscle Portfolio 1965-1970
High Performance Mustang IIs 1974-1978
High Performance Mustangs 1982-1988
Nash-Austin Metropolitan Gold P. 1954-1962
Oldsmobile Automobiles 1955-1963
Oldsmobile Muscle Cars 1964-1971
Oldsmobile Toronado 1966-1978
Opel GT Gold Portfolio 1968-1973
Packard Gold Portfolio 1946-1958
Pantera Gold Portfolio 1970-1989
Panther Gold Portfolio 1972-1990
Plymouth Muscle Cars 1966-1971
Pontiac Tempest & GTO 1961-1965
Pontiac Muscle Cars 1966-1972
Pontiac Firebird & Trans-Am 1973-1981
High Performance Firebirds 1982-1988
Pontiac Fiero 1984-1988
Porsche 356 Gold Portfolio 1953-1965
Porsche 911 1965-1969
Porsche 911 1970-1972
Porsche 911 1973-1977
Porsche 911 Carrera 1973-1977
Porsche 911 Turbo 1975-1984
Porsche 911 SC & Turbo Gold Portfolio 1978-1983
Porsche 911 Carrera & Turbo Gold P. 1984-1989
Porsche 914 Gold Portfolio 1969-1976
Porsche 924 Gold Portfolio 1975-1988
Porsche 928 Performance Portfolio 1977-1994
Porsche 944 Gold Portfolio 1981-1991
Range Rover Gold Portfolio 1970-1985
Range Rover Gold Portfolio 1986-1995
Reliant Scimitar 1964-1986
Riley Gold Portfolio 1924-1939
Riley 1.5 & 2.5 Litre Gold Portfolio 1945-1955
Rolls Royce Silver Cloud & Bentley 'S' Series
Gold Portfolio 1955-1965
Rolls Royce Silver Shadow Gold P. 1965-1980
Rolls Royce & Bentley Gold P. 1980-1989
Rover P4 1949-1959
Rover P4 1955-1964
Rover 3 & 3.5 Litre Gold Portfolio 1958-1973
Rover 2000 & 2200 1963-1977
Rover 3500 1968-1977
Rover 3500 & Vitesse 1976-1986
Saab Sonett Collection No.1 1966-1974
Saab Turbo 1976-1983
Studebaker Gold Portfolio 1947-1966
Studebaker Hawks & Larks 1956-1963
Avanti 1962-1990
Sunbeam Tiger & Alpine Gold P. 1959-1967
Toyota MR2 1984-1988
Toyota Land Cruiser 1956-1984
Triumph Dolomite Sprint Limited Edition
Triumph TR2 & TR3 Gold Portfolio 1952-1961
Triumph TR4, TR5, TR250 1961-1968
Triumph TR6 Gold Portfolio 1969-1976
Triumph TR7 & TR8 Gold Portfolio 1975-1982
Triumph Herald 1959-1971
Triumph Vitesse 1962-1971
Triumph Spitfire Gold Portfolio 1962-1980
Triumph 2000, 2.5, 2500 1963-1977
Triumph GT6 Gold Portfolio 1966-1974
Triumph Stag Gold Portfolio 1970-1977
TVR Gold Portfolio 1959-1986
TVR Performance Portfolio 1986-1994
VW Beetle Gold Portfolio 1935-1967
VW Beetle Gold Portfolio 1968-1991
VW Beetle Collection No.1 1970-1982
VW Karmann Ghia 1955-1982
VW Bus, Camper, Van 1954-1967
VW Bus, Camper, Van 1968-1979
VW Bus, Camper, Van 1979-1989
VW Scirocco 1974-1981
VW Golf GTI 1976-1986
Volvo PV444 & PV544 1945-1965
Volvo Amazon-120 Gold Portfolio 1956-1970
Volvo 1800 Gold Portfolio 1960-1973
Volvo 140 & 160 Series Gold Portfolio 1966-1975

Forty Years of Selling Volvo

### BROOKLANDS ROAD & TRACK SERIES

Road & Track on Alfa Romeo 1949-1963
Road & Track on Alfa Romeo 1964-1970
Road & Track on Alfa Romeo 1971-1976
Road & Track on Alfa Romeo 1977-1989
Road & Track on Aston Martin 1962-1990
R & T on Auburn Cord and Duesenburg 1952-84
Road & Track on Audi & Auto Union 1952-1980
Road & Track on Audi & Auto Union 1980-1986
Road & Track on Austin Healey 1953-1970
Road & Track on BMW Cars 1966-1974
Road & Track on BMW Cars 1975-1978
Road & Track on BMW Cars 1979-1983
R & T on Cobra, Shelby & Ford GT40 1962-1992
Road & Track on Corvette 1953-1967
Road & Track on Corvette 1968-1982
Road & Track on Corvette 1982-1986
Road & Track on Corvette 1986-1990
Road & Track on Datsun Z 1970-1983
Road & Track on Ferrari 1975-1981
Road & Track on Ferrari 1981-1984
Road & Track on Ferrari 1984-1988
Road & Track on Fiat Sports Cars 1968-1987
Road & Track on Jaguar 1950-1960
Road & Track on Jaguar 1961-1968
Road & Track on Jaguar 1968-1974
Road & Track on Jaguar 1974-1982
Road & Track on Jaguar 1983-1989
Road & Track on Lamborghini 1964-1985
Road & Track on Lotus 1972-1981
Road & Track on Maserati 1952-1974
Road & Track on Maserati 1975-1983
R & T on Mazda RX7 & MX5 Miata 1986-1991
Road & Track on Mercedes 1952-1962
Road & Track on Mercedes 1963-1970
Road & Track on Mercedes 1971-1979
Road & Track on Mercedes 1980-1987
Road & Track on MG Sports Cars 1949-1961
Road & Track on MG Sports Cars 1962-1980
Road & Track on Mustang 1964-1977
R & T on Nissan 300-ZX & Turbo 1984-1989
Road & Track on Pontiac 1960-1983
Road & Track on Porsche 1951-1967
Road & Track on Porsche 1968-1971
Road & Track on Porsche 1972-1975
Road & Track on Porsche 1975-1978
Road & Track on Porsche 1985-1988
R & T on Rolls Royce & Bentley 1950-1965
R & T on Rolls Royce & Bentley 1966-1984
Road & Track on Saab 1972-1992
R & T on Toyota Sports & GT Cars 1966-1984
R & T on Triumph Sports Cars 1953-1967
R & T on Triumph Sports Cars 1967-1974
R & T on Triumph Sports Cars 1974-1982
Road & Track on Volkswagen 1951-1968
Road & Track on Volkswagen 1968-1978
Road & Track on Volkswagen 1978-1985
Road & Track on Volvo 1957-1974
Road & Track on Volvo 1977-1994
R&T - Henry Manney at Large & Abroad
R&T - Peter Egan's "Side Glances"

### BROOKLANDS CAR AND DRIVER SERIES

Car and Driver on BMW 1955-1977
Car and Driver on BMW 1977-1985
C and D on Cobra, Shelby & Ford GT40 1963-84
Car and Driver on Corvette 1956-1967
Car and Driver on Corvette 1968-1977
Car and Driver on Corvette 1978-1982
Car and Driver on Corvette 1983-1988
C and D on Datsun Z 1600 & 2000 1966-1984
Car and Driver on Ferrari 1955-1962
Car and Driver on Ferrari 1963-1975
Car and Driver on Ferrari 1976-1983
Car and Driver on Mopar 1956-1967
Car and Driver on Mopar 1968-1975
Car and Driver on Mustang 1964-1972
Car and Driver on Pontiac 1961-1975
Car and Driver on Porsche 1955-1962
Car and Driver on Porsche 1963-1970
Car and Driver on Porsche 1970-1976
Car and Driver on Porsche 1977-1981
Car and Driver on Porsche 1982-1986
Car and Driver on Saab 1956-1985
Car and Driver on Volvo 1955-1986

### BROOKLANDS PRACTICAL CLASSICS SERIES

PC on Austin A40 Restoration
PC on Land Rover Restoration
PC on Metalworking in Restoration
PC on Midget/Sprite Restoration
PC on Mini Cooper Restoration
PC on MGB Restoration
PC on Morris Minor Restoration
PC on Sunbeam Rapier Restoration
PC on Triumph Herald/Vitesse
PC on Spitfire Restoration
PC on Beetle Restoration
PC on 1930s Car Restoration

### BROOKLANDS HOT ROD 'MUSCLECAR & HI-PO ENGINES' SERIES

Chevy 265 & 283
Chevy 302 & 327
Chevy 348 & 409
Chevy 350 & 400
Chevy 396 & 427
Chevy 454 thru 512
Chrysler Hemi
Chrysler 273, 318, 340 & 360
Chrysler 361, 383, 400, 413, 426, 440
Ford 289, 302, Boss 302 & 351W
Ford 351C & Boss 351
Ford Big Block

### BROOKLANDS RESTORATION SERIES

Auto Restoration Tips & Techniques
Basic Bodywork Tips & Techniques
Camaro Restoration Tips & Techniques
Chevrolet High Performance Tips & Techniques
Chevy Engine Swapping Tips & Techniques
Chevy-GMC Pickup Repair
Chrysler Engine Swapping Tips & Techniques
Engine Swapping Tips & Techniques
Ford Pickup Repair
How to Build a Street Rod
Land Rover Restoration Tips & Techniques
MG 'T' Series Restoration Guide
MGA Restoration Guide
Mustang Restoration Tips & Techniques
Performance Tuning - Chevrolets of the '60s
Performance Tuning - Pontiacs of the '60's

## MOTORCYCLING

### BROOKLANDS ROAD TEST SERIES

BSA Twins A7 & A10 Gold Portfolio 1946-1962
BSA Twins A50 & A65 Gold Portfolio 1962-1973
Norton Commando Gold Portfolio 1968-1977
Triumph Bonneville Gold Portfolio 1959-1983

### BROOKLANDS CYCLE WORLD SERIES

Cycle World on BMW 1974-1980
Cycle World on BMW 1981-1986
Cycle World on Ducati 1982-1991
Cycle World on Harley-Davidson 1962-1968
Cycle World on Harley-Davidson 19781-1983
Cycle World on Harley-Davidson 1983-1987
Cycle World on Harley-Davidson 1987-1990
Cycle World on Harley-Davidson 1990-1992
Cycle World on Honda 1962-1967
Cycle World on Honda 1968-1971
Cycle World on Honda 1971-1974
Cycle World on Husqvarna 1966-1976
Cycle World on Husqvarna 1977-1984
Cycle World on Kawasaki 1966-1971
Cycle World on Kawasaki Off-Road Bikes 1972-1979
Cycle World on Kawasaki Street Bikes 1972-1976
Cycle World on Norton 1962-1971
Cycle World on Suzuki 1962-1970
Cycle World on Suzuki Off-Road Bikes 1971-1976
Cycle World on Suzuki Street Bikes 1971-1976
Cycle World on Triumph 1967-1972
Cycle World on Yamaha 1962-1969
Cycle World on Yamaha Off-Road Bikes 1970-1974
Cycle World on Yamaha Street Bikes 1970-1974

## MILITARY

### BROOKLANDS MILITARY VEHICLES SERIES

Allied Military Vehicles No.1 1942-1945
Allied Military Vehicles No.2 1941-1946
Complete WW2 Military Jeep Manual
Dodge Military Vehicles No.1 1940-1945
Hail To The Jeep
Land Rovers in Military Service
Military & Civilian Amphibians 1940-1990
Off Road Jeeps: Civ. & Mil. 1944-1971
US Military Vehicles 1941-1945
US Army Military Vehicles WW2-TM9-2800
VW Kubelwagen Military Portfolio 1940-1990
WW2 Jeep Military Portfolio 1941-1945

CONTENTS

**Brooklands Books**

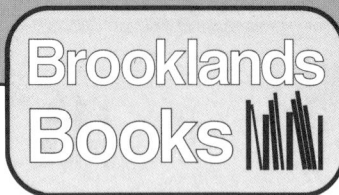

## ACKNOWLEDGEMENTS

We published our first Road Test book on the Triumph Stag many years ago, and quickly discovered that this car was going to be one of the more popular classics when enthusiasts demanded a second book on the subject! That became our Triumph Stag, Collection No.1. These titles have now sold right out, and so we decided to trawl through our archives to find some additional material and to publish the whole lot in one bumper edition in our Gold Portfolio series.

Brooklands Books are a living archive service for motoring enthusiasts, republishing valuable material which is no longer available about cars of interest. For volumes like this one, we depend on the generosity and understanding of those who originally published the copyright material we reproduce. So we are pleased to acknowledge our debt for material in the present volume to the owners of *Autocar, Autosport, Car, Car and Driver, Classic and Sportscar, Competition Car, Custom Car, Modern Motor, Motor, Motor Sport, Popular Classics, Popular Imported Cars, Road & Track, Road Test, Sports Car Graphic, Triumph World, Wheels* and *World Car Guide*. Lastly, our thanks go to motoring author James Taylor for his few words of introduction. We thoroughly recommend as further reading the definitive book he wrote with Stag Owners' Club historian Dave Jell, *Triumph Stag - the complete history*.

R M Clarke

Mention the Triumph Stag in enthusiast circles and everyone immediately thinks of the problems associated with its V8 engine. Expensive faults caused by poor build quality did untold harm to the Stag's reputation when the car was new, and more than a quarter of a century later their legend still persists

Yet the strength of the owners' clubs today makes quite clear that the Stag had and still has a large number of virtues, and painstaking development work by aftermarket specialists has made the car into the reliable and enjoyable machine it always should have been. This further development has shown how good the Triumph 3-litre V8 engine always was at heart, and has underlined how great was the tragedy of British Leyland's mismanagement in the early 1970s was. With a little more money put into development and a little less emphasis put on impossible launch deadlines, the whole sad story of the Stag would have been quite different.

As announced, the Stag was a genuine all-weather 2+2, snug as a small saloon in winter with its hard-top on or open to the sun and wind in summer with its convertible top folded. It accelerated and handled more like a saloon than a pure sports car, but that suited its intended customers down to the ground. With plenty of disposable income and - at most - two small children, they saw in it a chance to retain the sports-car fun of their younger days while adding the prestige and practicality of a smart, powerful and refined executive saloon. There was, quite simply, nothing else like it on the market.

The road tests and other articles reproduced in this book show just how exciting the Stag seemed on its introduction, and they also reveal how disillusionment with its faults gradually set in. The final articles show how the car has been reappraised by modern enthusiasts - and the irony of it is that the Stag is far better understood and appreciated now than it ever was during its production lifetime!

James Taylor

*The new Triumph Stag's 2 plus 2 body is available either as a soft top or with a detachable hard top.*

## JOHN BOLSTER tries

# The new Triumph Stag

The Triumph Stag is an entirely new model for which there must be an immense demand. In brief, British Leyland have produced a 2 plus 2 GT which has all the looks, refinement and performance of continental cars like the Mercedes-Benz 280 SL at just over half the price. In fact, it can probably see off the opposition fairly comfortably in at least two of these three respects, and how Lord Stokes proposes to satisfy the potential demand one would be interested to know.

The heart of the Stag is a very interesting new engine. It is a V8 with a single overhead-camshaft for each head, with chain drive. It is, in effect, the "other half" added to the inclined four-cylinder Triumph engine used in the Saab 99. The in-line valves are operated directly through inverted bucket tappets, seating in wedge-shaped combustion chambers in aluminium cylinder heads. The cast-iron 90 deg block carries the forged alloy steel crankshaft, which has integral balance weights, on five main bearings. A bonded rubber torsion vibration damper is on the front end.

An excellent vintage feature is the positive drive to the water pump, through skew-gears from a jack-shaft which also drives the distributor. The fan is belt-driven separately and has a viscous coupling to reduce noise at high revs. The engine is very much over-square with dimensions of 86 mm by 64.5 mm (2997 cc), and on the moderate compression ratio of 8.8 to 1 it develops a net output of 145 bhp at 5500 rpm. It is interesting that fuel injection is not at present standardised, an electric fuel pump at the rear of the car delivering to two horizontal Stromberg carburetters within the vee. An inertia switch cuts off the fuel supply in the event of an accident, a valuable safety feature.

The V8 engine can be mounted in unit with a manual gearbox, similar to that of the Triumph 2.5 PI. Alternatively, a Borg-Warner automatic transmission has been developed for this power unit. The hypoid final drive housing is secured to the rubber-mounted sub-

frame that also carries the rear suspension.

No separate chassis frame is employed, the body being a pressed-steel monocoque. It is a two-door 2 plus 2 and is available either as a soft-top or with a detachable hard-top. In either case, a padded roll-over bar gives the structural rigidity that a roof normally confers. It is an extension of the door pillars and has a central member that projects forward to join the screen. This overhead construction is immensely rigid and reinforces the whole body, acting like a roof in eliminating scuttle shake.

Independent front suspension of the MacPherson type is used, with semi-trailing arms behind. The rack and pinion steering has power assistance as standard. To give an

idea of the size of this new GT, it has the same track a the 2.5 saloon and is 6½ ins shorter in the wheelbase, but still 1 ft longer than the TR6. The very delightful styling of the body is unmistakably the work of Giovanni Michelotti.

### A brief test

I was able to try the Stag in Belgium on roads ranging from fast, smooth straights to the most appalling pavé, with plenty of corners thrown in, plus some heavy traffic. I chose an open car because the weather was superb, and I was not at all disappointed that it had automatic transmission. Triumph and Borg-Warner had already demonstrated such an extraordinary *rapport* over the 2.5 PI that I wanted to see if they could do it again with the 3-litre V8. They could!

The Stag is as far as it could possibly be from a hairy sports car. It is very quiet indeed and the engine is especially smooth, with hardly a suspicion of the V8 exhaust

**The interior is well planned with a nicely laid out dash.**

beat. There is a lot of torque in the lower and middle ranges, the car feeling very lively in consequence, and one is very seldom tempted to select the ratios manually, though there is provision for so doing. The transmission works very smoothly and there is no need to use high revs on the lower gears, though the engine spins very freely when required to do so.

The car rides astonishingly well over bad roads but does not feel excessively "soft" and is steady at its maximum speed, which appears to be just below 120 mph. The steering is extremely light, which is a little disconcerning at first, but it is found with experience to give adequate feel of the road. The insulation of road and tyre noises is of a very high standard indeed. There is enough understeer to give high-speed stability in side-winds but the Stag is well-balanced for taking fast curves; the amount of roll is rather less than one expects with such comfortable suspension.

Everything about the Stag is luxurious, but it accelerates like a sports car and seems very happy to cruise at 100 mph. The new Triumph 3-litre V8 engine is delightfully willing and plays with the relatively light 2 plus 2 body. This is a very important addition to the British Leyland range which will bring prestige to its makers and pleasure to its owners. We hope to publish a full-length road test soon.

**Car Reviewed:** Triumph Stag open 2 plus 2, price £1,996 or £2,042 with hard top, including tax.
**Engine:** Eight cylinders, 86 mm by 64.5 mm, 2997 cc. Single chain-driven overhead camshaft to each bank. Compression ratio 8.8 to 1. 145 bhp at 5,500 rpm. Two horizontal Stromberg carburetters.
**Transmission:** Single dry plate diaphragm-spring clutch. 4-speed all-synchromesh gearbox with central change, ratios 1.0, 1.386, 2.10, and 2.995 to 1. Chassis-mounted hypoid unit, ratio 3.7 to 1. Optional extra: Borg-Warner automatic transmission.
**Chassis:** Combined steel body and chassis. MacPherson independent front suspension with anti-roll bar. Power-assisted rack and pinion steering. Independent rear suspension by semi-trailing arms with coil springs and telescopic dampers. Disc front and drum rear brakes. Bolt-on steel wheels fitted Michelin 185 HR 14 radial ply tyres.
**Equipment:** 12-volt lighting and starting. Speedometer. Rev-counter. Water temperature, battery condition and fuel gauges. Heating, demisting and ventilation system. Two-speed windscreen wipers and washers. Flashing direction indicators. Reversing lamps. Optional extra: Refrigerated air conditioning.
**Dimensions:** Wheelbase 8 ft 4 ins. Track, front 4 ft 4½ ins; rear 4 ft 4⅞ ins. Overall length 14 ft 5¾ ins. Width 5 ft 3½ ins. Weight 1 ton 5 cwt.
**Performance:** (approximate) Maximum speed 118 mph. Speeds in gears: third 79 mph, second 52 mph, first 36 mph. Acceleration: 0-30 mph, 3.5 s; 0-50 mph, 7 s; 0-60 mph, 9.5 s; 0-80 mph, 16 s; 0-100 mph, 29 s.

*The rear of the car is unmistakably Triumph (above). Cutaway drawing of the new 3-litre V8 engine, which gives 145 bhp (below).*

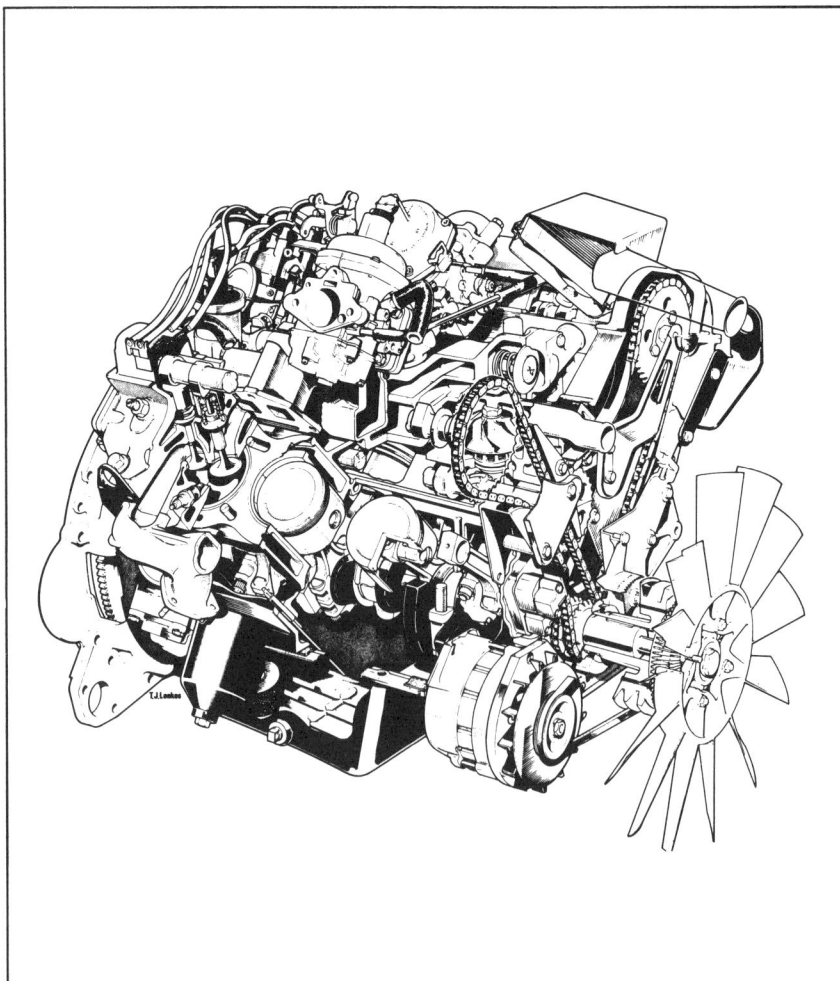

6

# TRIUMPH STAG V-8

### BY CYRIL POSTHUMUS

A HIGH-PERFORMANCE luxury grand tourer from a marque which, in its time, has produced a whining baby called the Super Seven, the knife-edged Mayflower and Renown series, and the early spartan sports TRs of the '50s, shows how far Triumph has progressed. Their new Stag is a fine, fast, exceptionally smooth performer in the currently somewhat neglected 3-liter class and its qualities should ensure a big demand for it from world markets. The U.S. version, which will come with wire wheels, stainless steel rocker panels and tinted glass as standard equipment, will have a list price of about $5300.

Its evolution is logical yet fascinating. Having designed and developed a 45-degree 1.7-liter slant-4 single overhead cam engine for Saab of Sweden to power their 99 model, what could be more inviting than to pair up two blocks at 90 degrees on a common crankcase, juggle the bore and stroke and produce a 3-liter V-8? The Saab's cylinder dimensions are 83.5 x 78 mm; the Stag has a slightly bigger bore and shorter stroke at 86 x 64.5 mm, and the 2-plane crankshaft runs in five plain bearings. Like Austin, Triumph has hedged on the toothed belt drive system which bolder makers like Fiat and Ford employ, putting their trust in time-proven single chains. There are two side-draft Stromberg carburetors

in the vee, Lucas alternator and ballasted coil ignition, and a no-loss cooling system with expansion chamber and pressure cap. Output is 145 bhp at 5500 rpm, with typical V-8 torque of 170 lb-ft at 3500 rpm, the overall result being velvet smoothness all along the line.

The excellent Triumph 2.0/2.5 PI Mk 2 sedan contributes both the independent front and rear suspension by coil spring/shock units and the all-synchromesh 4-speed gearbox to the Stag, while the front grille and twin quartz-halogen headlight units and stainless steel wheel trims also owe some allegiance to the Mk 2. The steel-panelled unitary body is a 2-plus-2, available with a soft top, a detachable hard top, or both, and the structure embodies a fully padded roll-over bar of T-form, running between the two door pillars and the center of the screen header rail and serving as an excellent safety feature. The soft top can be raised or lowered by one person in under a minute and when down lies concealed in a covered well behind the rear seats. There is a TR-6-style zip-out rear window panel for extra ventilation in hot weather and fixed quarter windows.

Electrically operated windows are standard, while the hard top has forward-hinged anti-draft quarter windows and heated rear window as standard, electrical circuits for the

7

latter automatically connecting or disconnecting when the top is fitted or removed. The standard of luxury in the Stag is indeed very high; it wouldn't be British without a bit of wood, of course, although the non-reflective walnut veneer on the facia and central console has a synthetic look about it. The facia has a matte-black padded surround and the instruments—all that a keen, conscientious driver would need—are set in a compact curved area ahead of him, as on the PI Mk 2. The controls are very well laid out; the separate, fully reclining, vinyl-trimmed seats adjust every way—fore, aft, height and rake—and the steering column is adjustable both axially and vertically. With near 120 mph and V-8 pickup, the Stag has extra large disc front brakes of 10.6-in. diameter and 9.0-in. drum brakes at the rear, with servo assistance.

## Driving Impressions

SUMPTUOUS COMFORT is the first impression on getting into Triumph's new 3-liter V-8. You can juggle the seat and the steering column around until they suit *you* and the control layout is the work of serious sports car designers—handbrake in the proper place in the middle behind the gearbox hump, gear lever just right, lights and winkers through one fingertip lever, wiper and washer on the other and good, big pedals. A leather-clad ignition key fitting into a steering column lock is a lot less awkward than most, once one is used to it, and a tweak of it brings pleasant 8-cylinder noises, subdued save when you stab the accelerator hard. Acceleration is highly impressive on the Stag and when happy that everything was warm we got a 0 to 50 mph in 7.3 sec and 0 to 70 in 12.7 with quite a wind blowing, whereas Triumph themselves claim 7.0s and 12.5s. The gearchange, with a nice short lever, has quite a wide movement but is satisfyingly free and not "notchy" (a particular hate of mine after many Maxi miles); the speedometer is set right ahead, the tachometer next to the right, somewhat concealed by one of the steering wheel spokes. A cluster of telltale lights mollycoddle the driver with information about oil pressure, winkers, headlight beam, choke, fuel and handbrake, adding to the feeling of confident well being which the Stag imparts.

Is it churlish to complain that the power steering is, if anything, too light? It happens with other cars too, that one takes too big a bite at the wheel initially, and in reaction a few miles further I found I tended to under-steer for a corner, then had to take a little more. Obviously familiarity acclimatizes one to this, which cannot fairly be called a fault, although

# TRIUMPH STAG V-8

one has to watch it when cornering really fast, and lack of feel can be a handicap. Here's where the Citroen SM's progressive power steering will come in eventually, one presumes.

My main impression of the Stag over a fine choice of Belgian roads ranging from *pavée* to pleasant tree-lined B roads and short, sharp stretches of autoroute, was of its almost soft smoothness; there was nothing taut or "sports car" about its handling, save for its precision. The ride was impeccable, the sudden tire patter over a patch of *pavée* reassuringly stable with no lurch or pitch while one could corner very fast without the tire squeal eloquent of forces in conflict. Effortlessness is a strong characteristic; at 1000

rpm, road speed in top gear was an indicated 19.9 mph, but engine flexibility enabled me to come down to 16 mph without snatch or niggle, before accelerating hard, still in top gear, with equal lack of fuss. On the other hand, a grabbed lower gear, then into top again at about 5300 rpm produced most intoxicating pick-up, and one could find oneself in the upper 90s very quickly, all without fuss other than wind noise. This gay deception caught me out at about 90 mph through a fast bend, but when I had to brake suddenly for oncoming traffic while still in the bend, the car wavered from its line not at all. Abundant power is always pleasant and passing in the Stag was a mere matter of aim and fire, with truly superb braking to look after the deceleration side of things.

It is in essence a fast, luxury grand tourer in the old meaning of the word, i.e., ideal for a rapid, comfortable drive for two or three plus luggage across France to the Riviera, or to Germany, Switzerland or Italy, without tiring its occupants. Obviously that V-8 engine has more potential, and its performance in the 2-plus-2 Stag makes one think immediately of its use in an angrier form with double-double Webers or fuel injection, in a shorter, lighter 2-seater sports car with firmer suspension—a TR-7, say—but for anything like that, we must wait and see. British Leyland has chosen for the moment to make maximum use of existing tooling and pieces from the 2.5 sedan and Saab engine for a luxury tourer. As it is, the Triumph Stag must rate highly for easy, untiring high performance with all mod cons. There are options with Laycock overdrive and Borg-Warner Type 35 automatic transmission to make driving it even easier and in such a form the Stag should appeal greatly to many Americans. The discerning few who want a real sports car must hope for a V-8 TR-7 some time in the future. 

## TRIUMPH STAG SPECIFICATIONS

| | |
|---|---|
| Engine..................sohc 90° V-8 | Tires...Michelin XAS radial 185HR14 |
| Bore x stroke, mm........86.0 x 64.5 | Chassis/body...............unit steel |
| Equivalent in..........3.385 x 2.539 | Front suspension: MacPherson struts, coil springs, tube shocks |
| Displacement, cc/cu in.....2997/182.9 | |
| Compression ratio..............8.8:1 | Rear suspension: independent with semi-trailing arms, coil springs, tube shocks |
| Bhp @ rpm................145 @ 5500 | |
| Torque @ rpm, lb-ft........170 @ 3500 | |
| Transmission: 4-spd manual, 3-spd automatic, 4-spd manual w/o'drive | Curb weight, lb.................2800 |
| | Wheelbase, in..................100.0 |
| Final drive ratio................3.7:1 | Track, front/rear...........52.5/52.9 |
| Steering: power assisted rack & pinion | Length.......................173.8 |
| Turns, lock to lock.............4.0 | Width.........................63.5 |
| Brakes: 10.6-in. dia. discs front, 9.0-dia. drums rear, servo-assisted | Height (soft top)...............49.5 |
| | Fuel tank capacity, U.S. gal.......16.8 |
| Wheels.............steel disc 14 x 5J | Maximum speed, mph (mfr).......118 |

# TRIUMPH STAG

## Vee-8 power and a sporting image

By Jeffrey Daniels

"Why the Stag?" is not an easy question to answer. It is not a replacement for anything, and falls in the middle of a range already well-endowed with sporting vehicles. If anything, it is a classic gap-filler in the Triumph tradition, without any direct competition.

THE Stag had its origins four years ago (in other words, well before British Leyland came into being). Michelotti, who had done most of Triumph's styling under contract for a long time, was asked to produce a design study for a sports coupé using the existing Triumph 2000 floor pan and mechanical units.

This original study was progressively developed, with one major mechanical change when the in-line engine was abandoned in favour of the new vee-8 unit, and several changes in the body concept, including a shortening of the wheelbase. A new nose and tail treatment was devised, which was thought so successful that it was adapted to the 2000 saloon when it emerged in Mark 2 form. It is interesting to note, when looking at the undoubted family resemblance between the Stag and the saloons, that the Stag was first on the drawing board.

This is just one aspect of the design which shows how much thought has gone into making the whole unit easy to service. The camshafts are chain-driven, and the drive sprockets can be undone and left *in situ* while the camshafts are removed, avoiding the necessity to reset the timing afterwards. Incidentally, the choice of chain drive was taken quite deliberately, despite the availability of toothed belts, in the interests of reduced engine length and better oil-tightness.

Another neat feature is the ease with which the entire inlet manifold assembly can be removed. The two Strombergs sit facing towards one another, in the manner of the SUs in the Rover 3500. Their good anti-pollution characteristics mean that the 1971 California regulations can be met with the aid of an inlet hot-spot and a thermostatically controlled air inlet valve. This was one reason why carburettors rather than fuel injection were chosen for the Stag. Triumph make no secret of the fact that they are looking at the possibility of injection, but stress that any system they use will have to be capable of meeting the American pollution requirements.

### Engine

The vee-8 engine is of course the real centre of interest in the Stag. It is by no means entirely new, for Saab have been using half of it in the 99 for some time. It was a simple matter (in mind from the earliest days of the design) to take two of the 45deg canted, single overhead cam Saab engines and make a vee of them.

The Stag engine is not, however, twice the capacity. When siamesing two four-cylinder engines in this way, one runs into trouble with balance weights fouling pistons unless the stroke is shortened. Originally, indeed, the vee-8 had a capacity of 2.5 litres, but was stretched to three litres rather more than a year ago. It now has a slightly larger bore than the Saab unit, but a much shorter stroke, giving a bore/stroke ratio of 1.33. There is still a little stretch potential left in the engine, but not a great deal.

The large bore makes it possible to achieve very respectable valve sizes within the limits of the in-line layout, and the combustion chamber shape is an efficient wedge, reminiscent of that used by Fiat in the highly successful 128 engine. Power output is modest at 145bhp (net), but turning at the moment is for flexibility, as the 8.8 compression ratio and 16-56-56-16 timing bear witness. Even so, there is more power here than comes from the 2.5PI in-line unit, and rather more torque as well.

Chrome-iron is used for the block, and aluminium alloy for the heads (and inlet manifold). The installed weight of the engine comes very close to that of the smaller 2.5PI. The head design itself is ingenious. The use of slanted bolts to secure it, allows the camshafts to be taken off without disturbing anything else.

The crankshaft is a two-plane, forged-steel unit running in five main bearings. Engines have been run with single-plane cranks, but their vibration characteristics are inevitably far worse. Each crankshaft throw accommodates two big ends, with their own oil supplies from the adjacent main bearing.

Triumph quite rightly stress that the tremendous amount of work done on the Saab engine, especially in the reliability and durability fields, can in large measure be read across to the vee-8. Since Saab are now apparently satisfied that the four is the most durable engine in Europe, this bodes well for the life of the Stag.

All the most important features of the smaller engine are retained, including the distributor and water pump drive from a jackshaft (driven by one of the camshaft drive chains) running the length of the engine in the vee. This has enabled the layout of the front end of the engine to be kept very simple.

A viscous-drive fan of no less than 16.5in. dia. is used, thus reducing power losses and noise generation at high engine speeds. Electric fan operation was seriously considered, but it was felt that a unit able to produce an adequate output in the most difficult cooling conditions would be bulky and difficult to install. There are separate vee-belt drives to the alternator and power steering pump and, when fitted, to the air conditioning pump. This use of separate belts means that belt length and width can be kept to a minimum, at the expense of having a rather wide, multi-channel pulley at the front of the crankshaft.

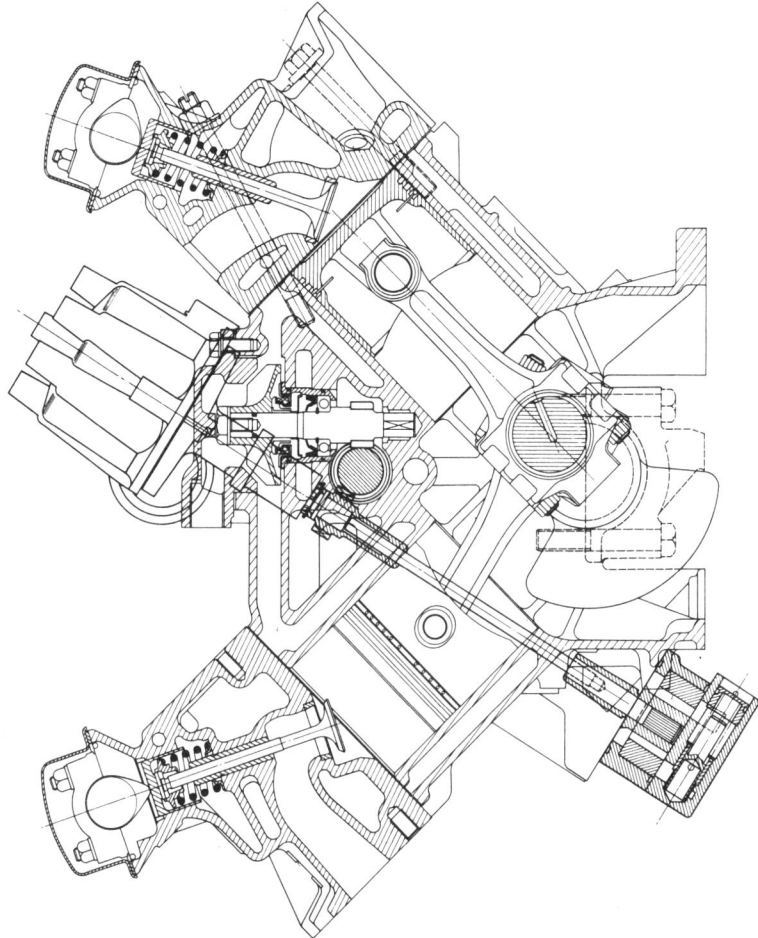

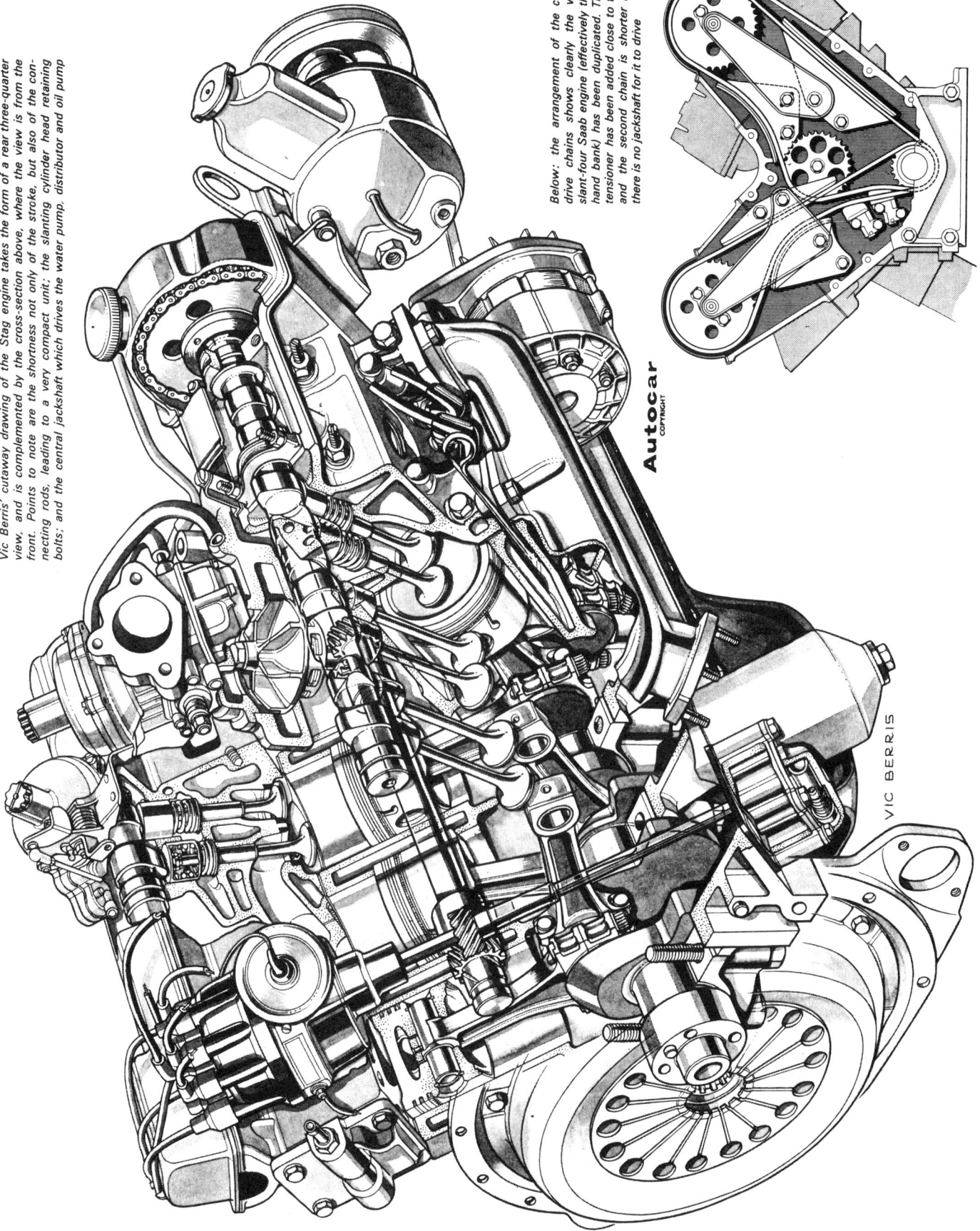

Vic Berris' cutaway drawing of the Stag engine takes the form of a rear three-quarter view, and is complemented by the cross-section above, where the view is from the front. Points to note are the shortness not only of the stroke, but also of the connecting rods, leading to a very compact unit; the slanting cylinder head retaining bolts; and the central jackshaft which drives the water pump, distributor and oil pump

Below: the arrangement of the camshaft drive chains shows clearly the way the slant-four Saab engine (effectively the right-hand bank) has been duplicated. The extra tensioner has been added close to the first, and the second chain is shorter because there is no jackshaft for it to drive

Autocar
COPYRIGHT

VIC BERRIS

11

# TRIUMPH STAG...

## Body

There are two points to bear in mind where the Stag body is concerned. One is that it bears a resemblance to the 2000 saloon, especially in the nose and tail; the other is that it is open.

Because it is open (neglecting for the time being the optional hard top), there is clearly a torsional stiffness problem. This has been got over partly by conventional stiffening of the floor area—a glance at our cut-away drawing shows the massive sills and two large transverse stiffeners—and also by using a 'Targa-bar' arrangement. This is a very strong, permanent structural member which absorbs a good deal of torsional strain between the front and rear halves of the car (the doors, of course, being the basic sources of weakness), and also adds to the beam-stiffness of the body shell as a whole. Apart from this, it acts as a valuable safety feature should the car be involved in a roll-over accident.

Compared with the 2000 saloon, the Stag sits on a wheelbase which is 6in. less. The overall length is 8in. less, which underlines the fact that the front and rear overhangs are very similar to those of the saloons. The vee-8 engine sits neatly under the bonnet, and not too far back, despite the encroachment of the MacPherson strut housings into the available width. Not surprisingly, the engine will fit the saloons with equal ease and, in fact, much of the development running was done with saloons so equipped, for obvious reasons.

The back end of the body follows conventional lines, with both the spare wheel and the large (14-gallon) fuel tank housed beneath the boot floor. A complex tank-venting system will be installed in cars for the American market, to comply with the latest regulations. Most of the rear suspension loads are taken out into the area of the rear bulkhead.

Usable length in the interior of the car is naturally limited by the shorter wheelbase to considerably less than that in the saloon. A few more inches are subtracted by the need to accommodate a folding hood. This is very neatly stowed in a well beside and behind the back seats, covered by a padded, rear-hinged panel. A good deal of effort was expended trying to arrive at some means of making the hood power-operated, which was thought important for the American market, but this proved not to be possible. However, raising and lowering the hood is a very simple, single-handed operation.

There is an alternative hard top; the car can be had with this *and* the soft top, or with either on its own, although if it is bought with the hard top only, the well behind the rear seats remains, so it is not a way of coming by some more room in the back. The hard top can be removed readily, and carries little stress (other than those involved with its own aerodynamic loads). Sealing is by rubber beading, resulting in the edges standing clear of the bodywork, so that there is no risk to the paint. Both the soft and hard tops are glazed at the sides and the back, preserving a good field of view for the driver. The rear window of the hard top is electrically heated as standard, and its rear quarter windows are hinged at the front, so that they can be opened to provide through-flow ventilation (saloon-type face level ventilation is provided).

A good deal of the body interior has been borrowed from the saloon, including the excellent dashboard and minor control layout. An innovation for Triumph is the use of electrically operated side windows, the switches for which are mounted on the massive central console.

Common features between the Stag and the saloons are emphasized by the fact that 23 of the body panels (not all external, of course) are interchangeable. In a way, it is surprising that the final result can have such different dimensions, one of the most significant being the Stag's much reduced height—it is over 6in. lower. Some of this is accounted for in reduced ground clearance, but the rest is a matter of body design.

## Suspension

Suspension design closely follows that of the 2000 saloon (or, more strictly, of the 2.5PI), with the same MacPherson struts at the front and semi-trailing arms at the back. Again, it is interesting to note that the wider rear wheel track introduced with the Mark 2 saloons last October was first mooted for the Stag.

At the front, there are a number of changes for the Stag. The front spring rates have been softened to 110 lb/in. in the interests of ride comfort, and the roll stiffness brought back up to the mark with the aid of an anti-roll bar. Front suspension travel between full bump and full rebound is 6.47in. Suspension loads are fed into the body in the conventional way, making use of the widely-spaced attachment points of this suspension layout. Braking loads are absorbed by semi-leading lower links, straight into the front bulkhead.

Rear suspension design follows that of the saloons very closely indeed, using the same principle of a massive sub-frame attached to the body by four widely spaced, rubber-insulated attachment points. The coil springs, with a rate of 322 lb./in., are placed well forward of the axle line, the mechanical advantage of roughly two-to-one reducing the equivalent stiffness at the wheel. The telescopic dampers are mounted aft of the axle line. Full travel between bump and rebound is slightly larger than at the front, at 6.55in.

## Transmission

There are no surprises here. The gearbox is taken from the 2.5PI, save for a higher first gear, and overdrive is an option (as is Borg-Warner 35 automatic transmission). Final drive ratio is 3.7-to-1, midway between the 4.1 of the 2000 and the 3.45 of the 2.5PI, clearly intended to match the potential performance as closely as possible. Gearbox and final drive are joined by a single-piece propeller shaft; the length involved is short enough to ensure that no whirling problems occur. Clutch diameter is 9in., which is half an inch more than that in the 2.5PI.

## Brakes

The Stag's brakes are the biggest yet from Triumph, notably larger than the 2.5PI brakes, even though these were uprated considerably last October. The front discs are no less than 10.6in. in diameter, and while the rear drums are still 9in. diameter,

# SPECIFICATION
## FRONT ENGINE, REAR-WHEEL DRIVE

### ENGINE

| | |
|---|---|
| Cylinders | 8, in 90 deg vee |
| Main bearings | 5 |
| Cooling system | Water, pump, fan and thermostat |
| Bore | 86.0mm (3.39in.) |
| Stroke | 64.5mm (2.54in.) |
| Displacement | 2,997cc (182.9cu.in.) |
| Valve gear | Single overhead camshaft per bank |
| Compression ratio | 8.8 to 1 Min. octane rating : 97RM |
| Carburettors | 2 Zenith-Stromberg 1.75CD |
| Fuel pump | SU electric |
| Oil filter | Full flow, replaceable element |
| Max. power | 145 bhp (net) at 5,500 rpm |
| Max. torque | 170 lb.ft (net) at 3,500 rpm |
| Max. bmep | 140 psi at 3,500 rpm |

### TRANSMISSION

| | | |
|---|---|---|
| Clutch | Laycock diaphragm spring, 9.0in. dia | |
| Gearbox | 4-speed, all synchromesh or Borg-Warner 35 three speed automatic with torque converter | |
| Gear ratios | Top 1.0 | |
| | OD top 0.82 | |
| | Top (Auto) 1.0 | |
| | Third 1.386 | Inter 1.45 |
| | Second 2.10 | |
| | First 2.995 | Low 2.39 |
| | Reverse 3.369 | Reverse 2.09 |
| Final drive | Hypoid bevel, 3.7 to 1 | |

### CHASSIS AND BODY

| | |
|---|---|
| Construction | Integral, with steel body |

### SUSPENSION

| | |
|---|---|
| Front | Independent, MacPherson struts, lower links, coil springs, telescopic dampers, anti-roll bar |
| Rear | Independent, semi-trailing arms, coil springs, telescopic dampers |

### STEERING

| | |
|---|---|
| Type | Power-assisted rack and pinion |
| Wheel dia. | 16.0in. |

### BRAKES

| | |
|---|---|
| Make and type | Lockheed disc front, drum rear |
| Servo | Lockheed vacuum |
| Dimensions | F 10.6in. dia. |
| | R 9.0in. dia. 2.25in. wide shoes |
| Swept area | F 220sq.in. R 127 sq.in. |
| | Total 347sq.in. (208sq.in./ton laden) |

### WHEELS

| | |
|---|---|
| Type | Pressed steel disc, 4-stud fixing 5.5in. wide rim |
| Tyres—make | Michelin |
| —type | XAS radial ply tubed |
| —size | 185-14in |

### EQUIPMENT

| | |
|---|---|
| Battery | 12 volt 56 Ah |
| Alternator | Lucas 11AC, 45 amp a.c. |
| Headlamps | Lucas four-lamp tungsten-halogen 110/220 watt (total) |
| Reversing lamp | Standard |
| Electric fuses | 8 |
| Screen wipers | Two-speed |
| Screen washer | Standard, electric |
| Interior heater | Standard, air-mixing type |
| Heated backlight | Standard with hardtop |
| Safety belts | Extra, mounting points standard |
| Interior trim | pvc seats and headlining |
| Floor covering | Carpet |
| Jack | Scissor-type |
| Jacking points | 2 each side under body |
| Windscreen | Toughened |
| Underbody | Phosphor treatment prior to painting |

### MAINTENANCE

| | |
|---|---|
| Fuel tank | 14 Imp. gallons (64 litres) |
| Cooling system | 18.5 pints (including heater) |
| Engine sump | 8 pints (4.5 litres) SAE 10W/40 Change oil every 6,000 miles. Change filter element every 12,000 miles. |
| Gearbox and OD | 3.75 pints SAE 90EP. Check oil every 6,000 miles |
| Final drive | 2 pints SAE 90EP. Check oil every 6,000 miles |
| Grease | No points |
| Tyre pressures | F 26 ; R 26psi (normal driving) F 26 ; R 30 psi (full load) |
| Max. payload | 728 lb (330kg) |

### DIMENSIONS

| | |
|---|---|
| Wheelbase | 8ft 4in. (254cm) |
| Track : front | 4ft 4.5in. (133cm) |
| Track : rear | 4ft 4.9in. (134cm) |
| Overall length | 14ft 5.8in. (442cm) |
| Overall width | 5ft 3.5in. (161cm) |
| Overall height (unladen) | 4ft 1½in. (125.8 cm) |
| Ground clearance (laden) | 4in. (10cm) |
| Turning circle | 30ft (1040 cm) |
| Kerb weight (with all optional extras) | 3,020 lb (1,375 kg) |
| (basic) | 2,807 lb (1,275 kg) |

### PERFORMANCE DATA

| | |
|---|---|
| Top gear mph per 1,000 rpm | 19.0 |
| Overdrive top mph per 1,000 rpm | 23.1 |
| Mean piston speed at max. power | 2,330 ft/min |
| Bhp per ton laden | 86.8 |

The Stag's kinship with the Mark 2 saloon shows up here in the similarity of the nose treatment. Actually the Stag is a smaller car all round, with a wheelbase six inches shorter and considerably less height. Tyres on this car, as on all Stags initially, are Michelin XAS

A glance into the cockpit would convince many people that the interior was a straight transplant from the saloons. In fact there are notable differences, including the door interior trim and the provision of switches in the centre console for the electrically operated windows

# TRIUMPH STAG...

*Above : Top view of the Stag emphasizes the form of roll-bar arrangement, which plays an important part in keeping the car rigid—always a problem with an open design of unitary construction. The rigid, padded cover for the hood recess can also be seen here, as can the length of the boot ; its relative shallowness is not so apparent*

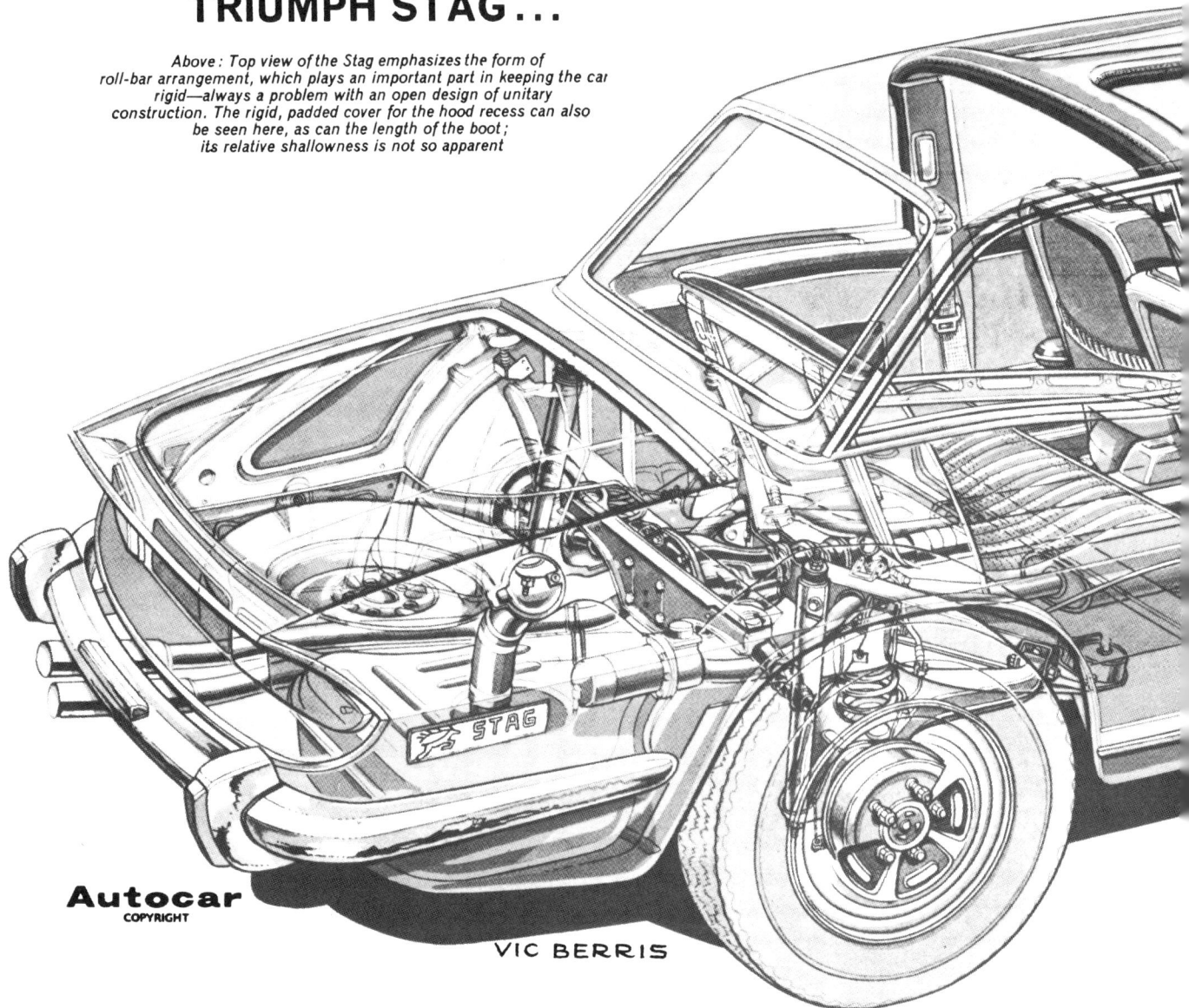

**Autocar**
COPYRIGHT

VIC BERRIS

*Above: Detail views show door interior treatment, with
combined door pull armrests; the Stag badge in the front grille centre;
the facia, very reminiscent of the 2000 Mk2 saloon; the other
Stag badge, on the rear wing (note the locking fuel filler cap); the
lamp cluster treatment at the front, again very like the saloon;
and the twin exhaust pipes*

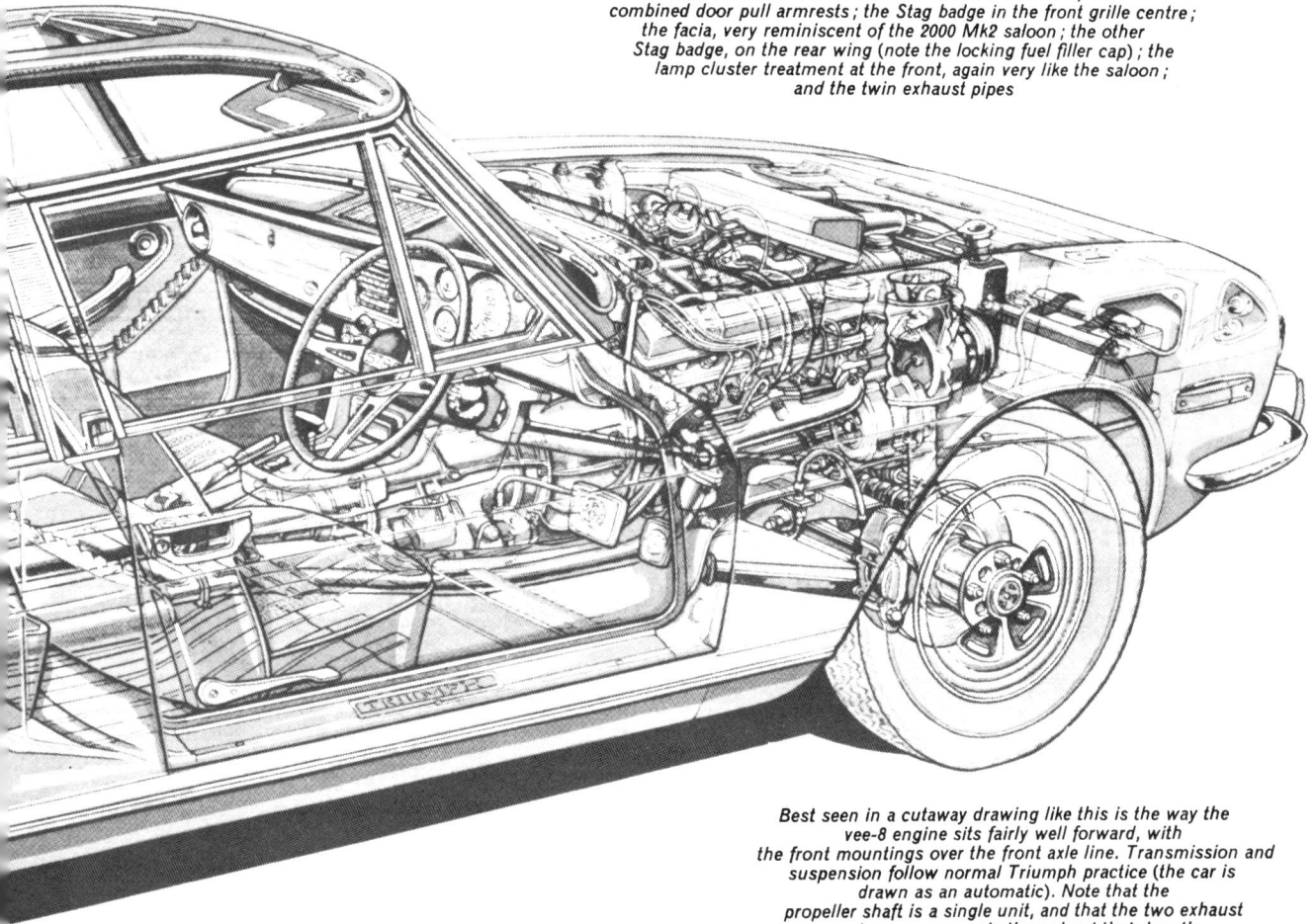

*Best seen in a cutaway drawing like this is the way the
vee-8 engine sits fairly well forward, with
the front mountings over the front axle line. Transmission and
suspension follow normal Triumph practice (the car is
drawn as an automatic). Note that the
propeller shaft is a single unit, and that the two exhaust
systems are separate throughout their length*

**continued . . .**

# TRIUMPH
# STAG...

their width has been increased to 2.5in. Total swept area is up to 347 sq.in., which more than compensates for the fact that the Stag is rather heavier than the 2.5PI. A split-circuit safety braking system is employed, of course, and a 3:1 brake servo is standard.

## Steering

Power steering is standard on the Stag. The decision that this should be so enabled the rest of the system to be designed with less regard to the need for keeping steering effort low. The pump is situated high on the right front of the engine, and powers an Adwest rack and pinion system. The steering wheel is relatively large at 16in. diameter, and the column adjusts for length and rake, as in the Triumph 1300. There is also a column lock.

Standard wheels are 5.5in. steel-discs, carrying 70-series 185-14 radials. Initially, Michelin XAS will be the tyre equipment on all cars. The effect of the lower final drive and the lower tyre profile in reducing the overall gearing is offset to some extent by the use of 14in. wheels, but the mph per 1,000rpm figure in direct top is still only 19 as compared with 20.2 for the 2.5PI.

## Equipment

The aim has been to cut the number of options to a minimum by having a high basic standard of equipment. Much of this (radial-ply tyres, brake servo, power steering, electric windows) has already been discussed. The interior bears a strong resemblance to the 2.5PI, with the same instrumentation and equipment; the front seats naturally have adjustable squabs, although lack of space means that they do not recline fully.

A Lucas 11AC45-amp alternator powers the electrical system, which carries considerable loads through its fully fused circuits. The starter is of the pre-engaged type, and the battery is of 56 Ah capacity.

An air-mixing heater and comprehensive ventilation system is virtually inherited from the Mark 2 saloons.

## Production

The design of the Stag engine means that it can be built on the same line as the Saab engine, and both units are now flowing off the same Coventry production line in large numbers. The body is produced in the new Triumph Merseyside plant, and sent fully finished and trimmed to Coventry for final assembly. Initial production target is 10,000 a year, with the capacity to expand if the demand proves high enough. The Mark 2 saloon production rate, for the sake of comparison, is currently about three times this figure. The first few months' production is earmarked for the home market, with introduction into overseas markets, including the all-important American and Common Market ones, when experience and stocks have been built up. □

*Above: The three-quarter rear view, with the hardtop in position, again betrays a relationship with the saloons. The hardtop is well endowed with window area; the rear quarter windows hinge open at their leading edges, and the rear screen is electrically heated*

*Above: The view into the engine compartment, looking forwards. Accessibility to most components is extremely good, indeed better than the picture suggests. The fan, not visible here, is of impressive diameter, and is of the viscous-drive type*

*Left: The wiper system has been carefully engineered to clear the greatest possible area of screen, especially on the driver's side*

*Below left: The rear view looks very saloon-like with the hardtop in place. There is no Stag badge on the back panel, but the twin tailpipes are a give-away*

*Below: The front seats are released by a catch high up on the door side, tipping slightly sideways as well as forwards to permit access to the back seat*

# DRIVING THE VEE 8 STAG

## First impressions from a trip in Belgium

SO far, the only Stag driving we have managed was a relatively short trip in northern Belgium. This is not ideal driving country, for the road system mainly consists of fast, open straights joined by very slow corners through little villages. On the other hand the celebrated Belgian *pavé* is still very much in evidence, so that one is well placed to learn a good deal about ride and road noise.

To some extent, the visual impression of the Stag as a "cut-and-shut" 2000 Mk2 saloon disappear when one enters the car, despite the close similarity of the instrument panel and minor control layout. One sits fairly low, with the result that the most comfortable driving position involves leaning back slightly; the steering column is adjustable for rake and reach, so that it is possible to settle really comfortably. Visibility is good without being outstanding. The nose is quite long and wide, but the front corners are well defined so that it is easy enough to place the car accurately in a traffic stream. The tail is visible for reversing. One excellent feature is the use of a parallelogram system for the driver's wiper, so that it clears the largest possible area of the big, steeply raked screen without leaving any blind spots.

Thoughts of the saloon recede even farther when the engine is started. A single turn of the key produces a subdued but characteristic vee-8 rumble, with a slightly uneven idle at about 700 rpm. Throttle response is smooth and progressive. But the pedal movement feels rather long. Indeed, all the control movements are on the long side, not least those of clutch and gearbox; this seems to have been preferred in order to keep the controls as light as possible —which they certainly are.

Moving away from rest, the smoothness and urge of the new engine becomes obvious. There is no impression of the car running away with the driver, but the low-speed punch is certainly there if required. As one moves higher up the speed scale it becomes obvious that the engine is going to stay with you all the way. There is no hint of low-speed torque having been obtained at the expense of high-speed strangulation. A yellow sector on the tachometer advises one not exceed 5,500 rpm—at which maximum power is developed—for cruising; the red line is at 6,500 rpm, which is a high limit for a relatively large unit like this. The engine is perfectly willing to rev right up to the red line, and in direct top gear it is possible to cruise well into the yellow. Triumph's claimed maximum is 118 mph in either direct or overdrive top, corresponding to 4,900 and 5,960 rpm respectively: in other words, about equally spaced either side of the maximum power point.

The gearbox is adapted from the 2.5PI, using a higher first gear, both to suit the engine characteristics and to reduce the torque loading on the final drive. By modern standards it is not a particularly good change, but the synchromesh is strong and there is no baulking. In any case, the lazy driver will find that with a bit of throttle-feathering, the Stag can be trundled away from 10 mph in direct top, and selecting overdrive is simply a matter of sliding the little inset switch in the top of the gear lever. The spacing of the gears and the choice of overdrive ratio means that overdrive third falls nicely between direct third and top, so that the Stag (if thus equipped) is effectively a six-gear car.

Considerable efforts have been made to make the Stag a quiet car to ride in and in large measure this has been achieved. The viscous-drive fan undoubtedly plays its part in keeping down noise from under the bonnet. The engine itself is certainly subdued up to about 5,000 rpm, after which some induction and exhaust noise starts to intrude. Wind noise is quite low, and the suppression of road noise has been very successful, so that even on steel-braced Michelins over dreadful Belgian surfaces, little noise penetrated the interior. In keeping with the general policy, other noises seem to have been suppressed as far as possible: the heater fan, wiper motor and electric windows are all quiet in operation, as is the comprehensive ventilation system, which works extremely well. It is even possible to close the doors quietly.

If the object of taking us to Belgium was to demonstrate that the ride was good even over very poor surfaces, then it was successful. Even though the Stag is no more than a half-way approach to the classic French theory of soft, long-stroke springs and lots of damping, it copes remarkably well with all but the very worst surfaces. The suspension travel of well over 6in. is sufficient to smooth out any sort of main-road imperfection up to and including deepish potholes. At the same time it is firm enough to avoid more than the slightest suggestion of float; there is no sign of pitch, and roll angles are small. The seat springing feels as though it complements the suspension well, and the design includes enough wrap-round to support occupants against sideways forces right up to shoulder level.

It was difficult to gain a real impression of the handling. Straight-line stability is excellent, and the power steering light: almost too light to begin with, but eventually feeling well harmonized with the lightness of the other controls. Even so, it is still not really good enough to compare with the very best systems available.

Where open corners could be found, the handling appeared to be basically slight understeer, with sufficient power always on tap to push the tail wider if needed. Beyond this, judgment will have to await our full Road Test, to be published as soon as possible.

Short drives of this nature are no place to start punishing the brakes systematically; but there was nothing to indicate that the Stag's massive disc-drum system, the largest yet fitted to a Triumph, is anything but entirely adequate. Pedal pressures are kept well down with the help of a large servo. The handbrake is situated between the seats, and is light and effective in use.

On the basis of this short meeting, it seems that the Stag might be a *genuine* example of that rare breed, the GT car. From the British Leyland point of view, it plugs the gap between the MGB and the Jaguar E-type. It is rather faster, much better-equipped and more comfortable than the former, but nothing like as fast, or as expensive, as the latter. Prices were settled after the main description in our colour section went to press: they are as follows. □

## PRICES

**Triumph Stag**
Soft top £1,995 17s 6d (£1,527 basic)
Hard top £2,041 11s 5d (£1,562 basic)
Hard + soft top £2,093 15s 10d (£1,602 basic)

# A TRIUMPH CALLED STAG

**All-new luxury £2000 2+2 coupe; open or closed body; 145 bhp V-8 engine based on the Triumph/Saab "four"; all independent suspension; power steering standard**

by Harold Hastings

The new Triumph Stag has no British counterpart—and few close rivals from the car factories of other countries if it comes to that. At a cost of £2000 plus or minus a few pounds according to equipment, it is distinguished by an entirely new 145 bhp, overhead camshaft 3-litre V-8 engine of Triumph design and manufacture; a striking but practical 2 + 2 open/closed body by Michelotti; all-independent suspension on well-tried principles; and furnishing and equipment that are both luxurious and forward-looking.

In introducing this new car Triumph have once again aimed at what they have already done so successfully with other models in the past—spotted a gap in the products of other British manufacturers and produced a car to fit it. The Stag is also planned with exports very much in mind—of the current Triumph range more than 50 per cent are sold abroad. Final assembly and engine manufacture will take place in Coventry, but the unitary-construction body will be made and trimmed in Standard-Triumph International's new £11-million plant at Speke, Liverpool, thus beginning a new phase in STI production plans.

Rear seat accommodation is much more practical than the token seating in the rear of many 2 + 2 bodies, and the car is planned for three-way open/closed motoring. As a pure soft-top model, it has a hood which stows away completely in a covered well but is easy to erect. For closed-car use there is a well-designed hardtop which can be regarded as a permanency for those who prefer a fixed, close-coupled coupe. And for the motorist who likes to make the best of all worlds, both hood and hardtop can be specified, the former remaining always available and the hard top discarded for summer or holiday occasions. However, the hardtop is not all that easy to put on and take off—this operation is intended for seasonal rather than for daily use.

Equipment and furnishing include all the items normally expected on a car of this standard, plus a number of special items some others haven't got. Among these are twin quartz-halogen headlamps, ergonomically planned instruments and minor controls; a special wiper arm on the driver's side to reduce the unswept area to a minimum; electrically operated windows; a heated rear screen in the hard top; power-assisted steering as standard. There are also a number of safety features such as lamps in the armrests which show a

red light to the rear when the doors are open; dual hydraulic brake operation; and a stout, padded roll-over bar to protect the occupants if the car should overturn in an accident. This adds quite a bit to the structural stiffness, although the exact amount is not known.

As for performance, the following manufacturer's figures with pre-production models give a good idea of the standard of performance: maximum speed 118 mph; acceleration through the gears from a standstill to 50 mph, 7sec; and to 70 mph, 12.5 sec. Acceleration in top; 30-50 mph, 7.5 sec.; and 80-100 mph, 13 sec.

## Engine

The engine is not only the only British V8 in real volume production for a British car but is also the full development of the four-cylinder PE 104S single ohc engine that Triumph produced for Saab (*Motor* June 15, 1968).

This four-cylinder unit was the result of an investigation instigated in 1963 by Harry Webster (then engineering director of STI and now executive chief engineer of the Austin-Morris Division of British-Leyland) into what sort of engines might be needed when the existing ranges eventually came to the end of their useful lives. After a hard look at fancy designs, it was decided that conventional principles would be best and that an in-line four would still be the most suitable arrangement for the basic version.

"For larger and more refined editions which might be required at a later stage but which had to be allowed for in initial production planning," I wrote, "the choice then lay with an in-line six or a V8. The very smooth and quiet six-cylinder 2-litre Triumph engine suggested there was a lot to be said for this type, but in-line sixes lack the advantage of short overall length

### In Brief

**Engine:** V8, 86mm x 64.5mm., 2997 cc; single oh camshaft per bank; two Stromberg 175 CDS carburetters; 145 bhp net at 5500 rpm; 170 lb. ft. at 3500 rpm.
**Transmission:** Laycock 9in. diaphragm spring clutch; 4-speed gearbox with synchromesh on all forward gears; optional extras: Laycock overdrive on top and third or Borg-Warner fully automatic transmission.
**Running gear:** Lockheed hydraulic brakes, disc front/drum rear, with servo assistance and dual operation; strut-type independent front suspension with anti-roll bar; independent rear suspension with coil springs and semi-trailing arms; power-assisted rack-and-pinion steering; Michelin 185 HR 14 radial-ply XAS tubeless tyres.
**Dimensions:** Length, 14ft. 5½in.; width, 5ft. 3¼in.; turning circle, 34ft.; basic kerb weight, 25cwt.
**Performance** (maker's figures): max. speed, 118 mph; 0-50 mph, 7 sec.; 0-100 mph, 29sec.
**Prices:** Soft-top, £1527 basic (£1995 17s. 6d. with purchase tax); hard top, £1562 (£2041 11s 5d. with PT); hard top with soft top, £1602 (£2093 15s. 10d. with PT).

which is becoming a progressively more important virtue in making roomy but compact cars for the crowded roads of the Sixties and Seventies. The vote therefore went to the V8, which could be produced if and when required. With this in view, the in-line four was designed from the outset as one half of a possible 90° V-8 and this is the prime reason why the block is set at an angle of 45° to the vertical. . . ."

The 3-litre V8 has a bigger bore and much shorter stroke than the four (86 x 64.5 as against 83.5 x 78). It was designed with a smaller capacity and increased to its present size by Spencer King when he took over as technical director. Because of limitations of space around the balance weights of the two-plane crank, the stroke cannot be increased without radical redesign. The two blocks and the top half of the crankcase, which is carried well below the crankshaft centre line, are a chromium-iron casting. A forged, alloy-steel crankshaft is used and runs in five steel-backed, lead bronze main bearings with lead indium overlay, and the same type of bearing is used for the big ends. A torsional vibration damper of the bonded rubber type is mounted on the nose of the crankshaft where it is combined with the belt pulleys for driving the alternator and hydraulic pump for the power-assisted steering. The forged steel connecting rods are split at right angles to the rods, which have fully floating gudgeon pins in the aluminium alloy pistons which have two compression and a scraper ring.

An unusual feature of the aluminium-alloy crossflow cylinder heads is that the holding-down nuts and studs are all accessible without removing the valve cover. This is achieved by a unique arrangement in which the studs on the lower side of the heads are parallel with the axes of the bores in the usual way, but the upper studs are angled slightly and have slotted heads so that, when their nuts have been loosened, the studs can be screwed out completely and the head then removed in the normal way.

Sintered iron inserts are used for both the valve seats and guides and the in-line valves operate at an angle in wedge-shaped combustion chambers which were chosen because they offer a good combination of smoothness and power and are also helpful from an exhaust-pollution standpoint. The shallow depressions in the pistons are incidental to combustion shape and are there to provide a ready means for changes in compression ratio (the standard

Versatility: the Stag topless, above, showing the substantial roll-over bar. The hardtop, right, is intended for seasonal rather than daily use. Below: the soft top version. Fingertip switchgear and well-planned instruments reflect ergonomic thinking on interior design, left

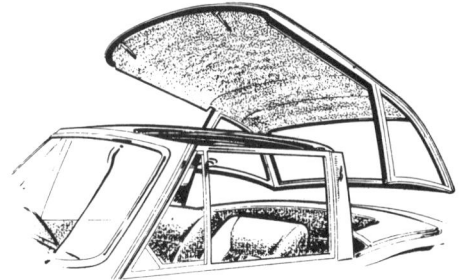

is 8.8:1) if required for special markets.

For each block of cylinders, a single overhead camshaft is used. The two camshafts are driven by separate single-roller chains from a pair of sprockets on the nose of the crankshaft. The forward chain drives the offside camshaft direct, with a guide and an hydraulic tensioner to look after the free run, while the rear chain drives the nearside camshaft with a third sprocket interposed in the otherwise free run to drive a jackshaft centrally disposed in relation to the two banks of cylinders. As with the first chain, an hydraulic tensioner is incorporated. Each camshaft runs directly in five bearings in the head and operates the valves (which have chromium-plated stems) via inverted bucket-type tappets with adjustment by hardened steel pallets. This system is well known for giving long periods between adjustment, although calling for removal of the camshaft to reset clearances; provision is, however, made for easy removal of the sprockets and for parking them in such a way that they can be replaced without disturbing the timing.

The offside bank of cylinders is numbered 1, 3, 5, 7 (from the front) and the nearside bank 2, 4, 6, 8, and the firing order is 1, 2, 7, 8, 4, 5, 6, 3. The timing is symmetrical and gives an opening period for both inlets and exhausts of 252°, with an overlap of 32°.

The purposes of the jackshaft mentioned earlier is to provide a skew-gear drive for the impeller-type water pump (see cut-away)

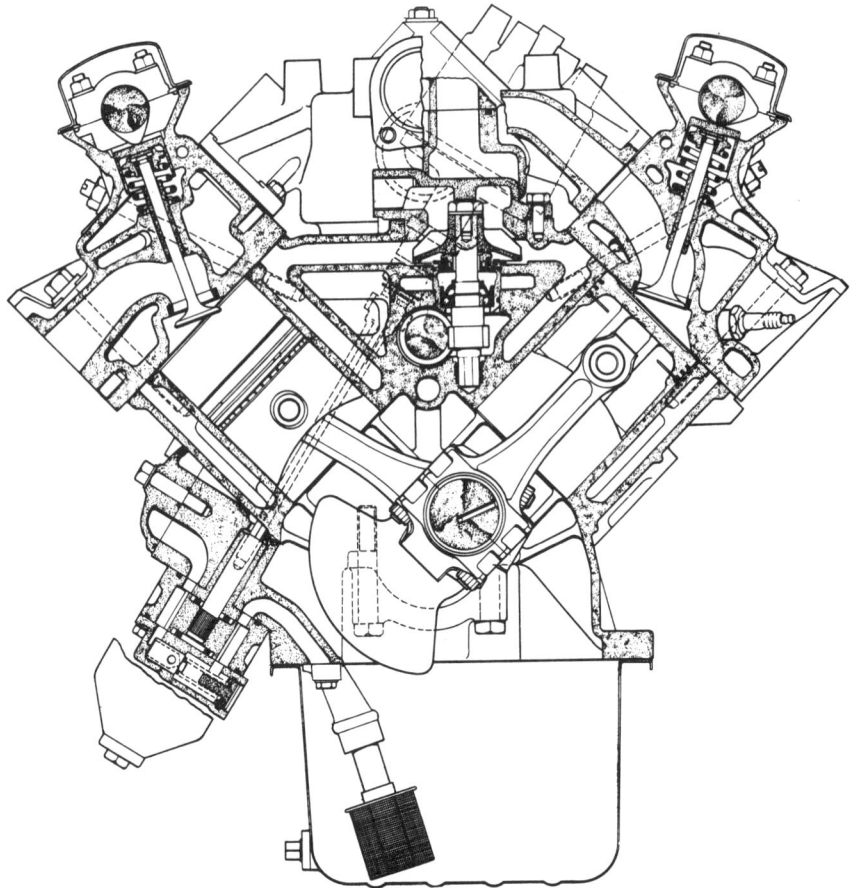

The new V-8 engine is in effect two Triumph/Saab blocks side by side. The bore is bigger and the stroke shorter than that of the "four". The drawing above shows the layout of the single overhead camshafts; pump/distributor drive; and light alloy block casting

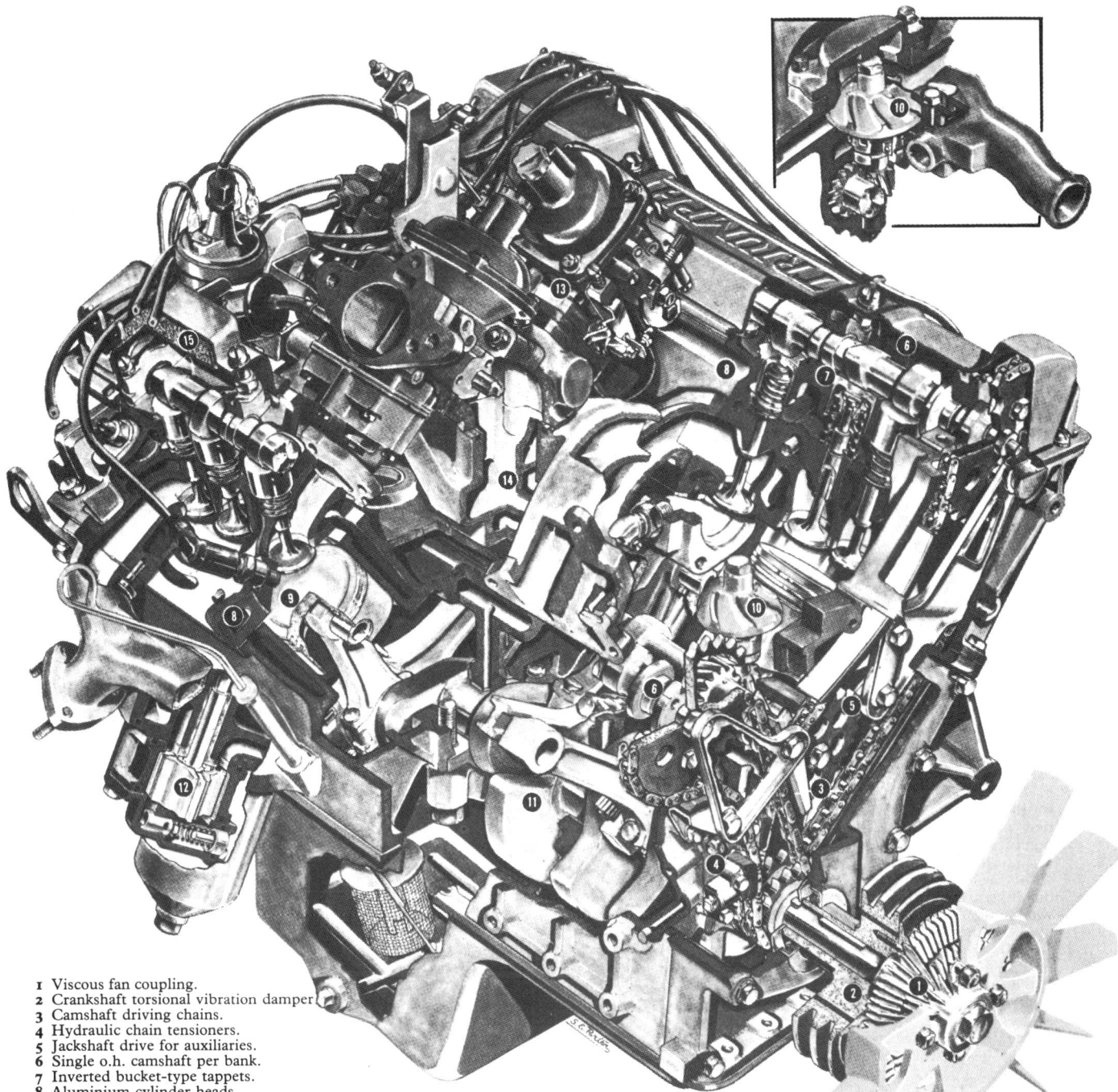

1 Viscous fan coupling.
2 Crankshaft torsional vibration damper.
3 Camshaft driving chains.
4 Hydraulic chain tensioners.
5 Jackshaft drive for auxiliaries.
6 Single o.h. camshaft per bank.
7 Inverted bucket-type tappets.
8 Aluminium cylinder heads.
9 Wedge-shaped combustion chambers.
10 Vertical water pump.
11 5-bearing 2-plane crankshaft.
12 4-lobe rotor-type oil pump.
13 Twin 175-CDS Stromberg carburetters.
14 Water-heated aluminium inlet manifold.
15 Crankcase ventilation oil trap.

which is vertically arranged between the cylinder banks, and for the distributor and oil pump via a second skew-gear, the two latter auxiliaries being linked by a long quill shaft.

To achieve a greater consistency in operating temperatures for the incoming mixture, water heating is provided for the cast aluminium-alloy inlet manifold (although it will be necessary to use exhaust heat on later smog-modified engines). This is of the four-branch type, with each branch incorporating two separate ports. Carburation is by a pair of sidedraught Stromberg 175-CDS instruments. They are fed from an electric diaphragm-type pump located in the luggage compartment and designed to deliver a pressure of 2.7 psi, with a relief valve to allow surplus fuel to return to the 14-gallon tank located under the luggage floor at the rear. An unusual safety feature is a fuel inertia switch which operates by means of a magnet and a steel ball located in a cone and so arranged that in the event of a violent impact, the ball rides up the cone and the electric supply to the pump is cut off. The same happens if the car overturns.

The cooling system, which incorporates a crossflow radiator, is arranged on the now-familiar "no loss" system and is pressurized at 13 psi. To reduce noise, the crankshaft-mounted plastic fan has 13 blades. A further detail that cuts down both noise and power loss is a viscous coupling, designed to slip above a predetermined speed.

Reduction of annoying noises is also in evidence in the exhaust system. From the three-branch, cast-iron manifold located on the outside of each bank of cylinders, separate pipe lines are led to the rear. Each one incorporates a silencer and an aluminium tail-pipe finisher which, by

means of internal perforations, provides further sound absorption; in addition, the two otherwise-separate systems are joined by a crossflow pipe just forward of the silencers with the object of eliminating the beat common with V-8 engines.

With its short-stroke design, moderate, rather than high, output of 145 bhp net at 5500 rpm and good torque of 170 lb. ft. at 3500 rpm, this new engine promises both good wearing qualities and effortless performance.

Transmission follows familiar lines with an hydraulically-operated 9-in. diaphragm-spring clutch and a strengthened version of the all-synchromesh gearbox used in the Triumph 2000 and 2.5 PI Mk. II. Options are a Laycock overdrive on top and third (and raises the road speed per 1000 rpm in top from 19.8 to 24.1 mph) or a Borg-Warner 35 fully-automatic transmission with a P-R-N-D-2-1 control giving manual selection of first and second gears when required. With this transmission, an oil cooler is fitted as standard. The rear final-drive—again a strengthened version of the saloon unit—is incorporated with the independent rear suspension sub-frame, which is insulated from the body on four rubber mountings similar to those used on the 2000 and 2.5 PI models.

Both the semi-trailing rear suspension and the MacPherson strut-type ifs are also of similar design to the 2000/2.5 PI. At both front and rear, the coil springs seat on noise-insulating washers and telescopic

hydraulic dampers are used. The steel disc wheels have 5J flat hump, safety-ledge rims and are shod with 185 HR 14 radial-ply Michelin XAS tubeless tyres. Stainless-steel trims are fitted as on some other Triumph models. Originally the car had 13-in. wheels but increasing the engine size and power made bigger brakes and hence bigger wheels necessary.

Notable points about the rack-and-pinion steering are that an Alford and Alder power-assisted system incorporating Adwest hydraulic valve mechanism is standard, and that the impact-absorbing column is adjustable for both rake and height, giving an axial movement of approximately 4 in. and a vertical range of about 2 in. Both adjustments are controlled by a single lever which can be reached from the driving seat.

The same accent on safety is noticeable in the servo-assisted, disc/drum Lockheed brakes, operated by a divided hydraulic system which incorporates a tandem master cylinder to give independent front and rear operation. The system also includes a pressure differential valve and a brake-failure warning light. Friction areas are large and give a total swept area of 347 sq. in., which should be adequate for a maximum laden weight of 33½ cwt. Basic kerb weight is 25 cwt.

Although the body is completely new, it bears the unmistakeable stamp of Michelotti and has a distinct family resemblance in such items as front and rear treatment to the most recent examples of the current range, the Mk. II versions to the 2000 and 2.5 PI models. Unitary construction is followed for the steel bodyshell and because this is basically an open car, a good deal of care and some

extra weight has been necessary to provide adequate rigidity. Evidence of this is to be found in the stout double-section body sills and the use of sturdy box sections and vestigal chassis members in the nose and tail. Another feature which adds rigidity as well as serving its primary purpose of protecting the occupants if the car overturns is the very sturdy roll-bar construction, which is also tied to the stout screen surround. This is of double-tubular construction and is attached to both the screen header rail and the B post, with stiffener plates in the angles. It remains permanently in position both when the car is open and when the pressed-steel hard top is in use.

The hardtop incorporates large quarter lights of toughened glass which are forward-hinged to provide anti-draught ventilation and a heated rear window as standard. Electrical connections for the latter are automatically connected and disconnected when the hard top is fitted or removed. With the permanent winding windows and hinged front quarter lights, the car offers all the amenities of a normal saloon, plus the extra refinement of electrically-operated door windows, and winders for the front quarter lights.

Wisely, the soft top does not have to be removed to fit the hardtop. It stows neatly into a well surrounding the rear seats and the usual untidiness of a button-on fabric cover is avoided by a rigid horseshoe-shaped cover. This is attached

1 Twin quartz halogen headlamps
2 Viscous fan coupling
3 3 litre V8 OHC engine
4 Power steering pump
5 Rack and pinion power steering box
6 Offset steering column
7 Strut type IFS
8 Fuel inertia switch
9 Electrically operated windows and winding quarter vents
10 Exhaust crossover pipe
11 Double boxed sidemember
12 Padded roll-over bar
13 Covered hood stowage
14 Semi-trailing IRS
15 High pressure fuel pump

*Hatton*

by fasteners at the front and a counter-balanced hinge at the rear so that when the fasteners are undone, it automatically swings upwards to allow the whole hood to be pulled out and secured. This can be done by one person without assistance in a minute or so; concise instructions are printed on a plate fixed to the driver's sun visor.

As with most soft tops vision is not quite up to the very high standard offered with the hardtop, but a reasonable approach is made by the provision of rear quarter lights as well as a large rear window. For maximum comfort when the hood is in use in hot weather, the plan first introduced on the TR6 is followed of making the rear window partially removable by means of a zip fastener which can

be reached from inside the cockpit.

Good access to both front and rear seats is provided by the two wide doors which offer a 41-in. aperture at waist level. They are forward-hinged and fitted with Wilmot-Breeden anti-burst locks, operated by the ignition and petrol filler car key. Both doors also lock internally and a separate key opens the glove box and boot. The boot offers 9 cu. ft. of unrestricted luggage space as the spare wheel lies horizontally alongside the fule tank in a well beneath the luggage floor. The boot lid is counter-balanced by torsion bars that also serve to counter-balance the hinged hood cover already mentioned.

The front seats not only have fully-reclining squabs, but are adjustable for

height as well as normal fore-and-aft movement, a particularly good detail being that all these adjustments can be carried out with the occupants seated. For rear compartment access, the squabs tip forward and self-locking, quick-release catches prevent accidental tipping on heavy braking. Each seat is 22 in. wide and the internal width from door to door is 52 in. Headroom is 34½ in. for both hard and soft-top conditions.

At the rear, the effective seat width is 40 in., the headroom 32½ in. and the distance from the squab to the back of

23

the front seat 22-28 in. according to front-seat adjustment. As at the front the trim is of expanded vinyl with a basket-weave pattern on the seat facings. The headlining of the hardtop is washable.

Odds and ends space is quite generous. There is a cubby locker on the passenger side of the facia with a parcel shelf below and, unusually, a neat, recessed section of the scuttle above in which odds and ends can be parked. In addition there are stretch pockets in both the doors and on the backs of the front seats, together with recessed compartments in the rear-seat panels.

The excellent ergonomic planning of instruments and minor controls introduced with the Mk. II 2000 and 2.5 PI models is continued in the Stag, the facia and controls of which are, in fact, closely modelled on the same plan. All the dials are concentrated in front of the driver in the non-reflecting walnut-veneer facia panel and comprise a speedometer and rev counter with large dials, flanked by further dials for a battery condition indicator, a coolant thermometer, a clock, and a fuel gauge. In addition the now-familiar Triumph "all-systems-go" dial is used to

group the warning lights for main beam, direction indicators, ignition, choke, hand brake, low oil pressure, low fuel level, and water temperature—all clearly identified. In addition, there is a brake failure warning lamp and, on left-hand-drive vehicles only, a hazard warning lamp switch enabling all direction indicators to be flashed simultaneously; this switch is deliberately tucked away in the glove compartment to avoid accidental operation.

Instrument lighting is rheostat-controlled and interior illumination is provided by neat lamps which are recessed into each side of the heater console and controlled by both courtesy and independent switches. For map reading, there is a small lamp in the glove locker lid which comes on when the latter is lowered so that it provides light where it is needed and out of sight of the driver's eyes.

For the minor controls which are required while the car is in motion, fingertip levers on the steering column are used. One controls the two-speed wiper and the screenwasher, while the other operates the direction indicators, headlamp flasher, dip switch and horn. The ignition switch and the main lighting

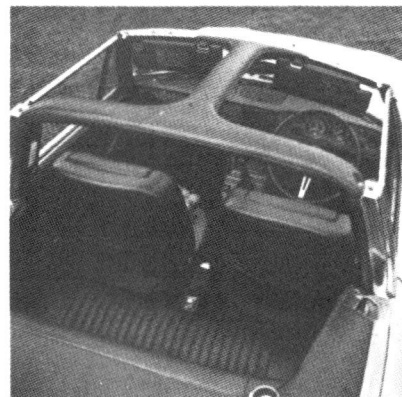

switch (which includes a parking-light position) are on the sides of the column nacelle. Switches for the electrically-operated side windows, interior lights and two-speed heater fan are mounted on the console, while the overdrive switch is incorporated in the gearlever knob as on the 2000/2.5 PI.

The heating and ventilation system includes adjustable face-level vents at

24

both the extremities of the facia board and in the centre; these supply cool, fresh air, the volume of which can be boosted by the fan. When required, a Delany-Galley air conditioning system can be fitted.

Other features of the luxurious equipment include a dipping interior driving mirror with a break-away support, padded sun visors with a vanity mirror for the passenger, attachments for safety harness at both front and rear; and twin quartz-halogen headlamps. With these items, plus all the usual fittings which are nowadays taken for granted, the Stag is equipped to a degree which very few cars can rival.

# DRIVING IMPRESSIONS

Charles Bulmer

Without any doubt this is going to be a very successful car and perhaps the source from which even more successful variants may spring. If one has any reservations at all, they arise from two causes—firstly that the Belgian roads over which we drove were unsuitable for making any judgment about its cornering behaviour; and secondly that it is fundamentally an open car. Presumably this is because Americans like open cars. So do many Europeans, but in return one must accept certain drawbacks—high weight for adequate structural rigidity; indifferent space utilization; and wind noise.

The Stag is open to criticism in these directions but there is very little else that arouses anything less than enthusiasm. It has, for example, that very desirable but rare quality in a GT car—a very good ride—better than its touring ancestors, the 2000 and 2.5PI saloons from which the suspension is adapted with changes of spring rates. This is matched by comfortable seats giving good lateral support and what one might describe as a typically Italian driving position with the steering wheel set at a rather flatter angle than usual. Since the column is adjustable both telescopically and for height, the seat is adjustable for height, rake and reach and the pedals are almost perfectly set, anyone should be able to find an ideal position in this car.

In fact the whole control layout and the way it works is something of an example to other manufacturers. It incorporates nearly all the features we have demanded for many years and the overall result fully justifies the effort. But many of these matters will be explored later in a proper road test. At present we are more concerned with broader issues like the new V8 engine which is undoubtedly a winner. You could pull it down to something like 300 rpm in top and still accelerate smoothly away and yet on the other hand it would run with complete smoothness and no feeling of mechanical strain to the red line at 6500 rpm.

With the optional overdrive there seemed to be little difference in maximum speed between overdrive and direct top. On the

## Engine

| | |
|---|---|
| Cylinders | 90° V8; five-bearing crankshaft with vibration damper. |
| Bore and stroke | 86 mm. (3.385 in.) x 64.5 mm. (2.539 in.) |
| Cubic capacity | 2997 cc (182.9 cu. in.) |
| Piston area | 465 sq. cm. (72 sq. in.) |
| Compression ratio | 8.8:1 |
| Valvegear | In-line, ohv operated by chain-driven single oh camshaft in each bank; wedge-shaped combustion chambers in aluminium-alloy cylinder heads. |
| Carburation | Two side-draught 175-CDS Stromberg carburetters mounted on water-heated, cast aluminium-alloy manifold between cylinder banks. Electric diaphragm-type pump in luggage compartment, with inertia cut-off; 14-gal. rear tank. |
| Ignition | Lucas 16C6, 6-volt ballast-resistor coil; centrifugal and vacuum timing control; Champion N-9Y sparking plugs. |
| Lubrication | Four-lobe rotor-type pump driven by skew gears from jack shaft; full-flow filter; 8-pint re-fill capacity (9 pt. when filter changed). |
| Cooling | Pump, thermostat and 13-blade plastic fan driven direct from crankshaft nose through viscous coupling; 18½-pint capacity. |
| Electrical system | 12-volt Lucas negative-earth; 56 amp. hr. battery; 540-watt alternator. |
| Maximum power | 145 bhp net at 5500 rpm equivalent to 2334 ft./min. piston speed and 2.0 bhp per sq. in. of piston area. |
| Maximum torque | 170 lb./ft. at 3500 rpm equivalent to 140 lb./sq. in. bmep at 1485 ft. min. piston speed. |

## Transmission—manual

| | |
|---|---|
| Clutch | Laycock 9-in. dia. hydraulically-operated diaphragm-spring type. |
| Gearbox | Four-speed with synchromesh on all forward gears. |
| Overall ratios | 3.70, 5.13, 7.77 and 11.08; reverse, 12.47. Laycock overdrive optional on top and third with gear-knob switch; overdrive ratios: top 3.04; 3rd 4.20. |
| Final drive | By open propeller shaft to hypoid-bevel final drive mounted on rear sub-frame and universally jointed halfshafts with low-friction splines to rear wheels. |

## Transmission—optional automatic

| | |
|---|---|
| Details | Borg-Warner Type 35 with torque converter and 3-speed epicyclic gearbox. P-R-N-D-2-1 floor-mounted control. Overall ratios: 3.70-8.50, 5.37-12.37 and 8.85-20.40; reverse, 7.75-17.80. |

## Running gear

| | |
|---|---|
| Brakes | Lockheed disc-front/drum-rear, with vacuum-servo assistance, divided hydraulic operation by tandem master cylinder, pressure-differential valve, and warning light. |
| Brake dimensions | Front discs, 10⅝ in. dia.; rear drums, 9 in. dia. x 2¼ in. wide. Lining areas: front, 24 sq. in.; rear 78 sq. in.; total, 102 sq. in. Swept areas: front, 220 sq. in.; rear, 127 sq. in.; total 347 sq. in. |
| Front suspension | Independent by coil springs; single lower transverse links braced by diagonal radius arms. Anti-roll bar. |
| Rear suspension | Independent by semi-trailing arms and coil springs; telescopic hydraulic dampers. |
| Wheels and tyres | Steel disc wheels with 5J flat-hump safety-ledge rims. Michelin 185 HR 14 radial-ply XAS tubeless tyres. |
| Steering | Power-assisted Alford and Alder rack-and-pinion type incorporating Adwest hydraulic valve system; 3 turns lock-to-lock; impact-absorbing column with driver adjustment for rake and height. |

## Dimensions

| | |
|---|---|
| Length | Overall, 14 ft. 5¾ in.; wheelbase. 8 ft. 4 in. |
| Width | Overall, 5 ft. 3½ in.; track: front, 4 ft. 4½ in.; rear, 4 ft. 4⅞ in. |
| Height | 4 ft. 1½ in. (soft top erected); ground clearance; 4 in. (4-up condition). |
| Turning circle | 34 ft. (between kerbs). |
| Weights | Dry, 23½ cwt.; basic kerb weight, 25 cwt.; kerb weight incl. opt. extras, 27 cwt. |

## Effective Gearing

| | |
|---|---|
| Top gear | 1000 rpm corresponds to 19.8 mph (in o'drive, 24.1 mph) |

motorway we held an indicated 128 mph in the former and 126 mph in the latter although the corresponding rev counter readings (5100 and 6100 rpm respectively) suggested that the true maximum was rather less—around 120 mph. Although these speeds were held for some distance in a strong cross-wind, the car felt completely stable as long as you left the very light power steering more or less to its own devices. The same was true on very bumpy surfaces—it felt like a very good and reasonably direct power steering layout but without driving the car hard on a circuit or on well-known winding roads it is difficult to say more.

There is no doubt that the weight added to make an open car structurally stiff has achieved its objective—both the open and hardtop versions feel stiff. They both suffer from wind noise though, particularly the

open car with the soft top raised in which it is difficult to converse above 80 mph although, surprisingly, the noise level does not increase proportionately at maximum speed. Nor does the hood flap or buffet which is quite an achievement at 120 mph.

At one time Triumphs suffered from noisy gearboxes. They seem to have overcome this and both the Stags we tried had quiet transmissions. The box felt very much like that of a 2.5 PI (which of course it is)—that is to say rather notchy. The change became much easier if the full clutch travel was used and in any case it would probably improve with use. Many people will elect to have the automatic instead—though we would prefer the manual for the sake of the optional overdrive which makes for such effortless high speed cruising—at least until Triumph produce a five-speed gearbox.

ANY STAG WORTHY OF attention is distinguished by brow, bez, trez and at least a three-point top, if not by a flurry on the Stock Exchange; but the creature emblazoned on the badge of Triumph's new V8 is to be noted for anatomic implausibility, if not for what Browning called Hornmadness. Never mind; from what I am told the car was named Stag (Triumph are careful to omit the article) mainly because their American salesmen thought the name just right for appealing to customers who wished or needed to have associations with masculinity, virility, and all that sort of thing. The word is an Icelandic one, referring not to the male red deer in particular but to any male animal, and it comes from *stiga* which means to mount. It is all laughably inappropriate, for there is nothing phallic or prurient about this car, none of what I recently read (in another motoring magazine and in another context) charmingly described as red-blooded foulbreathed hairy masculinity; this Stag performs with reasonable vigour, but its manner is suave and urbane, not just manly but gentlemanly.

Considering its parentage this is not surprising. Most of the car's suspension, running gear, transmission and minor assemblies come from the Triumph 2000, which is nothing if not smooth. The body is new and attractive, but has a strong flavour of current Michelotti 2000 lightly spiced with a hint of Pininfarina Fiat Dino. The engine, which is the most interesting part, is what we have been awaiting for some time—a V8 composed of two Saab slant fours.

Those slant fours are not pukka Saabs, of course, but were designed and built by Triumph in convenient time for the Swedes to buy them when they went looking for just such an engine. In fact they had asked Ricardo to do a design for them along similar lines; but Ricardo had been consulting with Triumph and knew that the British engine would suit the Swedes very well. Introductions were duly made and contracts eventually settled, and so the 1709cc Triumph engine found its way into the Saab 99. Something must have happened to it in the process, for the engine was originally designed as a 1500cc job; and it is probable though not certain that the process was reversed in making the 2997cc V8, for its bore and stroke are respectively 86 and

# STAG

**Is Triumph's £2000 personal car a logical upward step for Capri and 124S owners?**
**L J K SETRIGHT**

64.5mm, while those of the Saab engine are 83.5 and 78.

This does not quite explain how 2 × 80bhp = 145bhp. Yet the performance is quite satisfactory, brisk though not dramatic. Flexibility is beyond criticism, the transition from burble to bellow depending on throttle opening rather than on revs. Perhaps most important of all, there is little danger of the current version being superseded by a more powerful one in the near future, the fear of which often dissuades people from buying a new model lest its depreciation be artificially accelerated by planned obsolescence. For that matter there is little danger of the same engine finding its way into other cars soon (though the long-term likelihood of a V8 TR7 looks assured) for production limitations are imposed by the laziness, inefficiency and general incompetence of the British foundry trade, which has a lot to learn from its American counterpart. To put it another way, the Stag's iron block is not an easy one to cast.

Its aluminium alloy heads do not present the same problems. Much of their conformation is the product of manifolding requirements, for underbonnet space is severely limited. It is probably with the aid of non-standard heads that Triumph have experimentally run this V8 at considerably higher ratings—though they have naturally altered other things and even tried a single-plane crankshaft. Cam forms are easily changed, too: they are gentle enough at present, but there is plenty of room in the wedge-shaped combustion chambers above the mildly dished piston crowns for the valves to lift farther and faster. Alas, there is no room for making the valves and ports bigger. So the overhead camshaft makes no contribution to high performance; it is there merely as an aid to longevity, minimising the inertia loadings

throughout the valve train and keeping the noise down. Ample justification, of course—an overhead camshaft engine does not *have* to be racy. For that matter it does not have to be complex or difficult to work on, but this particular design was under way at a time when cogged belts were in their infancy, and when quick, easy tappet adjustment was just a spark of genius short-circuiting the conventions at GM, Fiat and MV Agusta. Not to put too fine a point on it, this Triumph design has been around for years.

So has the Triumph 2000, subscriber of so much to this new enterprise. When Triumph, in the persons of that polished salesman Mr Sims and that candid engineer Mr Lloyd, first presented the Stag to the motoring press and we all acted out our profundities of professed doubts and feigned enthusiasms, little was said of this background save that the suspensions were similar and that the gearbox was a stronger version

than that of the saloons. In conning the specification I later noticed that the internal ratios of the gearbox were the same as in the saloons save that bottom gear was higher, 2.995 to one instead of 3.281. Now this is not an age and Leyland's is not a business in which that sort of thing is done unless it is strictly necessary: what customer is likely to express concern, or even to notice, the fact that 5500rpm in bottom gear produces 36 mph rather than 33? What was that about making the gearbox stronger? One way of increasing the torque capacity of a gearbox (which is not quite the same thing) is to raise the bottom-gear ratio and thus reduce the torque multiplication; and this is what has been done for the Stag. It cannot be mere coincidence that the first-gear reductions in Stag and 2.5PI are inversely proportional to the maximum torque outputs of their engines. Make of this what you will: either Triumph are admirably punctilious about maintaining

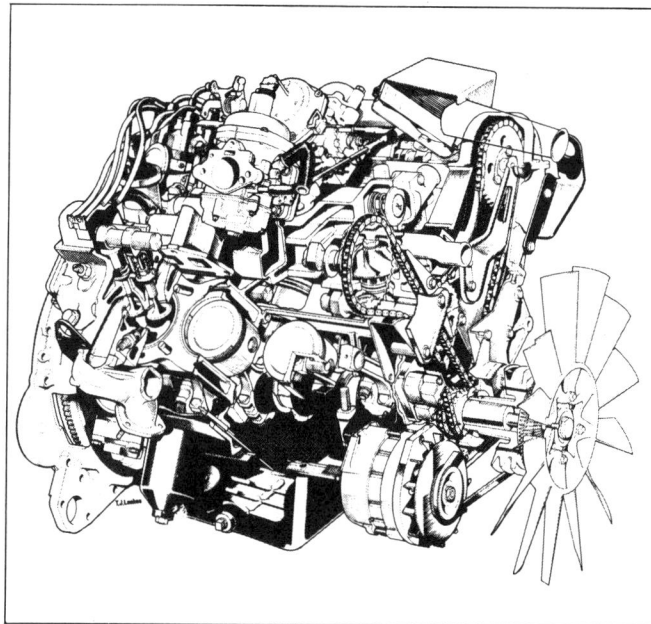

their safety factors, or that gearbox is teetering on the brink of expensive disaster.

While on the subject of teetering, I note with some alarm that the track of the Stag is quoted as 52.5in at the front and 52.875in at the rear, *four up*. Forsooth, is the track so very different when the car is otherwise laden? Are camber variations and tyre scrub so pronounced even within the range of normal static deflexion? Or is this new caution a product of the Trade Descriptions Act, under the provisions of which some petty monger of righteousness might sue the manufacturers because the rear track of his unladen Stag measures only 52.75in? Undoubtedly there are camber changes taking place as the suspensions flex, but there is nothing in the feel or behaviour of the car to suggest that in general driving they might be noticeable. Indeed they should be less so than in the saloons, for the Stag's bigger (14in) wheels are girt with Michelin XAs radials which have plenty of spare capacity for lopsided genuflexion. The car does not, it seems, depend on these tyres for its roadworthiness, for when sold to the USA it will wear G800GP70 Goodyears, which are also good but very different. Mr Engineer Lloyd said that the tyres were chosen merely because supplies happened to be conveniently available in the Coventry vicinity.

When it is in the right place at the right time the XAs is a superb tyre, and the Stag that I drove briefly during the press preview seemed to have large reserves of roadholding and safety. It was no epic drive, amounting to no more than 40 miles of Belgian roads (mostly major) with John Bolster navigating, followed by a further 40 as passenger while he drove. It was enough, however, to establish that the Stag is a car that can cover the ground quickly and without fuss, cruising happily at any speed up to its maximum and displaying no behavioural oddities even when cornering quite fast. The maximum speed is supposed to be 118.

I would not criticise anyone for choosing the car in this form. The Borg-Warner was very smooth in everything it did, and seemed admirably matched to the engine. It was even quicker and smoother to use the kick-down than to employ the PRND21 lever, and that is a rare experience. As for the topless order, on a fine sunny day it felt absolutely right to be driving this sort of car in wide-open trim. Aerodynamic backwash and uplift only began to affect the security of my hat at 80mph, while wind noise was actually rather nice—no roaring and buffeting but a subdued musical note, apparently generated by the roll-over hoop that is an integral and impressive feature of the design. When the Stag was being designed it seemed likely that legislation would call for roll-over protection, and having tried it the Triumph engineers were so taken with the idea that they kept it on. Apart from looking interesting and being very reassuring to the occupants, it also does sterling service as a body stiffener, restoring to the basically floppy open car some of the structural rigidity of the saloons. The hoop is made as an extension of the door pillars, with a central tee-piece linking it to the top rail of the windscreen; with its aid the car has a torsional stiffness exceeding 5000lb ft/deg. The soft hood fits over the top of it; so can the hard top, and somebody has had the bright idea of printing the instructions for deploying and dismantling the hood on the upper surface of the driver's sun visor.

There are other good ideas in the interior, though not all of them have been well carried out: release knobs for the backs of the front seats are very difficult to grasp if the seats are well back, though if anyone were wanting to sit in the rear occasionals the front seats would have to be moved forward anyway, so little space is there behind. On the other hand there is plenty of scope for steering-wheel adjustment, two inches vertically and four axially being at the disposal of a lever that can be unclamped and reclamped while driving. The steering itself is powered as in the 2.5 PI, with the same rather big wheel and the same rather slow gearing. The result is that the steering feels too light and dead, the four turns from lock to lock being at least one and perhaps 1.5 turns more than are wanted. The steering does not feel very accurate, either; in fact my main criticism of the car's handling is that it lacks precision. There is no sensation of lost motion, nor much sensation of anything else, but the overall effect is as if there were quanitities of rubber interposed between the steering wheel and the road wheels. The rubber mountings for the rear suspension subframe may have something to do with it. Whether they have or not, the car steers in a rubbery sort of way: in the type of fastish bend that one swings through at 60 or 70mph it does not feel possible to place the car precisely, and it may drift as much as a yard from the chosen line. On the other hand there is seldom any feeling of losing it, for the mild understeer is fairly consistent. Only once or twice did I have the suspicion that the tail might be getting a little wayward as roll angles increased or as power was poured on in a corner—and although Triumph speak of low-friction splines in the drive-shafts, I suspect that it was probably binding of these that produced slight lurching when trying a bit hard.

As I said, there was precious little opportunity for this sort of thing. Being in Belgium, we had plenty of humpy pavé that the Stag took very well at 60, and plenty of good modern roads where 110 was quite in order—though the cross-wind stability of the car (in open form, at least) is not all that it might be. Again it might be roll steering, for the suspension is quite soft, as is the whole car. It is emphatically not a sports car, but rather a good-looking sporting tourer, in the same vein as the big Peugeot coupé and convertible. It costs a mere £2000, so it should sell very well indeed, possibly taking some business from BMW and Reliant but generally creating a new demand. As the first car from Triumph with Liverpudlian parentage it augurs well for Speke's future. ●

Delightful drawing at left, done by an apprentice in Triumph's engineering department, shows salient points of the fourth but by no means the last modern V8 engine to emerge from the Leyland group in recent years. Michelotti-styled body was apparently conceived some time ago (as was the engine) but has been updated with the aid of recognition features borrowed from current 2000 saloons. There are many echoes of Pininfarina practice, exemplified by the Fiat Dino Spyder and the Peugeot 504 coupé and convertible, but the Triumph's clever price/power combination keeps it well apart from existing competition

# AUTOTEST

## TRIUMPH STAG
## (2,997 c.c.)

**AT-A-GLANCE:** New 2+2 sports car with vee-8 engine. Performance disappointing but above-average fuel consumption. Soft, well-damped ride and light, power-assisted steering. Good brakes. Fast, relaxed car on long journeys.

---

### MANUFACTURER

Standard-Triumph International Ltd, Coventry, England.

### PRICES (Hard-plus-soft top version)

| | | | |
|---|---|---|---|
| Basic | £1,602 | 0s | 0d |
| Purchase Tax | £491 | 15s | 10d |
| Seat belts (approx.) | £14 | 0s | 0d |
| Total (in G.B.) | £2,107 | 15s | 10d |

### EXTRAS (inc. P.T.)

| | | | |
|---|---|---|---|
| Automatic transmission | £104 | 8s | 11d |
| Overdrive * | £65 | 5s | 7d |

*Fitted to test car

**PRICE AS TESTED** .......£2,173 1s 5d

### PERFORMANCE SUMMARY

| | |
|---|---|
| Mean maximum speed | 115 mph |
| Standing start ¼-mile | 18.2 sec |
| 0-60 mph | 11.6 sec |
| 30-70 mph through gears | 11.2 sec |
| Typical fuel consumption | 22 mpg |
| Miles per tankful | 310 |

---

TRIUMPH obviously looked hard at their own line up as well as those of their competitors before deciding to design a new two-plus-two sporting car. Their intention was to provide a British model to compete with such Continental names as Alfa Romeo and Mercedes-Benz. What started out as a £1,500 project due for release in 1968 has turned out to be a £2,000 car introduced only this summer.

For those who missed our description and cutaway drawing published on 11 June, the Stag is a new model based on some of the Triumph 2000 Mk. II running gear and powered by a new 3-litre overhead camshaft vee-8 engine. The power unit in fact utilizes some of the tooling set up for the Saab 99 engine, sharing much of the top-end design. It is made mainly from iron castings and develops 145 bhp (net) at 5,500 rpm.

This is not a great deal more than that of the 2.5 PI engine used in the TR6 (142 bhp net), but one assumes that there is a lot of potential in it which is being saved for future years. It develops more torque than the TR6 engine though (170 instead of 149 lb. ft.) and is inherently a smoother and less temperamental unit.

Starting is always easy and there is a manual choke control for the twin side-draught Stromberg carburettors. It was seldom needed during the warm weather of the test, except for the first cold start of the day. Carburation on the test car was smooth, although we noticed traces of fuel starvation when making full throttle acceleration runs at MIRA. This may have accounted for the disappointing figures, which were not as good as those claimed by the manufacturers.

From a standing start 60 mph came up in 11.6sec and 100 mph in 36.9sec. These times are really no better than those measured on the 2.5 PI saloon we tested earlier this year and are 2.1sec and 7.9sec respectively more than those claimed by Triumph for the Stag. It should not be overlooked, however, that the Stag weighs about 2½cwt more than the 2.5 PI and is much higher geared, especially in bottom.

Overall gearing on the Stag is about the same as that of the 2.5 PI, a difference in final drive ratio offsetting slightly larger wheels. First gear ratio is 3.02 instead of 3.27 to 1 which permits it to run up to over 40 mph before the rev counter touches the red line at 6,500 rpm. There is an amber warning sector from 5,500 (peak power speed) to the red line. We found it best to change gear at about 6,000 rpm.

In second it is possible to reach an indicated 60 mph (true 58.5 mph) and in third over 90 mph. Unlike on other Triumphs, the optional overdrive (fitted to the test car and listed as a £65 extra) fills the gap between third and top, giving a useful maximum of about 100 mph and a theoretical one of 112 mph.

We found that the conditions affected top speed considerably, as did the trim of the car at

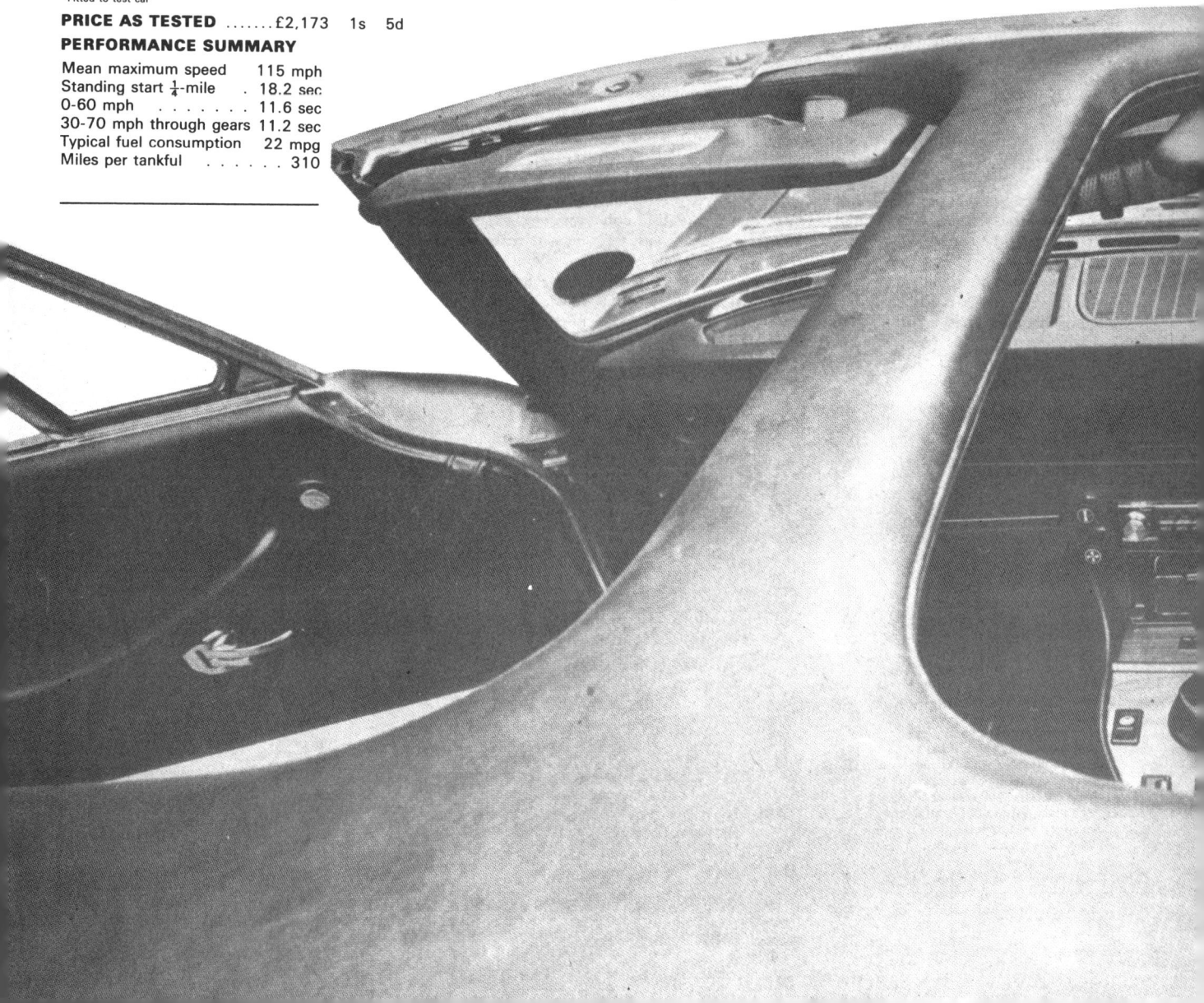

the time. In a straight line on a French *autoroute* with quite a brisk cross wind we recorded a mean maximum in overdrive top of 115 mph with the hardtop fitted. In direct top this speed fell to 113 mph and with the roof off to only 106 mph. Overdrive top is a very high gear indeed (24.2 mph) meant mainly for easy cruising, and the Stag is an effortless car to keep above the ton, with the rev counter reading no more than 4,500 rpm most of the time.

At our present legal motorway limit of 70 mph, only 2,900 rpm is required, which is below even the peak of the torque curve. At these high speeds there is a fair amount of wind roar with the hardtop in place or the hood erect, and with the roof down the roll-over frame causes a very audible whistle. In the main, however, the Stag is a quiet car, the engine and exhaust being particularly well silenced.

Our test car was a pre-production model and its gearbox appeared to have suffered from hard use. Second and top synchromesh were weak and the lever chattered noisily when accelerating in third gear. The clutch, too, on our car was heavy to operate and the action was not smooth. These points are probably not typical and we do not expect them to crop up on the cars being sold to the public.

Anyone coming to the Stag and expecting it to be a taut little sports car like the TR6 will be disappointed. It is much more a touring car and rides and handles in the same way as the Mark II saloons with which it shares front and rear suspensions.

Initially therefore the ride feels soft and the standard power-assisted steering needs learning. With longer acquaintance one begins to appreciate that the soft springs are well damped and that they soak up large and small bumps remarkably easily. Very little road noise is transmitted through from the wheels, which are shod with Michelin XAS radial-ply tyres as standard.

Roll angles are well controlled and once the lack of feel in the steering has become familiar the Stag can be driven very fast on dry roads. With power-on, cornering is stable, initial understeer being largely disguised by the power steering and any tail-out transition being finally killed by the inside rear wheel lifting and spinning. Cutting the power in a bend causes a progressive .and safe tucking-in effect at the front.

In the wet we experienced severe traction problems around country lanes as well as on greasy London streets, the tail sometimes letting go under acceleration without warning. In France also the front end slid out on very wet bends, making fast driving in these conditions a real test of skill and courage. At 100 mph the wipers lifted off the screen and above 110 mph on a badly drained *autoroute* the tyres could be felt aquaplaning.

Under better conditions the Stag is a fine touring car with a long-legged character which eats up miles very easily. On a run to the north-east and back we recorded 22.4 mpg and put 50 miles into every hour without effort.

For a 3-litre car of this weight the fuel consumption overall of 20.6 mpg is not at all

*Michelotti's styling is elegant and well in tune with the rest of the Triumph range. The wheeltrims are dummies. The roll over frame (below) contributes a lot to scuttle stiffness and also prevents collapse of the screen, should the car turn over in an accident*

# ACCELERATION

SECONDS

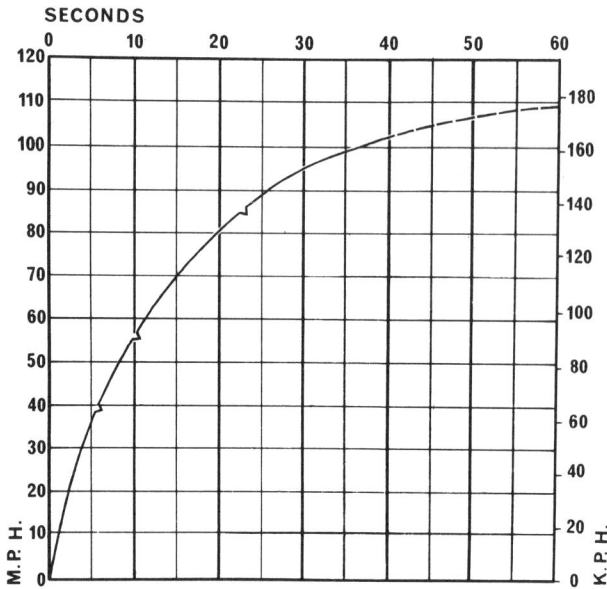

| SPEED MPH TRUE INDICATED | TIME IN SECS |
|---|---|
| **30** | 3.9 |
| 30 | |
| **40** | 5.8 |
| 40 | |
| **50** | 8.1 |
| 50 | |
| **60** | 11.6 |
| 61 | |
| **70** | 15.1 |
| 73 | |
| **80** | 19.6 |
| 85 | |
| **90** | 25.7 |
| 97 | |
| **100** | 36.9 |
| 108 | |

## SPEED RANGE, GEAR RATIOS AND TIME IN SECONDS

| mph | O.D. Top (3.04) | Top (3.70) | O.D. 3rd (4.20) | 3rd (5.13) | 2nd (7.77) | 1st (11.08) |
|---|---|---|---|---|---|---|
| 10-30 | — | 9.1 | 8.2 | 6.4 | 4.1 | 3.0 |
| 20-40 | 10.8 | 8.2 | 7.1 | 5.5 | 3.5 | 3.1 |
| 30-50 | 11.1 | 8.4 | 7.0 | 5.4 | 3.8 | — |
| 40-60 | 11.3 | 8.4 | 7.1 | 5.6 | 4.9 | — |
| 50-70 | 11.6 | 8.8 | 7.8 | 6.7 | — | — |
| 60-80 | 13.2 | 9.9 | 9.1 | 8.5 | — | — |
| 70-90 | 16.1 | 12.6 | 11.9 | 13.7 | — | — |
| 80 100 | 21.9 | 18.0 | 17.9 | — | — | — |

**Standing ¼-mile**
18.2 sec 75 mph

**Standing kilometre**
33.4 sec 98 mph
Test distance
1,762 miles
Mileage recorder
1.0 per cent
over-reading

# PERFORMANCE
## MAXIMUM SPEEDS

| Gear | mph | kph | rpm |
|---|---|---|---|
| O.D. Top (mean) | 115 | 185 | 4,770 |
| (best) | 117 | 188 | 4,850 |
| Top | 113 | 182 | 5,710 |
| O.D. 3rd | 100 | 161 | 5,790 |
| 3rd | 92 | 148 | 6,500 |
| 2nd | 61 | 98 | 6,500 |
| 1st | 42 | 68 | 6,500 |

# BRAKES
**(from 70 mph in neutral)**
**Pedal load for 0.5g stops in lb**

| | | | |
|---|---|---|---|
| 1 | 45-35 | 6 | 30 |
| 2 | 45-40 | 7 | 30-35 |
| 3 | 40-35 | 8 | 35-40 |
| 4 | 35 | 9 | 35-45 |
| 5 | 32 | 10 | 35-45 |

Retardation measured with Bowmonk decelerometer

## RESPONSE (from 30 mph in neutral)

| Load | g | Distance |
|---|---|---|
| 20lb | 0.24 | 125ft |
| 40lb | 0.53 | 57ft |
| 60lb | 0.77 | 39ft |
| 80lb | 0.93 | 32ft |
| 100lb | 0.95 | 31ft |
| 120lb | 1.0 | 30.1ft |
| Handbrake | 0.32 | 94ft |

Max. Gradient 1 in 5.

# CLUTCH
Pedal 40lb and 4¾in.

## MOTORWAY CRUISING
| | |
|---|---|
| Indicated speed at 70 mph | 73 mph |
| Engine (rpm at 70 mph) | 2,910 rpm |
| (mean piston speed) | 1,230 ft/min. |
| Fuel (mpg at 70 mph) | 26.1 mpg |
| Passing (50-70 mph) | 8.8 sec |

# COMPARISONS

### MAXIMUM SPEED MPH
| | | |
|---|---|---|
| Porsche 911 T | (£3,671) | 129 |
| Reliant Scimitar GTE | (£2,019) | 117 |
| Alfa Romeo 1750 GTV | (£2,431) | 116 |
| **Triumph Stag** | **(£2,042)** | **115** |
| Ford Capri 3000 GT | (£1,422) | 113 |

### 0-60 MPH, SEC
| | |
|---|---|
| Porsche 911 T | 8.1 |
| Ford Capri 3000 GT | 10.3 |
| Reliant Scimitar GTE | 10.7 |
| Alfa Romeo 1750 GTV | 11.2 |
| **Triumph Stag** | **11.6** |

### STANDING ¼-MILE, SEC
| | |
|---|---|
| Porsche 911 T | 16.0 |
| Reliant Scimitar GTE | 17.4 |
| Ford Capri 3000 GT | 17.6 |
| Alfa Romeo 1750 GTV | 18.0 |
| **Triumph Stag** | **18.2** |

### OVERALL MPG
| | |
|---|---|
| Alfa Romeo 1750 GTV | 23.9 |
| **Triumph Stag** | **20.6** |
| Ford Capri 3000 GT | 19.3 |
| Reliant Scimitar GTE | 18.5 |
| Porsche 911 T | 17.9 |

### GEARING (with 185-14in. tyres)
| | |
|---|---|
| O.D. Top | 24.1 mph per 1,000 rpm |
| Top | 19.8 mph per 1,000 rpm |
| O.D. 3rd | 17.3 mph per 1,000 rpm |
| 3rd | 14.2 mph per 1,000 rpm |
| 2nd | 9.3 mph per 1,000 rpm |
| 1st | 6.5 mph per 1,000 rpm |

**TEST CONDITIONS:**
Weather: Overcast. Wind: 10-15 mph. Temperature: 13 deg. C. (56 deg. F). Barometer: 29.35 in. hg. Humidity: 50 per cent. Surfaces: Dry concrete and asphalt.

**WEIGHT:**
Kerb weight: 25.1 cwt (2,805lb—1,273kg) (with oil, water and half full fuel tank). Distribution, per cent F. 55.7: R. 44.3. Laden as tested: 28.4 cwt (3,185lb—1,445kg).

**TURNING CIRCLES:**
Between kerbs L. 33ft 10in.: R. 34ft 3in. Between Walls L. 36ft 1in.; R. 36ft 6in., steering wheel turns, lock to lock 2.8.

Figures taken at unknown mileage by our own staff at the Motor Industry Research Association proving ground at Nuneaton and on the Continent.

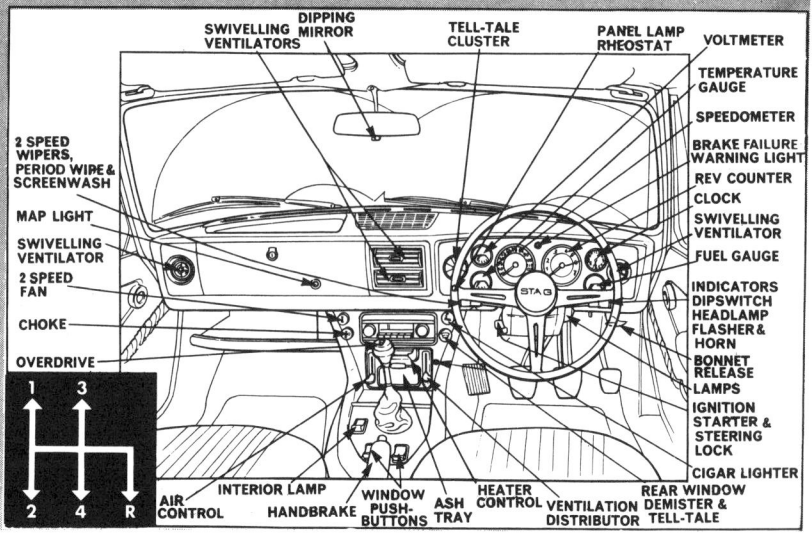

# CONSUMPTION

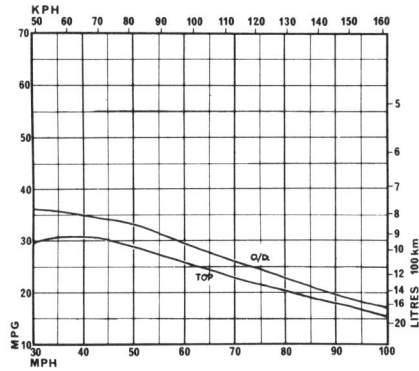

## FUEL
### (At constant speeds—mpg)

| | | Direct | Overdrive |
|---|---|---|---|
| 30 mph | . . . . . . . . . . | 29.8 | 36.0 |
| 40 mph | . . . . . . . . . . | 30.5 | 34.8 |
| 50 mph | . . . . . . . . . . | 28.8 | 33.3 |
| 60 mph | . . . . . . . . . . | 25.8 | 29.4 |
| 70 mph | . . . . . . . . . . | 23.0 | 26.1 |
| 80 mph | . . . . . . . . . . | 20.3 | 22.9 |
| 90 mph | . . . . . . . . . . | 17.8 | 19.8 |
| 100 mph | . . . . . . . . . . | 15.6 | 17.0 |

**Typical mpg** . . . . 22 (12.8 litres/100km)
Calculated (DIN) mpg  23.7 (11.9 litres/100km)
Overall mpg . . . 20.6 (13.7 litres/100km)
Grade of fuel . . Premium, 4-star (min. 97RM)

## OIL
Miles per pint (SAE 10W/40) . . . . . 1,500

# SPECIFICATION FRONT ENGINE, REAR WHEEL DRIVE

## ENGINE
| | |
|---|---|
| Cylinders . . . | 8, in 90-deg vee |
| Main bearings . | 5 |
| Cooling system . | Water; pump, fan and thermostat |
| Bore . . . . | 86.0 mm (3.39 in.) |
| Stroke . . . . | 64.5 mm (2.54 in.) |
| Displacement. . | 2,997 c.c. (182.9 cu.in.) |
| Valve gear . . . | Single overhead camshaft per bank |
| Compression ratio | 8.8-to-1. Min. octane rating: 97RM |
| Carburettors . . | 2 Zenith-Stromberg 1.75CD |
| Fuel pump . . . | SU electric |
| Oil filter . . . . | Full flow, renewable element |
| Max. power . . | 145 bhp (net) at 5,500 rpm |
| Max. torque . . | 170 lb.ft. (net) at 3,500 rpm |
| Max. bmep. . . | 140 psi at 3,500 rpm |

## TRANSMISSION
| | |
|---|---|
| Clutch . . . . | Laycock diaphragm spring, 9.0 in. dia. |
| Gearbox . . . . | 4-speed, all-synchromesh |
| Gear ratios . . | Top 1.0 OD Top 0.82 |
| | Third 1.386 OD Third 1.135 |
| | Second 2.100 |
| | First 2.995 |
| | Reverse 3.369 |
| Final drive . . . | Hypoid bevel, ratio 3.7-to-1 |

## CHASSIS and BODY
| | |
|---|---|
| Construction . . | Integral, with steel body |

## SUSPENSION
| | |
|---|---|
| Front . . . . . | Independent, MacPherson struts, lower, links, coil springs, telescopic dampers, anti-roll bar |
| Rear . . . . . | Independent, semi-trailing arms, coil springs, telescopic dampers |

## STEERING
| | |
|---|---|
| Type . . . . . | Power-assisted rack and pinion |
| Wheel dia. . . . | 15¾ in. |

## BRAKES
| | |
|---|---|
| Make and type . | Lockheed disc front, drum rear |
| Servo . . . . . | Lockheed vacuum |
| Dimensions . . | F 10.6 in. dia. R 9.0 in. dia. 2.25in. wide shoes |
| Swept area. . . | F 220 sq. in., R 127 sq. in. Total 347 sq. in. (245 sq. in./ton laden) |

## WHEELS
| | |
|---|---|
| Type . . . . . | Pressed steel disc, 4-stud fixing, 5.5in. wide rim |
| Tyres—make . . | Michelin |
| —type . . | XAS |
| —size . . | 815-14in. |

## EQUIPMENT
| | |
|---|---|
| Battery . . . . | 12 Volt 56 Ah |
| Alternator . . . | Lucas IIAC, 45 amp a.c. |
| Headlamps. . . | Lucas 4-lamp tungsten-halogen, 110/220 watt (total) |
| Reversing lamp . | Standard |
| Electric fuses . . | 8 |
| Screen wipers . | Two-speed |
| Screen washer . | Standard, electric |
| Interior heater . | Standard, air-mixing type |
| Heated backlight | Standard with hardtop |
| Safety belts . . | Extra, mounting points standard |
| Interior trim . . | Pvc seats and headlining |
| Floor covering . | Carpet |
| Jack . . . . . | Scissor type |
| Jacking points . | 2 each side under body |
| Windscreen . . | Toughened |
| Underbody protection . | Phosphate treatment prior to painting |

## MAINTENANCE
| | |
|---|---|
| Fuel tank . . . | 14 Imp. gallons (64 litres) |
| Cooling system . | 18.5 pints (including heater) |
| Engine sump . . | 8 pints (4.5 litres) SAE 10W/40. Change oil every 6,000 miles. Change filter element every 12,000 miles |
| Gearbox and overdrive . . | 3.75 pints SAE 90EP. Change oil every 6,000 miles |
| Final drive . . . | 2 pints SAE 90EP. Change oil every 6,000 miles |
| Grease . . . . | No points |
| Tyre pressures . | F 26; R 26 psi (normal driving); F 26; R 30 psi (full load) |
| Max. payload . . | 728 lb (330 kg) |

## PERFORMANCE DATA
| | |
|---|---|
| Top gear mph per 1,000 rpm . . . . . | 19.8 |
| Overdrive top mph per 1,000 rpm . . . | 24.1 |
| Mean piston speed at max. power . . . | 2,300 ft/min. |
| Bhp per ton laden . . . . . . . . . | 102 |

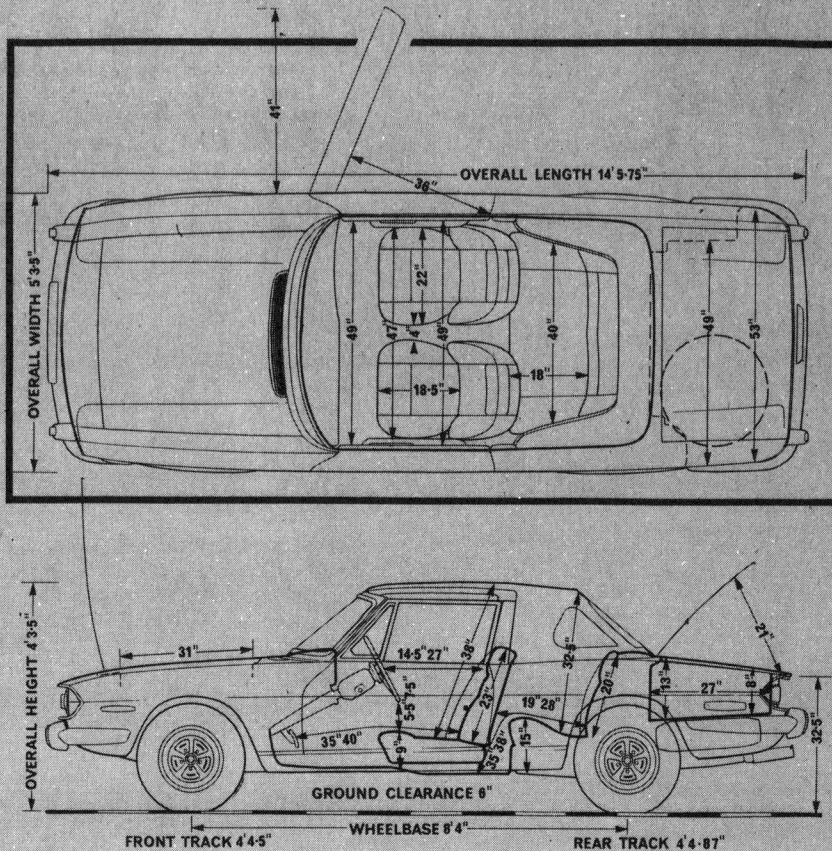

STANDARD GARAGE 16ft x 8ft 6in.

OVERALL LENGTH 14' 5·75"

OVERALL WIDTH 5'3·5"

OVERALL HEIGHT 4'3·5"

GROUND CLEARANCE 6"

SCALE 0.3in. to 1ft
Cushions uncompressed

FRONT TRACK 4'4·5"      WHEELBASE 8'4"      REAR TRACK 4'4·87"

bad and even when squirting the Stag about in heavy traffic we never got less than 18 mpg. At a steady 70 mph in overdrive top it will cover 26 miles on a gallon of 4-star premium, and at the same speed in direct it covers 23 miles.

Disc front brakes with servo assistance and drums at the rear worked efficiently and showed no real signs of fade either on the road or during our 10 stops in rapid succession at MIRA. Pedal effort for most check braking is reasonable, but it took a hefty 120lb shove to get the ultimate 1g stop. A large red warning light right in front of the driver warns of any fluid loss or any pressure difference between the divided front and rear hydraulic circuits.

The driving position is easy to tailor, with a neat tilt and telescope mechanism for the steering column and seatbacks adjustable for rake. There is enough legroom for a 6ft driver but we would have preferred a wheel smaller than the 15¾in. dia. one fitted. It has a nicely stitched leather rim and satin finished spokes.

In the back there is room for two adults if those in the front compromise on legroom, but it is meant for occasional use or for children. With the roof off, backseat passengers get an odd view of the roll-over frame, although it can be detached if required. We did not try taking it off because it must contribute considerably to the body stiffness.

The hardtop attachment and complete hood mechanism is a very fair copy of that used on the Mercedes-Benz 280 SL two-seater. A hinged metal tonneau panel hides the folded roof completely out of sight and putting it up is a quick and simple operation. Removing the hardtop is a two-man job as it is heavy and calls for care in avoiding the padded pvc of the roll-over frame where it rakes back behind the doors. The central rear pin is an electric connection for the standard heated backlight.

Two-speed wipers are standard and they are operated together with the powerful electric washers by a stalk on the left of the steering column. The wiper on the driver's side has a parallelogram action which drastically improves the area it sweeps. Most of the other controls are like those of the 2000, with similar rotary lighting switch, the same (ex 1300) "all systems go" combination warning dial and the same tasteful matt veneer woodwork. A battery condition voltmeter is fitted instead of an ammeter but there is not, surprisingly, an oil pressure gauge. The overdrive switch is in the top of the gearlever knob, surely the most logical place of all.

One key works the ignition and steering lock, together with both doors and the fuel filler flap. A second key is needed for the boot and yet a third key for the glove box. The boot is lined with carpet and takes a good volume of luggage although it is not very deep. Quartz iodine headlamp bulbs give bright illumination but their small diameter restricts their performance overall.

Heater controls are the same as on the 2000, with an independent cold air ventilation system and a two-speed fan which boosts both systems. Temperature control was sensitive but the swivelling eyeball nozzles at the ends of the facia did not pass enough air to be really effective in hot weather and with the roof on we noticed rather too much engine heat permeating through the bulkhead and floor for comfort in the cockpit.

The Stag is one of those cars which you appreciate the more you drive it. It has an easy and relaxing way of packing many miles into each hour and it is a satisfying and spirited car to drive fast. For a touring car it proves comfortable and we climbed out after 15 hours of travelling, tired but not fatigued and in no way stiff. Our test mileage was longer than usual because we took it abroad for maximum speed measurements and because we all enjoyed driving it. We liked it so much in fact that we shall be adding one to our long term test fleet as soon as we can get delivery. □

*Top: The soft top folds away completely out of sight under the padded metal tonneau panel. Instructions for hood stowage are on the back of the sun visor*

*Below left: Accessibility on the ohc vee-8 engine is good, with all the fillers nice and high. The battery is out of sight behind the right-hand headlamps*

*Below right: The rear seat room is generous for a sports car, but getting in and out is as tricky as ever*

*Above left: The spare wheel lies under the boot floor, which has been lifted out here together with the carpet. Above right: There is room for a reasonable amount of luggage and the lid is self-supporting*

*Below: The optional hardtop suits the body shape well. It costs £98 extra including a heated backlight and opening rear quarterlights*

# YEAR OF THE STAG

THE STAG SEASON IN THE UNITED STATES does not open until September, and it will be a skilled hunter who reckons to have a trophy in his garage before October. In order to do so he'll have to fire quite an impressive piece of armory at a Triumph dealer — like a check for about $5000.

Triumph (meaning British Leyland's Specialist-Car Division) is entering a completely new market with the Stag. In fact, it is hoping to *create* a market, for there is nothing quite like this 2+2 in the range of any rival manufacturer. It is a car which Triumph intends to sell at a minimum rate of 10,000 per year, but which it hopes will attract an annual demand of at least 15,000.

The Stag has a lot going for it — an interesting water-cooled, 3-liter V-8 engine, four-speed and overdrive (or three-speed automatic) transmission, independent suspension all around, three ingenious alternative roof specifications, optional Delaney Gallay air conditioning, electric window lifts and useful passenger space (though at the cost of some trunk capacity). It in no way replaces the TR series, which continues alongside the Stag, but it could well become a logical TR-replacement for the sports car buyer who did his courting in the TR and now finds himself with a wife and kids. It's also priced sufficiently below an XKE 2+2 not to worry one other member of the British Leyland Specialist-Car Division.

Like all new cars the Stag took a long time to gestate, and its final design wasn't conceived, it evolved. It started out to answer a demand (mainly from North America) for a coupe version of the Triumph 2000 sedan. In this form it would have had the same running gear (meaning the six-cylinder, in-line engine) and would

have been mainly a body change, for which the customer would have been expected to pay a few-hundred-dollar premium in return for a sexier line and less passenger space.

Then Triumph's customers started to ask for extras, like alternative hard and soft tops, and gradually the new car became visualized as something along the lines of the Mercedes-Benz 230SL, only with a lower price tag.

Coincident with these changes, Triumph was developing a new engine series to be produced initially in in-line, four-cylinder form, as a replacement for the 1300, and later as a V-8. At that stage no car had been earmarked for the V-8, although it was assumed that one day it might be dropped into the sedan.

Saab then came on the scene with a demand for a four-cylinder engine for its own new car, the 99, which was both good and bad news. Triumph was happy that its new four-cylinder engine was able to meet the performance requirements laid down by the Swedes, namely to achieve equal or better overall results than any other available engine, and expressly to give trouble-free starting down to temperatures of minus 30 degrees centigrade. But it meant redesigning the engine, and turning it back-to-front so that the flywheel was moved from the back of the block, as originally designed by Triumph, to the front. Also, the dimensional limitations imposed by Saab meant completely revamping the cooling system and moving the water pump from its conventional location at the front of the block to a position alongside the angled cylinders, where it could be gear-driven from a jackshaft above the crankshaft. In this way, valuable inches were carved off the length of the engine.

The engine had originally been conceived as a 1500, with a bore of 80 mm (3.15 in.) and a stroke of 76 mm (2.99 in.), but for a relatively heavy car like the Saab a larger displacement was clearly needed, and so the bore was increased to 83.5 mm (3.28 in.) and the stroke to 78 mm (3.07 in.) to give a figure of 1708 cc.

The encouraging performance of the new four-cylinder engines, particularly as regards long life, encouraged Triumph to think again about the power unit for the forthcoming Stag, and to speed up the development of a 90-degree V-8 version. Initially, the engine was built with the 80 mm (3.15 in.) bore and a short stroke of 64.2 mm (2.52 in.), to give a 2½-liter displacement, with the idea of bringing in a 3-liter version in the 1970s. (The Stag was originally scheduled to appear more than a year ago, but was delayed as a result of the big BLMC reshuffle.) However, when the Stag was given final approval for production, it was as a full 3-liter, an increase in cylinder bore giving a displacement of 2997 cc. Clearly, there is room for further increases in the future by lengthening the stroke.

In its production form the 3-liter V-8 delivers 145 bhp at 5500 rpm, and 170 lbs-ft of torque at 3500 rpm. American cars will have the Zenith Stromberg CDSE carbs to meet the emission regulations, along with an exhaust-heated inlet manifold, the temperature and flow being regulated by a thermostatically controlled valve.

The cylinder block is cast in chrome-iron and carries an alloy-steel forged crankshaft with its five main bearings and its big-end bearings in steelbacked lead-bronze with a lead-indium overlay. The eight aluminum alloy pistons are controlled by forged-steel con rods with fully floating gudgeon pins. Triumph design engineers were working to quite a strict weight target for this new engine series, and chose aluminum alloy cylinder heads from the start. Also, they were under instructions to design engines which would be as simple as possible to assemble on the production line. To meet this requirement they decided to do away with a separate camshaft carrier, and to cast the carrier integrally with the head, so that the head and valve gear could be assembled separately in one unit, then be bolted onto the block. In order to achieve this they had to experiment with angled head bolts, and it was some time before they found a layout of vertical and angled bolts which

IN 1962, TRIUMPH FAWNED THE TR4 . . . BUT IT'S 1970 NOW AND DEER SEASON IS OPENING / BY JOHN BLUNSDEN

prevented lateral movement of the head assembly at the block face. However, all is now well, and of course the engines show similar advantages in the event of an overhaul in that with the aid of a special key at the front end of the head assembly, which releases the head from the top timing sprocket and chain (which retains its tension), the head assembly can be removed, undisturbed, in a few minutes.

The camshafts of the V-8 engine run in five main bearings and operate in-line valves with chrome-plated stems through inverted bucket-type tappets. The valves are recessed into wedge-shaped combustion chambers and are timed at 16, 56, 56 and 16 degrees. The camshafts are chain-driven, and there is a jackshaft which, in addition to driving the water impeller through skew gears near the front end, drives a four-lobe, rotor-type oil pump through a similar pair of gears toward the rear of the shaft. The cooling system is served by a cross-flow radiator and a 16½-inch diameter 13-blade fan with a viscous coupling, the no-loss system being designed to operate at 13 psi.

The twin Zenith/Stromberg 175 CDSE sidedraft carbs are mounted on a four-branch cast-aluminum-alloy manifold between the cylinder banks, and have combined air cleaners and silencers with replaceable paper elements. The exhaust manifolds are three-branch iron castings.

A relatively lightweight, cast-iron flywheel with hardened-steel starter ring gear is matched to a 9-inch diameter Laycock single dry plate, diaphragm-spring hydraulic clutch. There is a choice of transmissions. The four-speed, all-synchro gearbox has ratios of 1.00, 1.39, 2.10 and 2.99 to 1 with a 3.37 to 1 reverse, which gives overall transmission ratios of 3.70, 5.13, 7.77, 11.08 and 12.47 to 1 with the standard axle. When the optional overdrive (by Laycock) is added, giving a step-up of 0.82 to 1, overdrive third becomes 4.20 and overdrive top becomes 3.04 to 1.

The third alternative is the Borg Warner Type 35 transmission giving three ranges: 1.00/2.30, 1.45/3.34 and 2.39/5.50 in the gearbox, or 3.70/8.50, 5.37/12.37, and 8.85/20.40 overall. An oil-cooler kit is standard with the automatic transmission option. In every case the prop shaft has U-joints with needle-roller bearings, and the iron casting for the hypoid-bevel final drive forms part of the rear subframe, and is insulated from the body by rubber mountings.

The Stag transmission is very similar to that of the Triumph 2000 and 2.5 series, but uprated where necessary. The suspension layout is also almost identical to that of the 70-series sedans,

which were given increased rear track last fall, which resulted in a notable improvement to the cars' already above-average handling qualities and ride.

The front suspension is by a strut system with coil springs and telescopic shock absorbers, single lower transverse links, leading radius rods to maintain fore-and-aft location, and an anti-sway bar. The springs are mounted on noise-insulating washers, the pivots are rubber-bushed, and the swivels are ball-jointed. The independent rear suspension is a semi-trailing-arm system based on the insulated subframe, again with noise-insulated coil springs and telescopic shock absorbers. United States-bound Stags are to be equipped with chrome-plated wire wheels as standard, whereas other markets are accepting pressed steel wheels with safety-ledge rims. The wire wheels have 5½J flat rims.

Triumph introduced a new power steering system for its sedans last fall, and this has been adopted for the Stag. It provides for four turns of the 16-inch, three-spoke wheel from lock to lock and a turn circle of 34 feet. The column is axially adjustable by 4 inches, and vertically by 2 inches.

The brakes have a divided system through a tandem master cylinder, operating on 10⅝-inch diameter front discs and 9 x 2¼-inch rear drums, with a 3-to-1 direct-acting servo. This gives a total frictional area of 347 square inches. Tire equipment specified for U.S.-bound cars is Goodyear G800s in the 185/70 HR x 14 size.

One of the interesting features of the unit construction body is the integral rollover hoop and roof spine, which remains in place regardless of which of the three top alternatives are chosen. The cheapest way to buy a Stag is to choose a soft top (this stows away neatly beneath a flush-fitting panel behind the rear seat). The next alternative is to select a hardtop with the idea of having this permanently in place. The third and most expensive alternative is to choose the hardtop, but with the soft top tucked away in its compartment, ready for use should it be decided to unbolt the hardtop and leave it at home. The soft top incorporates a partially removable rear window (it undoes on three sides) and the hardtop has a heated rear window as well as hinged quarter-windows.

The interior of the Stag is trimmed in expanded vinyl, with basket-weave finish on the seat facings. The headlining of the hardtop is washable. Each of the individual front seats is adjustable for height and rake as well as fore-and-aft movement. A bench seat is provided at the rear. The fresh-air heater has a two-speed booster fan, which can be used

to pump air through the intakes in the center and in the corners of the dashboard.

Trunk capacity has suffered somewhat from the need to provide reasonable rear-seat room plus stowage space for the soft top, a fairly wide spare wheel and tire, and a 16.5-gallon gas tank, and there's a meager 9-cubic-feet left over beneath the counterbalanced lid for the proverbial toothbrush. Odds and ends are better catered for in the cockpit itself with a glove locker, a parcels shelf below it, and pockets in each door and in the back of each front seat.

Triumph engineers invariably go to a lot of trouble when announcing a new car to provide a full set of performance figures; figures which have a history of fairness bordering on conservatism. The typical Stag which finds its way into private hands, therefore, should have little difficulty in matching the following acceleration times: 0-30 mph in 3.5 sec; 0-40 mph in 5.0 sec; 0-50 mph in 7.0 sec; 0-60 mph in 9.5 sec; 0-70 mph in 12.5 sec; 0-80 mph in 16.0 sec; 0-90 mph in 21.5 sec and 0-100 mph in 29.0 seconds. Maximum speed should be 118 mph, and the maximums in the indirect gears, using a moderate rev limit of 5500 rpm, are 34, 51 and 75 mph with the four-speed manual box. The gearing gives 19 mph per 1000 rpm in direct top, and 23.1 mph per 1000 rpm in overdrive top.

The Stag weighs in at 2460 lbs dry, or 3020 lbs with all extras, of which the new V-8 engine accounts for 446 lbs including the power steering equipment. Overall length is 173.8 in.; width 63.5 in.; height (soft top) 49.5 in.; height to top of screen 48 in.; wheelbase 100 in.; ground clearance 4 in.; front track (four-up) 52.5 in.; and rear track 52.8 in.

A fully detailed appraisal of the Stag with the aid of SCG's sophisticated test facilities must await the arrival of the new 2+2 in the States, but meanwhile a few hours in northern Belgium with one of the pre-production Stags was a valuable way of forming some preliminary impressions.

For a start, we are more than impressed with the V-8 engine, which is outstandingly smooth throughout the range, and does not feel at all overstressed at 6500 rpm, although the yellow band starts at 5500. Its other high point is its great gobs of middle-range torque, which means that the difference in acceleration times between this car and those quoted for the manual-shift version are minimal. A kick-down or a manual downshift on the B-W selector produces a momentary pause as the engine revs rush up to 5500, but then the transmission really bites and gets the car ac-

celerating strongly. Throttle response generally is very good, and mechanically the engine is extremely quiet. In fact, it seems to us to be an excellent power unit all around.

Turning to the suspension, the Stag is softly sprung to the point where it makes mincemeat of the notorious Belgian pavé. The cost is a certain amount of float, mainly noticeable during an initial change of direction. Cornered hard the Stag will give a slight lurch entering the turn, but then go through very quickly with progressively increasing understeer. Cornering at normal speeds is nearly level, but there is a considerable angle of lean if the car is pressed very hard. Adhesion in the dry on Michelin XAS tires is outstanding, even when the tires are soft enough to generate a lot of squeal.

The power steering still gives adequate front-wheel "feel," and it does not really seem as low geared as four turns lock to lock. Perhaps this is because it is fairly sensitive to steering wheel movement once the initial inch or so of "slack" is taken up. The brakes seemed powerful enough, although when we braked hard on the pavé we found it possible to lock up the left rear wheel.

Mechanical noise was very low from all areas, and wind noise was confined to a small amount of disturbance around the tops of the windows. Inevitably, there was a lot of roar at high speed if either vent pane was opened or the electric windows were partly lowered.

The cockpit is tastefully finished throughout, and the dashboard and control layout ergnomically very good. With fully adjustable front seats and adjustable steering column, there is a wide choice of driving positions. At first the driver's seat felt almost too well-shaped and close-fitting, but this was an illusion and, in fact, the Stag seems to be a great car for long-distance touring. Also, we were pleasantly surprised when we opened the trunk and found that 9 cubic feet can add up to quite a lot of very usable space.

The detail finish on the Stag is of the high quality one has come to expect from Triumph of late, and all in all it seems to be a very worthy addition to their range. The amount of interest shown in the car during the test-run suggests that Michelotti has produced an esthetically pleasing as well as practical shape, and that the car is going to be very well received. We have a feeling that two or three European manufacturers who have had the 2+2 GT market very much to themselves with higher-priced cars, are going to become very conscious of the Stag's existence during the coming year.

*You've seen the Toledo exterior on the cover — this is the Stag, a new two-plus-two with V8 power and luxury appointments. It is placed in Lotus Elan plus-two and E-type two-plus-two class, but sells for much less than these.*

Two new Triumphs for the open road — and Australian markets (eventually) . . .

# TOLEDO and STAG

Sloniger drives our cover car — the Triumph Toledo — plus its hairy-chested V8 big brother . . . the Stag.

THE Triumph (big boss' favorite) division of BLMC really spread itself this European spring, releasing new cars at the top and bottom ends of the price range which could hardly be less alike, though they do share a couple of curious specs.

One is called a **Toledo,** which is hoped to sell en masse. The other leaps into view as the **Stag,** and I would guess it will come closer to fulfilling Triumph hopes.

If you're wondering what a high-price GT and a cheap mid-range family car could have in common — both offer a 34-foot turning circle and the rated ability to carry slightly over 700 lb of people or packages, divided as you might prefer.

Taking U-turn ability first, this is a major disappointment in the Toledo. One of the ways Triumph always scored, alongside better finish and engineering for the class, was in manoeuvrability. It is hard

to like a 97 in. wheelbase machine which looks compact and needs as broad a boulevard as the stretched GT on its 100 in. base.

Triumph makes much of the fact that the Toledo is only 1.1 in. longer overall than the front-drive 1300 line with which it shares the shell. Yet the Toledo is conventional in having an oversquare engine up front driving the rear wheels.

In short, it is almost too conventional.

The point is not how well Triumph fitted normal bits into a one-time advance package (taking a good deal of footroom with the gearbox while at it) but that FWD was abandoned just when competitors in this middle sedan range are going for kinky engineering: the kind overseas markets expect from Triumph.

Oh, it does its 80 mph without fuss, pulls almost as well in fourth gear as third for lazy cruising, swallows four adults and decent luggage with no lip for senior citizens to hoist their cases over . . . in fact, I had few complaints. But equally few raves.

One definite slip is the steering.

There is little or no road feel to a point where you sometimes wonder where the wheels are pointed. Lining up for a fast bend is tricky because you have lost all slip feel or surface feedback.

Certain detail tricks work well — recesses over the rear wheel arches to widen elbow space in back, the proper dials and accurate gearbox action and above all the ride over really rough surfaces which is among the best in class to come out of England.

For once an English maker was well advised to release his new, inexpensive model in Belgium where they breed the worst cobbles known to civilised man.

Performance is not the Toledo's intended bag.

Triumph's Stag, on the other hand, is aimed right into the heart of the performance market. Not at the racers *per se,* but at the boulevarde barons who like a breath of youth, enough power to pass the peasants and the looks to identify their car immediately in a crowded parking lot.

To the firm's credit, Triumph went beyond these sales certainties to give the Stag first-rate brakes and plenty of handling for its 115 mph top, real safety engineering and one of the nicest small V8 engines I've encountered.

To make one thing very plain from the first — this new 3-litre V8 has no, repeat no, connection with the Rover pushrod eight. Triumph people turn puce at the mere thought, pointing out theirs was on the boards before Rover ever joined the team. They were a little less certain of how one big company like BLMC can justify three small V8s (including the Daimler) but considering the head-shed background I'd bet Triumph's will not be the first to feel an economy axe.

Nor should it be. This single-OHC mill is one of those fortunate powerplants which feels balanced, strong and *right* from the very first. It produces 145 bhp at 5500, which also happens to be the orange line but feels able to punch out far more revs any time you take an eye off the tach.

With 170 lb/ft of torque peaking 2000 rpm lower, the engine is more than able to shift a ton and a half of luxury 2 plus 2 without strain. Incidentally, the rather high kerb weight (that's with all options) explains a relatively low load factor of something over 700 lb.

On the other hand, you won't be carrying four overland plus luggage because two of them wouldn't be comfortable in the occasional rear

seats. Like most of this *genre*, the Stag is a two seater which can be pressed into service for a couple more briefly.

The pair in front have fine seat adjustment in all planes and a steering wheel which slides in and out on the column as well as lifting or lowering. Here they slipped, for my shape people at any rate. The answer is a smaller diameter rim to avoid clashing with the legs in lowered positions.

It says a good deal for the car that this was about the only item which seriously caught my negative attention. The pedals are somewhat more offset than I would choose, the transmission hump rather large, and the so-called one-man top took two of us many minutes, despite the clever feature of instructions on the sun visor.

Options include air conditioning — which must be somewhat of a first for volume-produced European cars — and either a Laycock overdrive for the top two ratios of the manual box or a Borg Warner 35 automatic. And a hard top you need a crane to fit or remove. The tin lid does have heated rear window though, plugging in automatically.

As for those box options, the BW fits this engine well enough but it's not precisely secret about full-throttle upshifts (at the 5500 orange line) nor kick-down lag. Also, the car is showing the ton at 5000 in automatic top and can only get

more flurried towards max speed.

With the manual and overdrive you can have 95 in OD/three and 115 in four or OD/four (if you can find a *very* long road) but cruising at the magic 100 is far less fraught. The switch is atop the shift knob.

The manual box could do with shorter lever movements from cog to cog but this is largely masked by the overdrive. Against usual inclinations, I think I'd take four on the tunnel plus overdrive.

The top folds out of sight under a metal panel in back (rather DB SL that) but whichever top you use, the roll cage never leaves.

This is a padded bar across the car behind the front seats with an arm reaching forward to meet the top of the windscreen. Entry or exit to the rear is no trickier than in a two-door coupe and the installation looks firm enough to do some good.

Further safety, if you should flip it, comes from an inertia switch to shut off fuel flow.

Alongside that, electric side windows (the rears pivot outwards too, but manually) are merely nice and classy. After all, while they didn't want to name a price I gather the Stag is going to cost at least $A800-900 more than a hairy chested TR6, so it wants a little plush.

Apart from too much rococo decoration on a basically trim enough form, the styling of the Stag works pretty well.

Triumph really seems to know what Grand Touring is all about. #

*Above: Stag styling is controversial, distinctive. Like sporty TR6, it is available in soft and hard top versions, but includes integrated roll bar on all cars.*

*Above right: Stag interior is built for ideal touring comfort for two up front. Two in back should be there for short hauls only — made quicker by an OHC V8 under the bonnet.*

*Toledo interior is basic and comfortable — prices lower than Triumph 1300, which has expensive FWD. Toledo is rear wheel drive.*

ABOVE: Interior is strongly
reminiscent of Mk. 2 2000
with curved dash, walnut
veneer, and padded wheel.
TOP: Stag in spider form with
Targa-type roll bar in place.
RIGHT: Shapely from all
angles, Stag offers space
for four adults and luggage.

*Triumph enters the GT
lists with a Targa-type
V8 with looks, comfort
— and performance,
reports Dev. Dvoretsky*

"STAG" may not be everyone's idea of a good name for a successful automobile, but this new two-plus-two from the Standard-Triumph division of British Leyland is the most exciting new model from any British factory for many a long day.

Stag is a good-looking, two-door, occasional four seater with room in the back for real people — not legless midgets.

Styling of the monocoque body is by Giovanni Michelotti who has been influencing the designs from Standard-Triumph since the days of the Herald in the late 50s. The heritage shows in the new car. The spider-coupe (and it is both) is **powered by a new 90-degree**

# STAG V8 BY TRIUMPH

short-stroke 3-litre V8 with chain-driven single ohc giving 145 bhp nett at 5,500 rpm.

It is nominally a 120 mph machine, which can get to 60 mph in under 10 secs, has a reasonable thirst, exceptionally long legs, and an all-independent suspension borrowed from the 2.5 PI and TR-6 models. Stag takes over where the Rover-Triumph concept (now more than seven years old) left off, and carries on into the sphere of real grand touring. The Rover/Triumph 2000's were oriented towards the family man with a sporting inclination. The Stag follows the same trend — only moreso.

It's BLMC's first real challenge in the highly competitive luxury grand touring class which has long been the domain of continental makers.

Price of only 1527 Stg. basic for the softtop version, 1562 Stg. for the hardtop or 1602 Stg. for the combined softtop/hardtop. This will make it one of the world's most-wanted cars. With automatic and air-conditioning, practically the only two extras, the price could rise to around 1800/1850 Stg. basic — say around $6000 - $6500 fully equipped in Australia. If it were assembled in New Zealand along with the XJ-6 and Rover 2000 the price could drop a thousand dollars. And at that price it would be a sure-fire winner in Australia.

But don't rush your dealers yet. For the first 12 months production will be restricted to the home market. A production rate of about 15,000 a year will be underway by September.

But even when exports do start, it's likely most will go Stateside.

## Hammering

Already U.S. dealers are hammering on the doors of the new 11-million Stg. factory at Speke just outside Liverpool where the bodies for Stag will be produced and at Coventry where the engines will be made and the car assembled.

It seems likely that something will have to make way for Stag on the now hard-pressed Coventry lines. No one will be surprised to see the end of the 11-year old Herald/Vitesse series to make way for the stylish newcomer.

Though production is bound to rise, I reckon this is going to be the Rover 2000/3500 and Jaguar XJ-6 all over again. (The former still three to six months lag on delivery and the latter 12 months and more in some markets)!

What makes the Stag such a certainty? It's a car that combines the good looks and sleekness of sports car and limousine. It is well appointed, extremely comfortable, has a high top speed and extremely good acceleration that few will use but which means so much at the traffic light G.P. It is easy to get into and out of — which means the not-so-young who'd like to feel they aren't out of it yet and who can afford it, won't hesitate. It's an easy car to drive (it comes equipped with Adwest power steering the same as the XJ-6) and even without automatic the torque is so good gear changing can be left to a minimum.

It's safe and though I wouldn't say the suspension is infallible, it will certainly be tops to most who buy it. Combine all this with the safety of a three-point roll-over bar, the air-conditioned comfort or the open-to-the-wind appeal plus the fact it can carry four at a pinch and you have answered the problem of hundreds of thousands of motorists.

## Engine

The basic design was originally dreamed up by Harry Webster (now BLMC chief engineer) and STI sales chief Lindon Mills four years ago. The design and concept hardly changed from the word go. Originally the idea was to use a vee made from two of the 1700 cc single ohc units built by Triumph for SAAB. But later this was reduced to 2.5 litres. When 'Spen' King was moved from Rover to take over from Harry Webster as STI engineer he immediately upped the

*ENGINE bay is a neat installation, with access to the important ancilliaries an easy matter. V8 is made up of two Triumph fours — developed originally for Saab — siamesed.*

capacity and increased the power output. The result was an engine that simply screamed "torque".

Bore and stroke of the original 1709 four cylinder unit is 83.5 by 78 mm and output 87 at 5500 rpm. The V8 has a bore of 86 and a stroke of 64.5 and an overall capacity of 2997 cc (182.9 cubic inches). Output is 145 bhp nett at 5,500 rpm. Torque is 170lb. ft. at 3500 rpm.

The single overhead camshaft on each bank is chain driven — the two-plane crankshaft has five main bearings as have the camshafts themselves.

Twin side-draught Stromberg carbs are used, though it's a pretty open secret that STI have been playing around with Bosch electronic fuel injection. This is unlikely to be seen for a couple of years, at least.

An alternator is standard equipment and there is a 'no-loss' cooling system with expansion chamber and 13 psi pressure cap. The fan has a viscous coupling which cuts down roar. Power gets down to the rear wheels through a nine inch diaphragm clutch. The four speed all-synchromesh gearbox is similar to that of the Triumph 2000/2.5 PI Mk 2. Overdrive is available on third and top and when fitted with the standard 3.7:1 rear axle gives a road speed at 1000 rpm in overdrive top of 24.1 mph — a real loper.

Standard ratio either with four speed box or Borg Warner overdrive is 19.8 mph per 1000 rpm. Tyres are Michelin 185 HR 14 in radial XAS tubeless on 5J rims.

Suspension is the fully-independent type of the 2.5 PI Mk. 2 with front struts with coil springs controlled by telescopic dampers and an anti-roll bar. At the rear there are semi-trailing arms, mounted on a rubber insulated steel sub-frame, with coil spring/damper units.

The servo assisted brakes — disc front, rear drum — are operated from a dual safety hydraulic system.

The monocoque body looks prettier in the flesh than some of its pictures suggest. The front treatment is very 2.5 PI, with quad quartz halogen headlamps — probably the first production car to be so fitted. The wraparound bumper treatment front and rear and the rear treatment with its gently sloping boot is neat and attractive.

The hard-top which — like the Mercedes 280SL type requires at least two men or a small hoist in the garage to lift — is a solid affair containing not only the two opening rear side vents but an electrically-heated rear screen as standard. The well for the soft top is neatly covered by a solid frame vinyl covered top. Putting the close-fitting soft-top up takes about two minutes. The hard-top is held in position at five points controlled by levers — two at the screen, two behind the roll over bars and one at the rear.

## Rollover

And its the roll-over bars in the Stag that are going to cause a lot of conversation. They are real roll-over bars and anticipate the possible criticism, on the grounds of safety, the sports car may come up against in safety legislation in the next few years. The softly padded triangulated bar is virtually an extension of the door pillars, and offers not only protection in case of roll-over but helps keep the monocoque stiff. It also acts as a sort of grab handle for the passenger if the

*Continued on page 42*

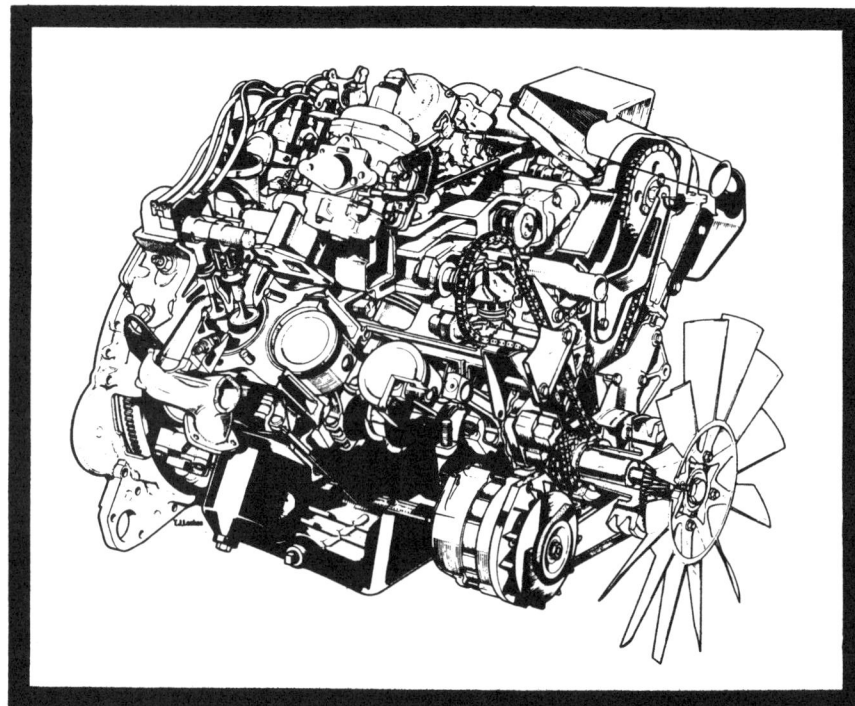

*CUTAWAY of 90 deg. V8 shows chain drive for camshafts, centrally-positioned Zenith carburettors, viscous-coupled cooling fan.*

# STAG V8

*Continued from page* **40**

driver is cornering fast. The framed windows of the wide opening doors are electrically controlled from the centre console. Internally the Stag is well finished, with multi-adjustable seats and aerated PVC coverings. The dashboard with its curved instrument panel taken from the 2.5 PI is well-laid-out and the instruments most easy to read. "Eye ball" socket ventilation on either side of the dashboard and central facial vents are included. The overdrive switch is on the top of the column. The left stalk from the adjustable-for-rake steering column controls the wipers and the (right) headlamps and horn. There's a three-spoke steering wheel with big central boss and leather covered rim.

Overall length is 14ft. 5¾in. and width 5ft 3½ in., height (with soft-top erect) is 4ft 1½in., wheelbase is 8ft 4in. and track front 4ft 4½in. and rear 4ft 4in. Ground clearance fully loaded — four in., turning circle (four turns lock to lock) is 34 ft.

The Stag is full of interesting ideas as you quickly notice once you step inside. A small light lower right of the glove box throws light down on to the floor (or the passengers feet) at night to become a perfect map reading light. 'Puddle' lights come on as the door opens and shine down from the armrests and form door warning lights for oncoming traffic. Stag has rear seat attachments for safety belts as well as the single attachment inertia-type lap/sash which are fitted as standard to the front seats.

The seats are extremely comfortable. They can be adjusted for height and rake and the squabs are also adjustable. The steering wheel has adjustment for rake only but this, combined with the seat adjustment makes for a driving positon to suit even the most fastidious.

With the windows up noise in the coupe even over rough Belgium pave road surfaces (where we took the cars for pre-release tryout) was very low. There's very little back wash turbulence when the hood is down even at 100 mph and the noise when the windows are lowered at high speed with the hard top in position is tolerable. However, for hotter climates — even the south of France — to get the best from this car would be to use the Delaney-Galley air-conditioning system.

## Ride

The ride over rough surfaces is very good, and the bumps are soaked up nicely by the all independent system. It's not the best I've met. You can feel it working on anything but a smooth road and under certain conditions (say braking in a fast corner) you upset the the balance and the tail will come out. The same applies under heavy braking when the front suspension seems to dive considerably. This has probably been brought about through compromise. Despite this, and the stupidity of four turns of lock with the good power-assisted rack and pinion steering, I found the car safe and relaxing.

If the tail does break away during a harsh manoeuvre correction is simply a matter of flicking the wheel. I've yet to try it in the wet.

I liked the overdrive car with its long loping gait: just over 4000 rpm and you cruise at the 'ton' in considerable silence. The automatic is the most responsively set Borg-Warner I've ever come across and selection of 2 or 1 on the P.R.N.D.2.1 quadrant is rapid. There's no waiting frightening seconds before the ratio is engaged. Kickdown is the same. The torque of the engine is eminently suited to the automatic.

I liked the map pockets in the doors, the fact that you can ask a couple of people to sit in the back with the hard top on and they can be reasonably comfortable.

The boot carries the spare wheel and the fuel tank under the flat floor. This leaves a maximum height of 13 in. and a minimum of eight in. Width between the wheelarches is 35 in. maximum width is 54 in. Capacity is 9 cu. ft. which means you've room for two 28 in. suitcases and perhaps a grip or two.

Performance is not to be sneezed at. The Stag weighs about 33.5 cwt gross — that's with everything aboard, so it is not light. Basic kerb weight is 25 cwt or with optional extras including the hard top about 27 cwt. But you can push all this up to 30 mph in 3.5 sec. to 40 in 5 sec. to 50 in 7 sec. to 60 in 9.5 sec. to 70 in 12.5 sec. to 80 in 16 sec. 90 in 21.5 sec. and to 100 in 29 sec. Top gear acceleration is good as can be expected with that torque. 20-40 mph, 30-50 mph and 40-60 mph all come up in 7.5 sec. 50-70 takes 8 sec. and 60-80 takes 8.5 sec. 70-90 takes 10 sec. and 80-100 13 sec.

Fuel consumption we didn't have time to take and the factory hadn't done their homework in this direction either. Judging by the way the fuel gauge registered on the 14 gallon tank I'd estimate an average 22 mpg.

This is a good car. It has its shortcomings but they are indeed few. The thinking is right and the result augers well for the first generation of the new BLMC giant. The Triumph Division certainly knows where it is going (next month I'll have some more news from there). The other specialists, Jaguar and Rover are doing things that will keep them to the forefront. I wonder what Harry Webster can come up with from the trailing Austin-Morris Division? Only time will tell. ●

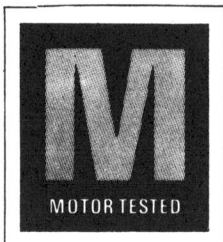

# Al fresco in style

**Very comfortable and well appointed; excellent performance, ride and roadholding; lavish hood and heavy hardtop difficult to erect; very easy to handle but manners and brakes imperfect; wind noise high, boot space modest**

Under the new code of permissiveness Triumph allow themselves to issue a discreet warning to Mercedes Benz and Alfa Romeo in their Stag advertisements. If they can produce the cars to back the threat, it's a warning that the opposition will not take lightly, even though it may be only image building. Although the new Triumph doesn't have quite the sporting spirit of the Alfa 1750 GTV, or perhaps the prestige of a Mercedes 280SL, it does have certain qualities—above all style and versatility—that no rival import can match at the price. It has no British counterpart, of course, so here at least it enjoys a monopoly in a sector of the market that has long been neglected. For this reason alone it can hardly fail to succeed.

Costs have risen so much in the past year or so that the Stag's £2000 price tag—more or less according to equipment—seems to us quite a modest outlay for such a striking, thoroughly modern V8 coupe that will exceed 120 mph, seat four adults—two of them in exceptional comfort—and successfully combine the roles of an *al fresco* fun car with that of a civilized town carriage.

In some respects—seating, ride, controls, finish, appointments and stability at speed, for instance—it exceeds expectations. In others it merely lives up to them: the acceleration, though good, is no better than you'd expect of a car which, by absolute

standards, has a modest power/weight ratio; and the roadholding is as tenacious as the fat tyres and all-independent suspension would suggest.

But there are also disappointments. Wind noise is high, even with the hardtop, and the engine is neither as quiet nor even as smooth as we hoped it might be—a relative observation for by no stretch of anyone's imagination could it be called rough and noisy. A tendency to twitch off course when changing gear and (for some drivers) feel-less power steering marred otherwise exemplary handling.

It could be argued that for a small sacrifice in versatility a fixed superstructure with detachable (and stowable) roof panels *à la* Porsche Targa, 914 and Corvette Sting Ray would have saved on cost, complication and weight. Thoughts along these lines—not shared by us all here—stemmed from the trouble we had in erecting (and removing) the Stag's two modes of headgear.

Despite some detail faults, though, we were certainly impressed by the Stag's concept and quality and completely hooked by its highly individualistic charm—at its greatest on these cars equipped (like ours) with soft and hardtop coverage, as well as the optional overdrive, equipment that raises the price to £2159. If that sounds a lot consider that the Mercedes 280SL coupe/convertible—the Stag's closest European rival in character and performance—costs over £4655, a Porsche Targa (left-hand drive only) even more. American V8 convertibles don't come much cheaper; nor will they do over 20 miles to the gallon like the Stag.

**PRICE: hard and soft top model as tested, £1602 plus £491 15s. 10d. purchase tax equals £2093 15s. 10d. Overdrive (as fitted) £65 5s. 7d.**

## Performance and economy

Triumph's completely new ohc V8 engine is in effect a doubled-up version of the PE104S "four" that helps make the Saab 99 such a good car. Both engines are likely to figure prominently in British Leyland's plans for the Seventies. Because the Triumph engine in the Saab feels so refined we had hoped the full V8, with twice as many power strokes to each revolution of the crankshaft, might set new standards in smoothness. It doesn't. Certainly it *is* very smooth, but no more so than some much larger American V8s. Despite the crossflow pipe linking the two exhausts just forward of the silencers, it has a rather throbby beat which you hear rather than feel—though there's also some vibration transmitted through the gearlever.

With 145 bhp (net) from 3 litres the engine is quite modestly rated when compared with, say, BMW's 2.8 "six" which develops 170 bhp. It also has quite a lot to do for the Stag is a heavy car—almost 26 cwt with the hardtop—so perhaps the acceleration isn't quite so vicious as you'd expect from a sporting 3-litre coupe. To get the best from the car, our performance tests were done with the soft-top up, saving about a hundredweight without increasing the drag.

The engine was always an instant starter and completely without temperament during the warm-up period. As you'd expect for a torquey V8, its usable pulling range spans the entire rev band, from idling at 800 rpm to the red-lined maximum of 6500 rpm, which we used during performance testing but seldom, if ever, on the road. The orange sector starts at 5500 rpm which the handbook warns should not be exceeded for long periods. On Continental motorways this restriction would impose a maximum speed of 109 mph in direct top (19.8 mph per 1000 rpm with the Michelin XAS tyres on our car) so the optional overdrive (£65 extra) is really essential if the car is to be used as a long-distance express—a role for which certain qualities make it particularly well suited.

In overdrive (24.1 mph/1000) 5500 rpm corresponds to over

## PERFORMANCE

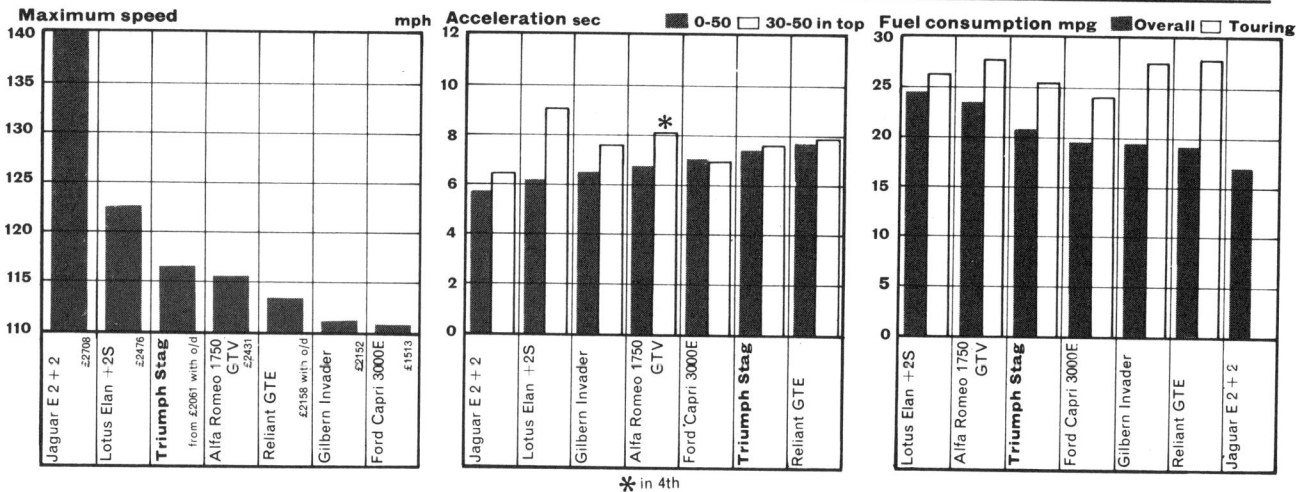

Maximum speed (mph) — Jaguar E 2+2 £2708; Lotus Elan +2S £2476; Triumph Stag (from £2061 with o/d); Alfa Romeo 1750 GTV £2431; Reliant GTE; Gilbern Invader £2152; Ford Capri 3000E £1513

Acceleration sec — 0-50 / 30-50 in top — Jaguar E 2+2; Lotus Elan +2S; Gilbern Invader; Alfa Romeo 1750 GTV *; Ford Capri 3000E; Triumph Stag; Reliant GTE — * in 4th

Fuel consumption mpg — Overall / Touring — Lotus Elan +2S; Alfa Romeo 1750 GTV; Triumph Stag; Ford Capri 3000E; Gilbern Invader; Reliant GTE; Jaguar E 2+2

### Conditions

Weather: Warm, dry, windy
Temperature: 56-62°F
Barometer 29.6 in. Hg.
Surface: Dry asphalt
Fuel: 97 octane(RM), 4 Star rating

### Maximum Speeds

| | mph | kph |
|---|---|---|
| Mean lap banked circuit (see text) | 116.5 | 187.5 |
| Best one-way ¼-mile (see text) | 123.2 | 198.5 |
| o/d 3rd gear | 112 | 180 |
| 3rd gear ⎫ | 92 | 148 |
| 2nd gear ⎬ at 6500 rpm | 60 | 97 |
| 1st gear ⎭ | 42 | 67 |

"Maximile" speed: (Timed quarter mile after 1 mile accelerating from rest)
Mean 109.8
Best 112.5

### Acceleration Times

| mph | sec |
|---|---|
| 0-30 | 3.5 |
| 0-40 | 5.0 |
| 0-50 | 7.4 |
| 0-60 | 9.7 |
| 0-70 | 13.2 |
| 0-80 | 17.0 |
| 0-90 | 22.2 |
| 0-100 | 29.6 |
| Standing quarter mile | 17.3 |
| Standing kilometre | 31.6 |

| | O/d | | |
|---|---|---|---|
| | Top | Top | 3rd |
| mph | sec | sec | sec |
| 10-30 | — | — | 5.8 |
| 20-40 | 10.2 | 7.8 | 5.4 |
| 30-50 | 10.1 | 7.6 | 5.1 |
| 40-60 | 10.2 | 7.5 | 5.1 |
| 50-70 | 10.6 | 7.7 | 5.7 |
| 60-80 | 11.6 | 8.5 | 6.6 |
| 70-90 | 13.2 | 10.0 | 9.1 |
| 80-100 | 16.3 | 13.4 | — |

### Fuel Consumption

Touring (consumption midway between 30 mph and maximum less 5 per cent allowance for acceleration)
25.5 mpg
Overall 20.9 mpg
(=13.5 litres/100km)
Total test distance 1520 miles

### Brakes

Pedal pressure, deceleration and equivalent stopping distance from 30 mph

| lb. | g. | ft. |
|---|---|---|
| 25 | 0.47 | 64 |
| 50 | 0.90 | 33 |
| 75 | 0.92 | 32½ |
| Handbrake | 0.38 | 79 |

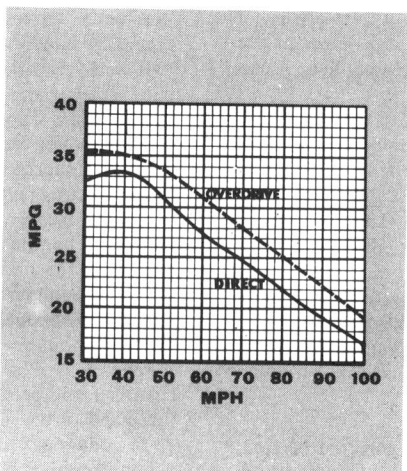

### Fade Test

20 stops at ½g deceleration at 1 min. intervals from a speed midway between 40 mph and maximum speed (=78 mph)

| | lb. |
|---|---|
| Pedal force at beginning | 25 |
| Pedal force at 10th stop | 35 |
| Pedal force at 20th stop | 40 |

### Steering

| | ft. |
|---|---|
| Turning circle between kerbs: | |
| Left | 33 |
| Right | 31 |
| Turns of steering wheel from lock to lock | 2.75 |
| Steering wheel deflection for 50ft. diameter circle | 0.95 turns |

### Clutch

| | |
|---|---|
| Free pedal movement | = ½in. |
| Additional movement to disengage clutch completely | =3½in. |
| Maximum pedal load | =35lb. |

### Speedometer

| Indicated | 20 | 30 | 40 | 50 | 60 | 70 | 80 | 90 | 100 |
|---|---|---|---|---|---|---|---|---|---|
| True | 20 | 30 | 40 | 50 | 59½ | 68½ | 77½ | 87 | 95½ |

Distance recorder 0.5% fast

### Weight

Kerb weight (unladen with hardtop and fuel for approximately 50 miles) 25.9cwt
Front/rear distribution 54/46
Weight laden as tested (without hardtop) approx. 28.1cwt

### Parkability

Gap needed to clear 6 ft. wide obstruction in front

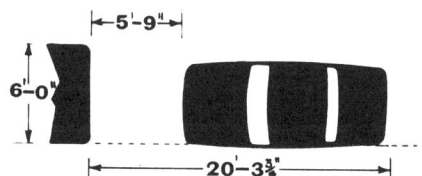

5'-9"
6'-0"
20'-3¾"

130 mph, a speed unattainable in normal conditions but we suspect possible under really favourable ones. The MIRA lap speed of 116.5 quoted in our tables is not a true maximum because at this speed the tyres are scrubbing away several mph on the banking. In any case our test driver confesses to lifting off momentarily for the turn at the end of the long wind-assisted straight. That the car was doing 125 mph here, still accelerating and feeling astonishingly stable, underlines its excellent high-speed cruising ability. Certainly the MIRA performance suggests a mean two-way maximum of well over 120 mph.

Despite its weight, the car surges away effortlessly and quite strongly in top gear from under 15 mph: 20-40 mph takes a smooth 8 sec., 50-70 mph no longer, indicating the consistency of the pull. Through the gears the acceleration is virtually identical to that of a Capri 3000GT: both cars reach 100 mph from rest in a fraction less than 30 sec.

More than with most cars, we suspect the Stag's performance must vary considerably according to its equipment. Taking obvious extremes, the acceleration times of a hard-top automatic with Michelin tyres must be much poorer than those of a lighter soft-top manual with the alternative Goodyear covers that effectively lower the gearing by 1 mph/1000 in top.

For such a fast and heavy car, the consumption of 4-star fuel is quite reasonable—21-26 mpg according to how and where you drive. At an average 23 mpg, the 14-gallon tank gives a useful maximum range of 320 miles.

## Transmission

The four-speed manual gearbox is a strengthened version of that in the 2000/2.5PI. It feels and sounds much the same, too. The gearlever moves in a precise and positive gate, but obstructive synchromesh makes the change notchy if you snatch at it. With easy, unhurried movements, though, it engages smoothly. Apart from baulking rushed changes, the synchromesh can sometimes be beaten. Gear whine is quite audible in the intermediates.

The clutch on our car was remarkably good—not only utterly

Above right: the front seats, adjustable for height, reach and rake, are superb; the rear bench, deeply troughed for thigh support, is passable, above. Five children, centre, is a squash but three will fit comfortably

RVC 435H
GB

smooth but also unusually progressive, so even indelicate footwork failed to jerk the car; ideal for learners. Our measurements show the pedal travel to be quite long (which we don't normally like) and the weight no better than average, but in use we were aware only of the smoothness of its action—more like a torque converter than a clutch.

The overdrive, operated by a handy switch in the gearlever knob (some drivers still prefer a column stalk), engaged with a gentle slurr, too, without help from the clutch; under light acceleration, disengagement was also smooth but on the overrun or hard power it was jerkier.

## Handling and brakes

Fat radial-ply tyres on a car with respectable all independent suspension and a well balanced weight distribution is a recipe that can hardly fail to give high cornering powers and a good hold on indifferent roads. The Stag's bond with the road is certainly excellent and to break it in the dry you've got to explore g forces that are well beyond what most people would regard as

Lifting the hardtop, above, is a two-person—preferably a two-man—job. Once in place it is easy to secure. The substantial hood unfurls from a covered well, left, and is rather more difficult to latch down

By popular request, our old test boxes have been replaced by real suitcases. See page 48 for further details. These were the only two that would fit the Stag's boot—a number 1 and 3—though there was some space left for squashable bags

the limit. Unfortunately (and unusually) we had little opportunity of assessing wet road adhesion.

In some respects then handling is every bit as good as the grip. Very light and responsive steering, seats that hold you in place and minimal body roll—not to mention the good roadholding—make hard cornering very safe and easy. But the Stag has its faults. Changing gear after a burst of hard acceleration through a corner, sometimes even on the straight, can make the car twitch off course. We suspect the reason for this irritating rather than alarming behaviour is the driveshaft splines locking up under heavy torque so that they cannot adjust in length with changes in wheel camber—most pronounced under cornering. As something must 'give' the suspension and thus the wheel twists fractionally on the rubber bushes. When the torque is released in a gearchange the splines slide, the wheel straightens and the car twitches.

To be fair, we were a lot more conscious of this inherent fault (the 2.5 PI suffers from it too) at the beginning of our test than at the end, partly because experience showed it to be more of a bad habit than a vice, partly because we found ourselves automatically avoiding mid-corner changes. For different reasons, lifting the throttle in mid corner—as opposed to breaking the drive altogether—will make the car tuck in, to tighten its line. In moderation—and it is not over pronounced in the Stag—such behaviour is not a bad thing as the car helps itself round any corner, rather than plough straight on, if you misjudge your speed. In the extreme we found that lifting off when right on the limit was the most likely way to cause a tail slide, though even then breakaway is gentle and easily checked.

It is during such untoward antics, though, that you are most aware of the absence of feel on the featherweight steering. All messages come through the seat of your pants, not through your hands, as there is little informative self-aligning tug with which to gauge adhesion and understeer. Searching for it will make for untidy, even lurchy cornering. In short, the Stag's steering is over-assisted, like that of the Jaguar XJ6, for really spirited cornering. If you don't indulge in this sort of thing, though, we'd say that it's a very good set-up, especially when judged purely on the amount of effort and twiddling it demands.

The strong winds we had for some of our test were unwelcome when taking performance figures but underlined the car's exceptional stability in adverse condition at high speed. Nothing seemed to deviate it from the straight and narrow, even at maximum speed.

The brakes on our car were less satisfactory: they juddered when applied hard at speed; faded a bit under really hard usage (though they *did* just pass our 20-stop test); and, like the steering, were too heavily assisted for our liking. Feathering off a stop was also difficult because of a rather unprogressive release action. The handbrake held the car in both directions on the 1-in-3 hill.

## Comfort and control

To say that the suspension feels quite firm, even a bit jolty on bad roads and a little jiggly on secondary ones, would be true but rather misleading. The overall impression is of great stability, of a very flat-controlled ride despite any underlying disturbances. There is little of the harshness often associated with a sports car, practically none of the wallowy float you get with a really soft saloon. The compromise felt to us exactly right. The stability of the ride is emphasized by the absence of exaggerated body roll on corners, by the rigidity of the structure (the overhead bracing certainly prevents scuttle shake and body tremors), and by well-muffled road noise; radial thump is particularly subdued.

In contrast wind noise is excessive. Detachable hardtops present special sealing problems and Triumph don't yet seem to have overcome those on the Stag. The top rail seal is the worst offender; that improvement is possible we demonstrated by reducing the hiss with a strip of masking tape. Even then, unpleasant wind rush dominates all other sounds above 60 mph and is getting pretty loud at 80, especially on a windy day. Almost inevitably, wind noise is even greater with the soft top: although the hood is taut, and free from draughts, the roar of the wind became almost unbearable during our maximum-speed

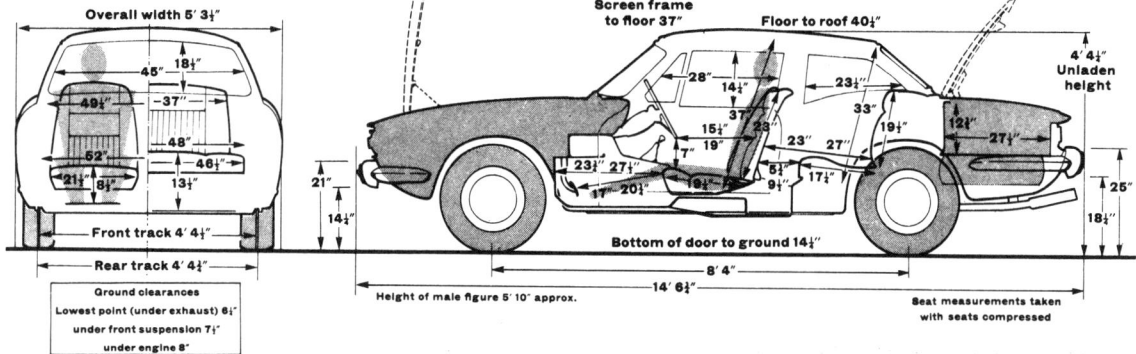

Overall width 5' 3½"
45° 18½"
49½" -37"-
52° 48"
-46½"
21½" 8½" 13½"
Front track 4' 4¼"
Rear track 4' 4¼"

Ground clearances
Lowest point (under exhaust) 6½"
under front suspension 7½"
under engine 8"

Screen frame to floor 37"   Floor to roof 40½"
28" 14½"
37¾"
15½" 19"
23" 27"
22"
7"
23½" 27½"
17" 20½"
19½"

23½"
33"
19½"
5½" 17½"
9½"

4' 4½" Unladen height
12½" 27½"
25"
18½"

21" 14½"
Bottom of door to ground 14½"
14' 6½"    8' 4"
Height of male figure 5' 10" approx.
Seat measurements taken with seats compressed

## Engine

| | | |
|---|---|---|
| Block material | . . . . . . . . . . | Chromium iron |
| Head material | . . . . . . . . . . | Aluminium alloy |
| Cylinders | . . . . . . . . . . . | 8—90° Vee |
| Cooling system | . water, no loss system, thermostatically controlled flow | |
| Bore and stroke | 86mm. (3.385in.) 64.5mm. (2.539in.) | |
| Cubic capacity | . . . . . . . | 2997 cc (182.9 cu. in.) |
| Main bearings | . 5—steel backed lead bronze with lead indium overlay | |
| Valves | . . . . . . . | Overhead camshafts |
| Compression artio | . . . . . . | 8.8:1 |
| Carburetter(s) | . 2—Stromberg side draught—175 CDS | |
| Fuel pump | SU electric. Diaphragm type. Inertia switch in engine bay | |
| Oil Filter | . . . . . . AC: Delco Full Flow—replaceable paper elements | |
| Max. power (net) | . . . . . | 145 bhp at 5,500 rpm |
| Max torque (net) | . . . . . | 170 lb.ft. at 3,500 rpm |

## Transmission

| | | |
|---|---|---|
| Clutch | . . . . . Laycock—single dry plate diaphragm spring type—9in. dia. hydraulically operated | |
| Internal gear box ratios | | |
| Top gear | . . . . . . | 1.00 |
| 3rd gear | . . . . . . | 1.386 |
| 2nd gear | . . . . . . | 2.10 |
| 1st gear | . . . . . . | 2.995 |
| overdrive top | . . . . . | 0.82 |
| overdrive third | . . . . . | 1.135 |
| Reverse | . . . . . . | 3.369 |
| Synchromesh | . . . . | All forward gears |
| Overdrive type | . . . . Laycock—electrically operated on top 2 gears | |
| Final drive (type and ratio) | . Hypoid bevel gears, 2 pinion differential—3.7:1 | |
| Mph at 1000 rpm in:– | | |
| O/d top gear | . . . . . | 24.1 |
| Top gear | . . . . . . | 19.8 |
| O/d 3rd gear | . . . . . | 17.4 |
| 3rd gear | . . . . . . | 14.3 |
| 2nd gear | . . . . . . | 9.45 |
| 1st gear | . . . . . . | 6.62 |

## Chassis and body

Construction   Integral construction—steel sub-frame at rear

## Brakes

Type . . Disc/drum, servo-assisted. Mechanically operated handbrake to rear wheels
Friction areas:
Front . . . . . . 24 sq.in. of lining operating on 220 sq.in. of disc
Rear . . . . . . 78 sq.in. of lining operating on 127 sq.in. of drum

## Suspension and steering

Front . . . . . . Independent by coil springs and wishbones; anti-roll bar

Rear . . . . . . Independent by coil springs and semi-trailing arms.
Shock absorbers:
Front and rear . . . . Telescopic hydraulic dampers
Steering type . . . . . Rack & pinion
Tyres . . . . . . Michelin 185 HR-14 radial ply tubeless
Wheels . . . . . . Steel disc
Rim size . . . . . 5J

## Coachwork and equipment

Starting handle . . No
Tool kit contents . . Combination tool: plug spanner wheel nut spanner; w/trim removal tool; tool pouch
Jack . . . . . . Scissor type
Jacking points . . 4, under sills adjacent to each wheel
Battery . . . . . 12 volt negative earth 56 amp hrs capacity @ 20 hr. rate.
Number of electrical fuses . . . . . 11
Headlamps . . . . 4 Quartz halogen
Indicators . . . . Front & rear & indicator repeaters in front wing side panels
Reversing lamp . . . Yes
Screen wipers. . . . 2 speed
Screen washers . . . Twin nozzle electrically operated
Sun visors . . . . 2—vanity mirror in passenger side
Locks:
With ignition key . Ignition & steering column lock, side doors & petrol filler cap
With other keys . Glove compartment & luggage compartment
Interior heater . . . . Smith's air/mix heater/demister
Upholstery . . . . PVC leathercloth with basket weave pattern on seat facings
Floor covering . . Moulded pile carpet with felt underlay
Alternative body styles Available with soft top or hard top or both
Maximum load . . . 943 lb (Difference between kerb & gross vehicle weight)
Maximum roof rack load . . . . . 112 lb. (hard top only)
Major extras available Hard top, air conditioning unit, Laycock overdrive, Borg-Warner automatic transmission

## Maintenance

Fuel tank capacity . . 14 galls
Sump . . . . . . 8 pints SAE 10W 20W/50
Gearbox . . . . . 2¼ pints SAE 90EP
Rear axle . . . . 2 pints SAE 90EP
Steering gear . . . 1¼ pints Type A (Power steering reservoir)
Coolant . . . . . 18½ pints (2 drain taps) including heater
Minimum service interval . . . . 6000 miles

Chassis lubrication . . None
Ignition timing . . . 14° btdc
Contact breaker gap . 0.015in.
Sparking plug gap . . 0.025in.
Sparking plug type . . Champion N–11Y
Tappet clearance (cold) Inlet 0.008in. Exhaust 0.018in.
Valve timing:
inlet opens . . . . 16° btdc
inlet closes . . . . 56° abdc
exhaust opens . . . 56° bbdc
exhaust closes . . . 16° atdc
Rear wheel toe-in . . 0—1/16"
Front wheel toe-in . . 1/16"—1/8"
Camber angle . . . . Front 1° Pos. ± 1°
Rear1½° Pos. ± 1°
Castor angle . . . . 2° ± 1°
King pin inclination . 10½° ± 1°
Tyre pressures:
Front . . . . . 26 p.s.i.
Rear . . . . 30 p.s.i.

## Safety Check List

**Steering Assembly**

| | |
|---|---|
| Steering box position | Forward and above engine bearer |
| Steering column collapsible | No |
| Steering wheel boss padded | Yes |
| Steering wheel dished | Slightly |

**Instrument Panel**

| | |
|---|---|
| Projecting switches | None seriously |
| Sharp cowls | No |
| Padding | Along top and bottom of facia |

**Windscreen and Visibility**

| | |
|---|---|
| Screen type | Zebra zone toughened glass |
| Pillars padded | Slightly |
| Standard driving mirrors | One inside |
| Interior mirror framed | Yes |
| Interior mirror collapsible | Yes |
| Sun visors | Two |

**Seats and Harness**

| | |
|---|---|
| Attachment to floor | By bolted slides |
| Do they tip forward? | Only after releasing side catch |
| Head rest attachment points | No |
| Back of front seats | Firmly padded |
| Safety Harness | No |
| Harness anchors at back | No |

**Doors**

| | |
|---|---|
| Projecting handles | No |
| Anti-burst latches | Yes |
| Child-proof locks | No |

1 map reading light. 2 fan. 3 ventilation grilles. 4 radio (extra). 5 warning lights. 6 battery condition. 7 temperature. 8 speedometer. 9 brake failure warning light. 10 rev counter. 11 clock. 12 fuel gauge. 13 fresh air vent. 14 choke. 15 heater controls. 16 interior lights. 17 and 19 windows. 18 handbrake. 20 rear window heater. 21 cigar lighter. 22 panel light rheostat. 23 washer/wiper. 24 indicators/horn/dip

runs—which also caused the Velcro seal above the passenger's window to come undone.

Fortunately, the weather was kind and we had the top down for much of the time. We loved the Stag like this. Buffetting can be kept in check, if not eliminated, by raising the side windows (rigidly held by slim frames) and, to a lesser extent, by tilting the sun visors to act as wind deflectors. With the exhaust burbling a discreetly sporting tune, we discovered in the Stag a new sort of open air motoring; nothing else at the price can do it in such style, comfort and luxury.

This, of course, is the car's special charm. But the excellence of the concept, of a versatile three-in-one car, is not quite matched by the means of achieving it. There are snags. The fully trimmed hardtop, for instance, is inordinately heavy and very much a two-man job to fit and remove—more a seasonal than a daily routine even though securing it, with five simple catches, is very easy. The contact for the heated back window is automatically made when you snap the top down.

With the soft top (which is invisibly stowed under a hinged cover that blends with the trim) the problem is one of tension rather than weight. Whichever way we arranged things the final task of securing the top rail catches proved a real strain—beyond the pull of most women—simply because there was not enough "give" in the frame initially to engage the pegs. The hood on our car was also invariably reluctant to hinge back again smoothly. It's a very substantial, elaborate and weatherproof piece of equipment—even partially trimmed on the inside—but it's by no means the easiest or quickest to erect and stow. Perhaps with further practice we'd have overcome the snags.

Open or closed it's hard to imagine anyone being uncomfortable in the front. The seats, deeply contoured for lateral, lumbar and side support, are adjustable for reach, rake and height. The steering wheel is also quickly adjustable in two planes. So except perhaps for very long legged drivers who might want an exta inch or so of rearward movement, everyone should be able to tailor themselves a perfect position.

The controls are particularly good. Within limits you can put the steering wheel where you want; and the pedals, though slightly offset to the right, are all arranged at the same comfortable height so they can easily be worked by pivoting your heel on the floor. The brake and throttle are reasonably placed for simultaneous heel and toe operation but the strong servo assistance makes the brake too flimsy a fulcrum unless you are braking hard. The handbrake is conveniently placed between the seats, the gearlever a hand's span away from the steering wheel.

1 coil. 2 distributor. 3 twin Stromberg carburetters. 4 fuel filter. 5 oil filler cap. 6 viscous coupled. 7 radiator overflow. 8 brake servo. 9 dip stick. 10 washer bottle. 11 power steering pump

The important minor controls are outstanding, a model example of how they should be arranged. The horn, dip, flasher and indicators are all controlled by the right-hand column stalk, the washers and two-speed wipers by a stalk on the left. Both are within fingertip reach, crisp in action and clearly identified with symbols. A large knurled knob just ahead of the right stalk operates the lights. Opposite on the left is the ignition key which, if you wriggle and twist, can also be enticed to lock the steering or release itself. Steering /ignition locks can be infuriating things.

The rest of the switchgear is on the central console, rather distant except for handy rocker switches in front of the gearlever operating the interior lights and the rather lethargic electric windows. Awkwardly placed behind the gearlever, and surrounded by the ash tray, are three slides. One regulates airflow to the excellent central ventilation grilles, supplemented by conventional eyeball vents at each end of the facia. The other two regulate the temperature and distribution of air from the heater—pretty effectively so far as we could judge during a warm spell—though output from the ventilation grilles is greatly reduced when the heater is on. With the powerful (but noisy) two-speed fan in operation, warm-up is quick and demisting efficient.

Good vision was obviously a design priority: the heated back window is a standard fitting and the driver's wiper has a parallelogram arm so that it sweeps with an efficient squeegee action right up to the pillar, leaving no murky wedges. With the hardtop in place you get a very good all-round view. With the soft top the side panels butting up to thick rear pillars can form an awkward blind spot. If you sit up, the back corners are just visible from the driver's seat and there's an excellent automatic reversing light. The headlights give an intense, uniform spread of light, free from spots and whirls: but their range was not particularly long on our car. Perhaps they needed raising a little.

Back seat accommodation is cramped though not impossible for six footers. Sensibly, the seat cushion is deeply troughed so that your thighs get some support even in the knees-up position forced on adults. With the front seats right back there is ample room for two or three small children.

## Fittings and furniture

As in the design of its switchgear so in decor and appointments, the Stag is a front runner. The clear and comprehensive instruments are set into an unpretentious matt walnut facia, concave on the driver's side to minimize parallax error. A clock, rev counter, battery charge indicator and all-systems-go warning light cluster are standard. Rich leatherette trim luxuriously covers the doors, the seats are richly upholstered in non slip basket-weave, and well fitted carpets cover the floor. As we've observed before, Triumph are also setting the standard in seat belt design.

We liked some of the details too: red warning lights in the door edges; powerful reversing lights; anti-thief quarter light knobs; neat door release levers and locks. There was something amiss with the lockup arrangements of our car though since all the keywork—for doors, facia locker, flush petrol filler and ignition—proved a fiddly and sometimes exasperating job. Smokers also complained about the inaccessibility of the central ash tray.

Stowage space inside is generous and well planned. Apart from the locker there's a shelf beneath the facia, pouches on the doors and seat backs, and recesses in the rear sidewalls. By saloon standards, the boot is not very big—wide but rather shallow because of the spare wheel and fuel tank beneath.

## Servicing and accessibility

As you'd expect, there's nothing very demanding about the servicing schedule, done on a 6000-mile cycle. All the ancillaries needing regular attention are quite easy to get at under the wide, front-hinged bonnet, supported by a self-locking stay. There's a modest tool kit in the boot. ∎

MAKE: Triumph. MODEL: Stag. MAKERS: British Leyland (Triumph) Ltd., Coventry, England

# a luxury V-8 sports car

## Triumph Stag

by Joseph Lowrey
(Photos by the Author)

The Triumph Division of British Leyland Motor Corporation sells mainly sports two-seaters in the U.S.A., mainly sedans in Europe. To bridge the gap it introduces the Stag "2 + 2" GT coupe, powered by a 3-litre V-8 of overhead camshaft layout, which will reach US dealerships late this year. At prices below Porsche and Mercedes Benz levels, the Stag offers a very high standard of comfort and road holding and a top speed of about 115 mph.

There is a strong family likeness between the Stag and the Triumph 2.5-litre 6-cylinder sedan. The wheelbase is cut 6 inches to 100 inches for the new GT, height is reduced 6 inches to 49.5 inches and width slenderized 1.5 inches to 63.5 inches. Beneath rather similar styling, the Stag has slightly larger disc-and-drum power brakes of similar design, the same telescopic strut front suspension, the same IRS by one semi-trailing link per wheel and a similar rubber–mounted sub-frame to keep rear suspension noise out of the car. Coil spring flexibility, anti-roll torsion bar diameter, damper settings and other details are of course tailor-made for the new car. Radial-ply tires are on all examples of the Stag, as is power assistance on the rack-and-pinion steering, so it has been possible to set up the handling without compromises to suit tire and steering options.

An integral steel body/chassis underframe has been reinforced cleverly so that it is as strong as a steel-topped job. There are, in effect, two roll-over hoops braced to the front and rear body structures, a hoop around the windshield and another behind the doors.

Along the car centerline, another steel member runs fore-and-aft to brace the two roll-over hoops integrating them in one rugged H-shaped superstructure of padded steel. This structure does not spoil the Stag as an open car, would give a great deal of protection in any crash, and holds the car firmly rattle-free over Belgian block torture surfaces.

Seating for "2 + 2" has its usual meaning. Two luxury seats in front, a roomy external-access trunk behind, and in between a pair of seats which are short of legroom and short of headroom too if the soft top is up or a hardtop is fitted.

The Stag is really a two seater which will carry kids or a very great deal of luggage. This is mainly a luxury car for one-time sports car lovers whose kids now have their own cars, and who have time to enjoy long-distance Grand Touring in a fine car. Emphatically the Stag does not replace the Triumph TR6, GT Six or Spitfire 4 sports two-seaters.

Power for the Stag comes from the second in a planned series of sohc Triumph engines. The first product of the automated transfer machines has been the 1.7-liter slant-four for the Saab 99 sedan, which has had to meet Swedish durability tests (see road test this issue).

Wedge combustion chambers have in-line valves mounted at an angle to the cylinder bores, and operated by piston tappets from one overhead camshaft per aluminum cylinder head. With its 5-bearing crankshaft in an iron Y-block, this engine will probably go up in rpm thru the 1970s as the ex-Buick (now Rover V-8) from the same group goes up in displacement. At present, maximum power is developed at 5500 rpm, and the tach is marked with a red line at 6500 rpm.

The 145 horsepower, net as installed in the car, is a catalogued figure which most production engines are claimed already to be bettering. A balancer hole between two separate halves of the cross-over induction manifold has been re-positioned to improve both torque and power.

At present, pistons with recessed crowns give a compression ratio of only 8.8 to 1, which seems to anticipate lead-free gas. A clean air package does not reduce this engine's power. Cams with 16-56-56-16 timing, in conjunction with two Zenith Stromberg CD carburetors, let the Stag accelerate cleanly from a mere 10 mph in 4th gear.

Uniquely, each aluminum cylinder head is held to the iron block by two angled lines of studs to keep the latter clear of an offset camshaft location. This also ties the head down firmly the way guy ropes secure a tent! One line of studs parallel to the cylinder bores is permanently screwed into the block: after a head has been lowered into place, the second line of slotted-end studs is put with a screwdriver, and bolts on the two lines of studs are tightened. Just don't plan on milling the head or block faces to increase that compression ratio!

A typically Triumph array of three transmissions is available behind this new motor. Four on the floor, with ratios closely spaced between 1-2 and between 3-4 but widely spaced between 2-3. An epicyclic overdrive which hot-shifts if you flick a switch (located in the gearshift knob) can be added to this transmission. Or, there is the usual Borg Warner automatic which permits manual selection of 2 or 1 if you don't like the way it behaves in D.

Weighing from 2800 to 3000 pounds according to how it is equipped (and likely to cost between $5,000 and $6,000), the Stag could use plenty more horsepower than it has. Anyone who does a swap should find a TR6 Triumph sports car owner eager to replace his "six" with the little V-8.

As it is however, the Stag is just the sort of car I'd like (with the optional air conditioning please!) for a transcontinental trip such as I've not made other than by jet since a dozen years back.

*stabhaz*]
**staff²** (-ahf), n. Mixture of plaster of Paris, cement, etc., as building-material. [orig. unkn.]
**stag,** n. **1.** Male of red deer or of other large kinds of deer; bull castrated when (nearly) full-grown. **2.** ‖ (St. Exch.) person who applies for allotments in new concerns with a view to selling at once at a profit; ‖ (sl.) irregular dealer in stocks. **3.** ~-*beetle* (with branched mandibles like ~'s antlers); ~-*evil*, lockjaw in horses; ~-*horn*, kinds of club-moss & coral; ~-

You might find this hard to believe, readers, but there are cynics amongst us who'd have you believe that the ConOxDic's definition might have been specifically determined for a certain new V8 smoothmobile from a certain specialist car division of BLMC. In fact, to labour the point a bit further, the reverse is more accurate since STI's new Stag, instead of being castrated when full grown, was actually endowed with a bit more fervour. But we're confusing the issue—muddled readers please begin here.

First thing which struck us about the Stag was that it looked like it should have come out two years ago. Talking later to Triumph's Director of Engineering Spen King and Sales Director Lyndon Mills we found out that this was more or less true. Development of the Stag started about five years ago with a scheduled three-year span before announcement. For various reasons, mainly connected with the amalgamation of STI into British Leyland but also connected with the Triumph-Saab engine deal, development slowed for some time, giving time to see where the new car would fit into BLMC's plan for the seventies and also giving time for a closer look at which sector of the market the Stag was aiming for. In this interim, several important factors were altered, especially on the engine side.

The exact chronological details of the Stag's birth are not easy to unravel due to big reshuffles—Spen King in fact came from Rover to STI about three years ago and picked up the Stag as a

half developed project. It seems though as if the car was conceived before the engine, mainly because North America seemed keen on a smooth coupe version of the strong-selling 2000 and later 2500PI series sedans.

Concurrently, STI were working on a new four-cylinder general-purpose engine which could later be developed as a 90deg pair in Vee-form for such upper crust ranges as may come along. Saab then made it known that they were wanting a good four banger for their impending 99 saloon and it was this stimulus that led STI to develop the short slant four unit, originally as a 1500 and later (for the heavy Saab) as a 1700 by increasing the bore and stroke. Triumph managed to surprise themselves by the performance of the new unit and turned again to thoughts of developing a 90deg V8 for the new Stag.

First time out it came as a 2½ litre with 80mm bore and 64·2mm stroke which was quite cute but lacking in the punch that the Great British Public would expect of the V8, nottermention the Great American Public who are averagely pro-Triumph for the sporting image. Ha,

# V8 STAG
# -THE ANSWER TO THE PONY?

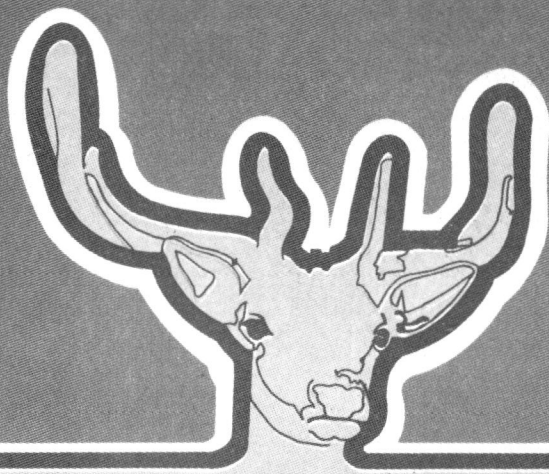

said Spen King, freshly arrived from new stablemate Rover and watching the new Stag trying to pull the skin off the test-bed rice pudding, more of bore and another five hundred cubes it must be. So that's how it came out at 2997cc.

Just in case you thought someone had rushed around at the last minute sticking bits of the Mk2 2000/2500 series on to the Stag to give it a lift, we can reveal (with apologies to Cardiac Basil) that the body styling for the Stag had been fixed with Michelotti from way back. Styling bits from the Stag were pinched for the saloons and only production delays on the Stag line made it appear otherwise.

You like the styling? Well we sort of do, except that we don't like the hard-top line too much. Styling is a bit cliche-ed Micho, but slick and smooth for the market it's intended for. And which market is that, sez you.

Well opinions differ on that one, even within Triumph. Their kick-off advertising warned Alfa and Mercedes to watch out, which meant that they were trying to identify with the sophisticated 2+2 scene, but certain high-ups in Triumph feel this makes them look like the underdog trying to nip the heels of the master.

In this country anyway it is probably closer to the Elan 2+2, Scimitar GTE and Capri 3000 market and indeed the new Stag might help to define this market more sharply as did the Rover and Triumph 2000s for the luxury medium performance saloon market. Nearest definition seems to be for the sporting minded family man, making a fair slice of bread, who can't now cram the new brood into his TR or BGT. Compared with most other +2s the Stag offers very generous rear seat accommodation, and if it wasn't contrary to the sporting bit to claim four seats, they could do so with more justification than a Capri.

So wadderyerget for your two grand?

Short answer is a lot of car, and a heavy car at that. Being designed from ground up as an open car, weight has to be sacrificed on the altar of structural rigidity if the beast isn't going to shake itself to bits after a couple of years. And yes you're right, that three-point roll cage does help keep it stiff as well as protecting your long and curlies should you flip.

Transmission, running gear and suspension are mostly beefed-up 2500PI which give a very familiar feel to the handling. But the big story of course is under the hood where lurks that nice fat V8. Three litres with twin side-draught Stromberg 175 CDS give a fairly modest 145brake at 5500rpm which is only a hair over the V6 Ford turning at 4750. Heads are aluminium alloy for weight saving, primarily, but also efficiency, with wedge-shaped combustion chambers; block is chrome iron and the engine has been designed so that complete head assembly can be quickly removed without disturbing the valve gear. Carbs are mounted on a cast ally four-branch manifold with combined air cleaners/silencers; exhaust disappears down three-branch cast iron manifolds. Cooling is revolutionary new $H_2O$ liquid (what will they think up next), the water being sealed and pressurised in a no-loss system. Cross flow rad is wafted by a 16½in 13-bladed plastic fan, driven by vee-belt off the crankshaft.

Triumph are very much aware of the potential of the engine and we can expect performance options, either bigger capacity or more developed, before too long. In its present form though, the engine is ultra smooth and willing, if lacking the big kick. We were only able to get our hands on the three-speed automatic (Borg Warner 35) in time for this feature though the four on the floor's following soon, and we'd be unfair if we criticised the Stag's performance on the auto change. Apart from the spaced-out ratios there is a big time lag between the boot being put in and the power coming on, so if it's wheelies you're after you're probably better off with the manual. But then there are undoubtedly a lot of tired ageing would-be racers around who'd think it was nice.

A well-driven manual will see 0–60 in 9·5sec, 0–70 in 12·5 and the ton in 29sec. Top speed is a modest 118mph (how about a five-speed box, Mister King). What the engine loses in performance it gains in smoothness and you can accelerate away from 500rpm in top without the engine rupturing itself—so who needs an automatic then?

One thing we really hated was the standard equipment power steering which is dead and vague for a performance car. Servo assistance on a heavy car yes, but complete control no. Yes you do get used to it but it rids the car of a lot of its sporting character.

One thing we really liked was the cockpit—the layout and general comfort of the driving position. Take the seats (no I'll take the seats and put them in the E-type)—they adjust fore and aft, the back adjusts finely for rake and in addition there's a little winder to raise and lower the seat squab. Seats are large but shaped for good location, trimmed in textured plastic. Couple that with a steering column adjustable for distance and rake and the Stag can accommodate your spastic hunchback of Notre Dame without even trying.

Wheel is large and flattish and does not obstruct the view of the instruments which are clearly laid out in the matt wood finish fascia which curves in front of the driver. Standard of trim and finish, in fact, is excellent within the price—at least on the car we tried. Carpet is thick and sybaritic, trim is in a choice of colours (black on ours) and fascia and other odd trims are matt-finished walnut.

Driving? Well it's a big car, and that must be remembered as you prepare to go all clever with the big V8. Remember that it isn't set up to be a rorty sportster, and within that limitation it handles well with a bias towards oversteer, especially with full power on. Body roll we expected to be more and were surprised how smoothly the Stag behaved when hammered. Don't compare it with the Elan +2, but give it credit for a sporting smoothie.

Not my choice for a two-grand motor car, but not a lot of money for the 15000 frustrated family motorists that Triumph expect to buy their new baby. **MH**

THE SECRET OF SURVIVAL FOR A small manufacturer in a world of giants is to find a niche in the market —to locate a gap that British Leyland, Ford and the rest have ignored, and to concentrate on filling it effectively.

Reliant, under the able guidance of Ray Wiggin, have been doing this for some time. Although their bread and butter business continues to come from three-wheelers, and from setting up package deal vehicle plants in developing countries, there is increasing emphasis on the jam of the sporting Scimitar range. We make no apologies for returning to the GTE—it is one of the most interesting cars produced in Britain and certainly one of the most thoroughly worthy. And it is unique in that it supplies genuine two-plus-two seating accommodation, plentiful luggage capacity, fine handling and the lazy, loping power of an underdeveloped three-litre engine at only just over £2000.

At least it *was* unique until a few months ago, when British Leyland's Triumph division introduced the Stag. The first few production examples are beginning to filter on to the roads now, and they are remarkably comparable with the GTE in specification as well as in price. This is not so surprising when you realise that the Triumph people themselves are dab hands at finding gaps in the market and plugging them under the very noses of their rivals. This habit started in the days before Leyland took over Triumph, when the Coventry company was indeed struggling for survival against Ford, BMC, Rootes and Vauxhall. More recently it has sheltered safely under the Leyland wing and now, as part of the British Leyland Motor Corporation's Specialist Car Division, is still charged with the same basic role.

Hence the Stag. It makes no pretence at being a sports car in the traditional sense of the phrase. Neither, although BLMC would have you think otherwise, does it quite qualify as a GT, either in the original or in the modern, devalued meaning. In fact it is a true, straightforward personal touring car—a throwback to the tourers that became a halfway house between saloons and two-seaters between the wars. As such, it has little direct competition but a lot of appeal, if BLMC have done their market research correctly, to the man who wants to move away from the staid dignity of a saloon, yet not *that* far away.

Its nearest rival, Reliant's GTE, achieves very much the same end although it does so by way of an entirely different route. It, too, is a real tourer, suited to long, laden journeys with two adults or a young family (neither car has much back seat space to offer anyone over early-teens size). But it has more of the old fashioned sports car in its makeup, and this shows immediately you get behind the wheel. It is obvious, too, in the design, for under the skin the Reliant is simplicity itself whereas the Stag is really rather complex.

Much of the Triumph's complexity, however, stems from the unusual and laudable degree of attention the designers have paid to details. By comparison much of this minutiae is missing from the GTE, which is all the more a pity when you realise that in hand-assembled, limited-output products like this there is much more opportunity to incorporate the thoughtful little touches that can really civilise a car.

## STYLE AND ENGINEERING

The debt the Stag owes to the 2000 and 2.5PI Triumphs is instantly visible. The styling around the front is almost identical to the 2000's. From that point back, however, the shape differs to suggest the Stag's sporting qualities with a slight drop in the waistline at the cockpit before humping up over the rear wheels to finish once more with the saloon's smiling, sheet-metal flourish. In fact this similarity is a case of chicken and egg, for the Stag was in fact evolved first.

With the hardtop and hood both out of the way, the Stag's built-in roll cage does nothing to detract from its appearance. Overall, the styling bears the unmistakable stamp of Giovanni Michelotti, Triumph's main consultant in this field for many years.

Equally, the GTE's lines are clearly the work of Tom Karen, head of the Ogle design consultancy and responsible for the external appearance of most of Reliant's more attractive products. The GTE has a higher, narrower look than its rival, emphasised by the way in which the roofline is carried straight back, breadvan-style, to achieve the estate-car-like rear end. At the back the GTE scores an aesthetic victory with its uncompromisingly vast rear window-cum-tailgate, a sharp contrast to the Stag's conventional boot and rear-end styling, weak echoes of the more mundane saloons.

The Reliant like most of its kind has glassfibre bodywork and the finish, although better than it used to be, is still not up to that achieved on the Stag's steel panels. On the other hand the metal is more expensive to repair, and corrosion-prone.

The Stag's chassis/body is the customary unitary construction affair, using a modified 2000 floorpan. The Reliant has a separate ladder-type chassis with cruciform bracing, welded up from stout steel tubes.

Suspension on the Reliant is by double wishbones and coil springs at the front. At the back the engineers have avoided the potential pitfalls of independence and stuck to a live axle located by trailing arms fore and aft and a Watt linkage in the lateral plane. The Stag has MacPherson struts in front, with semi-trailing arms and coil springs for the rear wheels. In fact, this and much of the rest of the running gear and other components come from the 2000/2.5PI.

The Stag's steering, too, is borrowed from the saloons. It has power-assistance as standard, by courtesy of Adwest. The Reliant also uses a rack and pinion system but for some reason cannot be obtained with servo.

In considering cars like these the details of design are of as much importance as the major facets. In this instance, it is highly relevant that the Stag as a result of its saloon background has plentiful compliance built into its suspension and steering whereas the Scimitar has relatively little. This has a considerable effect in determining their respective characters and behaviour.

GIANT TEST
**RELIANT SCIMITAR GTE v TRIUMPH STAG**

The wheels on both cars are ordinary pressed steel, thinly disguised by black and chrome trim (à la Vitesse, TR6 and others) on the Triumph and by equally unappealing glassfibre moulded covers on the GTE. The intention in both cases is to suggest that the wheels are rather more exotic than they in truth are. Braking in either case is by near identical power-assisted disc front/drum rear systems.

The Stag's V8 engine is a newcomer and one of uncertain future in view of the existence of the ex-GM alloy V8 which British Leyland acquired with Rover, not to mention the perennially expected efforts in this field from Jaguar. At all events the Triumph engine is a pleasing design and one that deserves wider application.

As a result of choosing the lusty Ford engine Reliant have also saddled themselves with the much less desirable gearbox that accompanies it. As ever, the second ratio is hopelessly low and the change action no more than fair. The Stag has a modified version of the much sweeter 2.5PI gearbox and can be ordered with overdrive on third and fourth at an extra cost of £65. Overdrive costs £72 extra on the GTE.

Both cars are also available with Borg Warner automatic transmission, at £104 more for the Stag and £139 for the GTE.

## USE OF SPACE

There's little to choose between these two in major dimensions. Both cars are quite generously proportioned and no attempt has been made to obtain compactness. Externally, in fact, they are of medium saloon size, with overall lengths near 14.5ft and widths nearing 5.5ft. The use made of the space so provided is another matter. Long engine compartments are necessary, partly for reasons of styling and partly because the power units themselves are set back from the front axle line in the interests of weight distribution. V-type engines, even a six like the Ford used by Reliant, never save as much space as they should. Obviously they are greedy of width, but length, too, goes by the board, especially when servo pumps, alternators and other ancillaries have to be hung on the front of the block because the already restricted side space has been hogged by the exhaust manifolds. Reliant, however, do manage to cram the spare wheel into the front of the engine compartment.

On wheelbases measuring not much over 8ft all this means that passenger space is at something of a premium. Both cars provide ample room around the front seats. Back seats are another story. The Triumph's is inviting because it is well upholstered and of reasonable width. Leg- and, particularly, knee-room are barely acceptable unless the front seats are well forward, while headroom is insufficient for anyone over 5ft 10in or so. The Reliant's back seat is divided into separate buckets by a folding armrest and although it locates the occupants better than the Triumph bench it gives them no more space for the extremities.

The Reliant's seat backs fold to extend the estate car-like luggage compartment that stretches back to the tailgate. The Stag, being basically an open car, has to make provision for the hood to fold into a compartment around the back seat, a prime reason for the restricted space in that area. Behind, there is a normal boot of none too generous capacity.

Space-saving rear suspension systems mean that room has been found in both cars for outstandingly generous fuel tanks.

## COMFORT & SAFETY

In contrast to the coupé bodywork of the Reliant, the Stag is basically a convertible for which a metal hardtop is an expensive and desirable option. It adds £98 to the price, though anyone who disdains open air motoring—or doesn't mind risking being caught in the rain with the top down—can save £56 by dispensing with the hood altogether and just having the hardtop. Because it has a comparatively long passenger compartment to cover, and because it contains a lot of glass, the hardtop is no lightweight. Undoing the quick-release catches that retain it is no problem; lifting it off, and finding somewhere to stow it, are not so easy.

The hardtop incorporates a heated rear window, for which the connector plugs-in automatically as the top is fitted on to the cockpit. The rear side windows are hinged at the front to act as vents. The top fits very snugly. In British weather, at least, there is little incentive to remove it. Those who do will find that the rather floppy and ugly hood, neatly hidden away under a panel, folds up and into place fairly easily. It has a zipped-in back window for extra ventilation when one keeps it erect to ward off Mediterranean sunshine. With the hood down the wind whistles irritatingly around the roll cage but doesn't buffet the occupants as

photography: Ivor Lewis

much as in an open two-seater.

The Stag's controversial cage is so positioned as to present no obstruction to headroom. The only time when it does interfere is if one turns to reach back over the very high squabs of the front seats into the rear compartment. Its padded vinyl covering can be unzipped for cleaning.

We preferred the front seats in the Stag, mainly because they have a simpler, more effective method of rake adjustment than the Reliant's as well as being more comfortable. The Reliant seats are firmer but do give much better lateral support. A minor disadvantage with the Reliant is that in front one sits closer to the floor, which can be tiring for the legs in addition to diminishing slightly the driver's forward view. We approved strongly of the Reliant's driving position, however, in which the controls are correctly laid out and the nicely angled steering wheel has a fairly thick rim of smallish

diameter. All round visibility is good in both cars.

The Stag uses the large-diameter wheel of the 2.5PI, adjustable over a limited range for both rake (2in) and reach (4in) once a clamping lever under the dash has been released. Even with all the travel used up the angle of the wheel remains much less vertical than our drivers like. This apart, we found nothing to complain of in the Stag's controls. They are almost identical to those of the 2.5PI, which is no bad thing.

The Stag facia looks like that of the saloons and is covered with matt-finished wood veneer. It houses a generous complement of instruments, plus switches that are simple enough to locate in a hurry. The lighting control is on a rotary switch projecting from the column. In fact the entire control system, even down to the single dial containing a multiplicity of warning lights, will be entirely familiar to

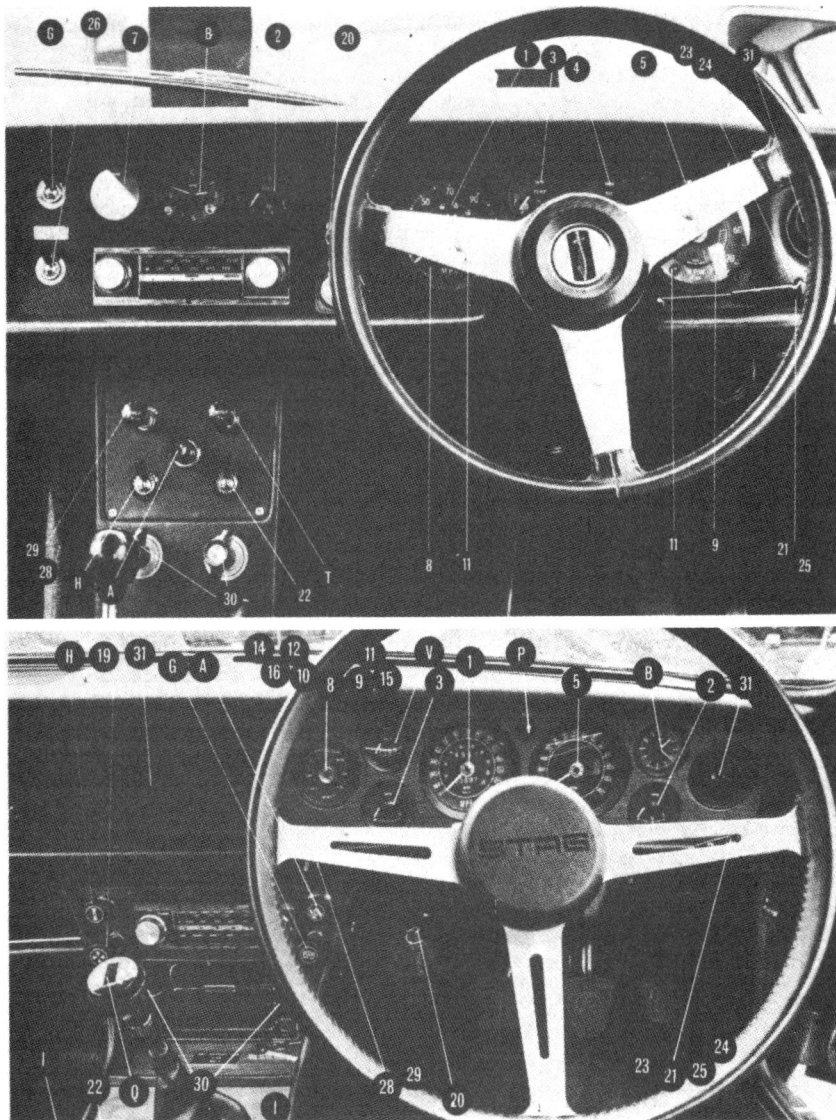

**Instruments: 1** Speedo **2** Fuel **3** Water Temp **4** Oil Pressure **5** Tacho **7** Ammeter **Warnings: 8** Ignition **9** Main Beam **10** Oil Pressure **11** Indicators **12** Water Temp **14** Handbrake **15** Fuel Low **16** Choke Control Warning **Controls: 19** Choke **20** Ign/Start **21** Indicators **22** Lights **23** Dip **24** Flash **25** Horn **26** Panel Lights **28** Wipers **29** Washers **30** Heater **31** Fresh Air Vent **Special Items: A** Cigar Lighter **B** Clock **G** Rear Window Demister Control **H** Blower **I** Electric Window Control **P** Brake Light Fail **Q** Overdrive Control **T** Rear Window Wiper **V** Voltmeter

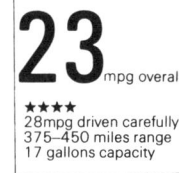

| | GTE | STAG |
|---|---|---|
| **PRICES** | At £2019 a mere £23 more than the basic Stag but £75 less than the more nearly comparable hardtop version. Adding £33 for a heated rear window to match the Stag's standard one makes the differential only £42 | At £1996, or £2094 when fitted with both hard and soft tops, the most expensive Triumph yet produced. Priced neatly to fit into a niche in the 2 plus 2 market occupied hitherto only by Reliant |
| **ACCELERATION** from standstill in seconds | 18.0 24.2 33.3 **13.5** 10.0 7.4 5.1 2.9 | 17.9 22.9 32.3 **13.6** 10.1 7.5 5.3 3.6 |
| **FUEL** | **23** mpg overall ★★★★ 28mpg driven carefully 375–450 miles range 17 gallons capacity | **21** mpg overall ★★★★ 25mpg driven carefully 280–335 miles range 14 gallons capacity |
| **SPEEDS IN GEARS** | mph 87 38 o/d 106 1 3 2 4 54 114 o/d 121 top speed | mph 91 41 o/d 101 1 3 2 4 60 112 o/d 118 top speed |
| **HANDLING** | Greater cornering power than Triumph on smooth roads, but more easily disturbed by bumps. Mild initial understeer on Pirelli Cinturato radials, changing progressively to oversteer and an impending slide. Grip maintained well in the wet | Excellent cornering power on Michelin XAs radials. Strong understeer emphasised by power-assisted steering, turning eventually to easily controlled oversteer. Adhesion remains good in wet and/or on poor surfaces |
| **LUGGAGE CAPACITY** cubic feet | | |

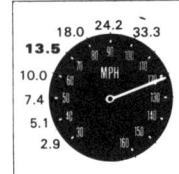

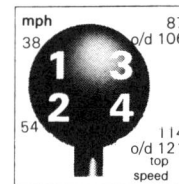

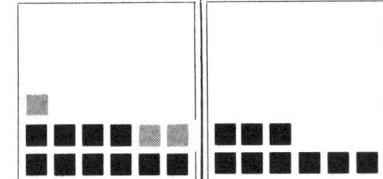

Tone indicates seats folded

**BRAKES**
RESPONSE in normal use. Deceleration (percent g) vs pedal load (lb)
A = Stag
B = GTE

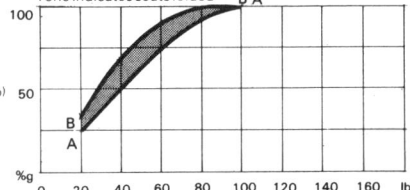

*FADE*
peak deceleration achieved in 10 crash stops from 60mph at one minute intervals
A = Stag
B = GTE

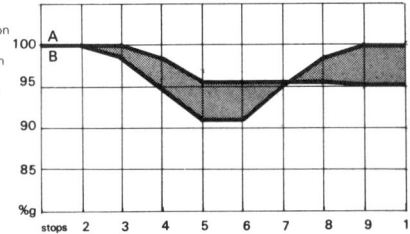

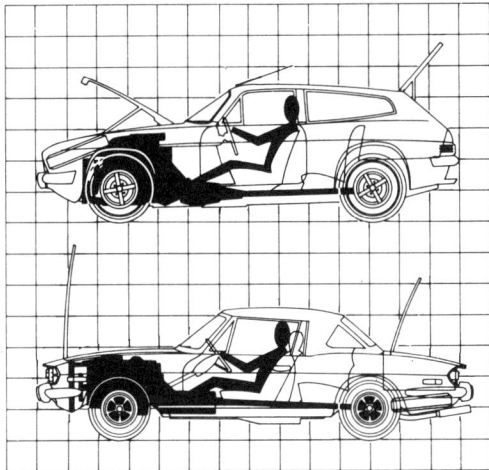

| | GTE | STAG |
|---|---|---|
| **DIMENSIONS** | inches | inches |
| wheelbase | 99.5 | 100 |
| front track | 55 | 52.5 |
| rear track | 53 | 52.8 |
| length | 171 | 173.75 |
| width | 64.5 | 63.5 |
| height | 52 | 49.5 |
| ground clearance | 5.5 | 5 |
| headroom, front | 33 | 34.5 |
| legroom, front | 28 | 30 |
| headroom, rear | 32 | 32.5 |
| kneeroom, rear | 6 | 5 |
| **ENGINE** | | |
| material | iron/iron | iron/alloy (heads) |
| bearings | 4 | 5 |
| cooling | water | water |
| valve gear | pushrod ohv | single ohc |
| carburettors | 1 Weber 40 DFAI | 2 Stromberg 175-CDS |
| capacity cc | 2994 | 2997 |
| bore mm | 93.67 | 86 |
| stroke mm | 72.42 | 64.5 |
| compression to 1 | 8.9 | 8.8 |
| net power bhp | 128 | 145 |
| rpm | 4500 | 5500 |
| net torque lb ft | 192 | 170 |
| rpm | 3000 | 3500 |
| **TRANSMISSION** | | |
| control | floor lever | floor lever |
| synchromesh | 1-2-3-4 | 1-2-3-4 |
| ratios to 1   1st | 3.16 | 2.99 |
| 2nd | 2.21 | 2.10 |
| 3rd | 1.41 | 1.39 |
| o/d 3rd | 1.16 | 1.13 |
| 4th | 1.00 | 1.00 |
| o/d 4th | 0.82 | 0.82 |
| final drive ratio | 3.58 | 3.70 |
| tyre size | 185/14 | 185/14 |
| rim size | 5.5 | 5 |
| **SUSPENSION** | | |
| front | double wishbones, coil springs and telescopic dampers | MacPherson struts, coil springs and telescopic dampers |
| rear | live axle, trailing arms, Watt linkage, coil springs and telescopic dampers | semi-trailing arms, coil springs and telescopic dampers |
| **LUBRICANT** | | |
| engine oil | 10W/30 | 20W/50 |
| sump, pints | 9.5 | 9 |
| change, miles | 6000 | 6000 |
| other lube points | 7 | 1 |
| lube intervals | 3000 | 6000 |

| | GTE | | STAG | |
|---|---|---|---|---|
| AIR | 24psi | rack and pinion | 26psi | rack and pinion |
| BRAKES | disc 10.6in | | disc 10.6in | |
| STEERING | 35ft turning circle | 3.6 turns lock to lock | 34ft turning circle | 4 turns lock to lock |
| AIR | | 24psi | | 26psi |
| BRAKES | drum 9in | | drum 9in | |
| WEIGHT | 2500lb | | 2632lb (3024 with optional extras, as tested) | |

anyone with experience of the related Triumph saloons.

The Reliant's dashboard layout is in the tradition of such cars—lots of gauges spread around and away from the matched speedo and rev counter. Down on the central console are the various switches. The setup takes a little puzzling out but works well enough.

Interior trim in the Reliant is entirely in black, regardless of the external colour. The Stag is available with a choice of trim colours. Both cars have vinyl upholstery; internal finish is to a high though not luxurious standard.

Borrowing again from the 2000, the Stag boasts a highly comprehensive and efficient heating/ventilating system that makes the front quarterlights (opened by turning a knob) quite superfluous. The Reliant's arrangements provide rather coarser adjustment between full heat and cold air but nonetheless work well enough. Power-operated windows are standard on the Stag.

Both cars differ from the norm in the matter of wipers. The GTE has a wiper and washer unit to clear the big back window—a job it does well. The Stag follows racing practice by fitting a pantograph arm to the windscreen wiper on the driver's side. It forms a parallelogram with the main arm and keeps the blade near-vertical.

Dual headlamps are common to the Stag and the GTE. Oddly, for these are fast, far from inexpensive cars, only on the Triumph has the opportunity been taken to fit quartz halogen bulbs.

The Stag also scores on interior illumination. A cunning light is let into the glove compartment lid so that it shines in and at the same time sends another beam downwards for map reading. Other lights illuminate the footwells and still more are built into the door armrests. When a door is opened the ground beneath is illuminated and a red light shines rearwards.

This is only one aspect of the Triumph's safety story. Like the Reliant it has a thickly padded scuttle surround and a collapsible steering column (the Reliant also has an impact-absorbing steering wheel). But there are many hidden details such as the inertia switch that cuts off the fuel supply if the petrol pipes are severed in an accident, or if the car turns over. An inversion is in any case made a safer affair than in most open cars by the provision of the stout roll cage hooped over the cockpit and linked to the top of the windscreen. Brought into the original design at a time when the US government was expected to insist on roll protection in open cars—something it has still to do—it was left in place, not least because it adds greatly to the stiffness of what would otherwise be a rather floppy body/chassis structure.

The Stag is much the quieter of the two. Its characteristically uneven V8 exhaust beat is audible but only just so, vibration negligible (as you would anticipate with eight small cylinders), the transmission is well insulated from the body structure, tyre thump and roar are equally well controlled by suspension insulation and wind noise

is at a low level that few comparable saloons can match. The Reliant suffers from an engine that is inherently much rougher running and is mechanically less refined. It makes its presence felt during acceleration but softens to a completely acceptable noise level once a steady speed has been attained. Much firmer springing than the Stag's, plus a good deal less rubber insulation and compliance, mean that road noise is a good deal more noticeable, as is wind roar over 100mph. Up to that point, oddly enough, it is almost non-existent.

Thanks to soft suspension, which allows a certain amount of body roll when cornering, the Stag has a smooth, even ride that makes a strong contrast with that of the Scimitar. The Reliant rides firmly, bumps are heard and sometimes felt and, on French roads, a well-laden Scimitar bottoms at the rear from time to time. But like most such sporting cars, the Reliant gets better as speed builds up; over 60mph it gives a smooth ride.

## PERFORMANCE, HANDLING, BRAKES

It's been a long time since we last tested a pair as evenly matched as this. A mere 3mph difference in maximum speed, within a tenth of a second of each other from 0 to 80mph and still only a second apart in reaching 100mph. All of which might come as something of a surprise, for the Stag has substantially more power (but less torque, admittedly), and the GTE, handsome as it is, looks considerably bulkier.

The Stag is held back to some extent by its high first gear, not to mention extra weight, though on the other hand the engine is a willing revver. The Reliant is aided

*Life inside both the Stag and the Scimitar is quite a comfortable affair if you're quick enough to get the front seats, but living in the back is a bit cramped for anyone apart from children. The Stag (top right) probably has the more expensive air about its living quarters since it has those little extras, such as electric windows, which make for real luxury motoring. Under the bonnet of the Scimitar is everything bar the kitchen sink, including the spare wheel and, of course, the Ford 3litre power unit, while under the Stag's bonnet (bottom right), there is just a great hunk of engine. The side view of the cars (below) shows up the Scimitar's box-like back end in comparison with the smoother Stag lines, while from the front you can see how two designers thought the front end of a grand tourer should look in 1970*

against the watch, though hindered otherwise, by the low Ford second gear. Up to a point, on the road, one can overcome this defect by the simple expedient of ignoring the rev counter and hanging on to 6000rpm. This is a full 25percent faster than the maximum net power speed and the engine feels rather distraught. Acceleration, as well as engine life, benefit by changing up much earlier.

Because of the improvements made in saloons in recent years, neither of these cars can truly be considered fast. A BMW 2800 will outrun them with ease and several other under-three-litre models would give them a gallop for their money. Their forte lies in providing reasonably high performance at what is nowadays little more than a medium price. Particularly when fitted with overdrive, they are both splendid machines for motorway cruising at 100mph and more. They are also gratifyingly economical. We have noted in the past the Reliant's frugal fuel consumption, especially when cruising on open roads. The Stag transpires to be nearly a rival for it in this respect. In both instances much of the economy can be traced to high gearing, reasonable mid-range torque and an ability to bowl along at a goodly pace in no more than the proverbial whiff of throttle.

The Stag is easily the more restful car to drive over long distances, being quiet and smooth by any standards, let alone the not very high ones of the Reliant. Similarly, the Stag is a much softer car in the suspension department. This gives rise to body roll during brisk cornering but pays off in facilitating much better adhesion than the Reliant's on bumpy surfaces. The Stag,

thanks to compliant springs and an independent rear end, has outstanding grip at the back, enabling one to be liberal with the throttle even when accelerating out of the tightest corners. In racing parlance, it puts the power down on the road. The Reliant, with much harder suspension and a live rear axle, rolls much less and still has very good traction but is more easily joggled off line by bumps, especially at low speeds.

Notwithstanding such shortcomings, the GTE is still the more enjoyable car to drive along winding, country roads. Primarily this is because it is altogether a more responsive, taut machine. In particular the steering is admirable—not unduly light, but precise and reasonably quick. The Stag is fitted as standard with Adwest power-assistance, a better steering servo system than any we know with the outstanding exception of the German ZF. Good as the Adwest is, it is excessively low geared and robs one of road feel, though it is not so ridiculously light and hypersensitive as some rival designs. Both cars have cumbersome turning circles.

Given a smooth surface the GTE displays better roadholding than the Stag by a small margin. Both cars offer excellent adhesion, attributable at least in part to the fact that in both cases the makers have standardised on a single make and type of tyre and can therefore develop their respective designs to get the best from them. The Stag is on Michelin XAs and the wider-rimmed GTE on Cinturato. Providing one keeps a light grip on the wheel the Scimitar feels entirely stable at maximum speed. The Stag is a little less reassuring in open form but came right up to scratch when the centre of pressure

was changed by fitting the hardtop.

Understeer is the dominant characteristic in both cars. It is more pronounced, and less easily overcome with the throttle, in the Triumph. The Reliant comes closer to neutral handling and can be brought closer still on a fast road by judicious application of torque in the bends.

Uncharacteristically, the Reliant proved more responsive to pedal pressure in our brake tests, needing less effort for a given rate of deceleration. Its braking, however, remained satisfactorily progressive, as did that of the Triumph. Under duress the Triumph's brakes faded fractionally and then, improving with heat, came back to their original, full power. Those of the Reliant showed a slight and steady fade pattern that would be too small to show up in normal, even arduous, driving.

## IN CONCLUSION

Anyone with £2000-odd to spend and a yen for semi-sporting transport could until recently have made a beeline for the Reliant. Now the advent of the Triumph has complicated the issue, and the choice becomes a difficult one. Evenly matched on performance and very nearly so on fuel economy and handling, these two differ—on paper—only in comparatively minor matters.

Yet the real difference lies in their natures. The Stag is in every way more sophisticated, a softer car leaning towards the saloon end of the spectrum. The GTE is simpler, harsher, more nearly a sports car that has been adorned with practical bodywork. It is the more satisfying of the two to drive hard, but the less relaxing. Its unique bodywork makes it particularly attractive. ●

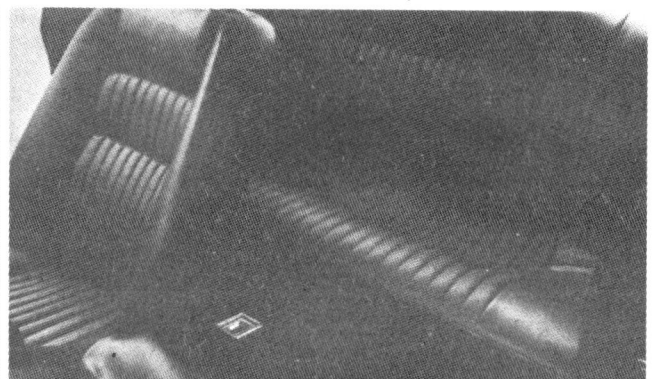

# THE STAG—
# AT LAST

**The Editor Belatedly Road-Tests
British Leyland's New V8 3-litre
Triumph Convertible**

*The Triumph Stag photographed outside the Buckley Arms Hotel at Dinas Mawddwy, not far from Bwlch-y-Groes, the once-popular 1-in-6 test hill. It was here that members of The Autocar's staff stayed 46 years ago, when reporting the RAC Small Car Trials. They naturally tried their 12/50 Alvis up the famous pass, ascending it in 1st, 2nd and 3rd gears. (The Stag went up fast in 2nd and 3rd gears.) This Welsh hotel still serves trout fresh from the Dovey river but its tennis court and private golf course have gone with the passing of the years.*

LAST MONTH I felt obliged to explain in an editorial why MOTOR SPORT had not published a report on the controversial new Triumph Stag convertible. I have since driven this new Coventry car, but it was touch and go, inasmuch as on the appointed Friday when I was to drive it, the telephone rang just before 1 p.m. and Simon Pearson of the S-T Press Office enquired when would I like my Stag? "Oh, by 3 p.m., to beat the traffic congestion, would be fine", I replied, "or 4 p.m. if that would be more convenient." "I'm afraid it will be more like 5.30", said the distant voice, "because there has been a hitch and we are still working on it." "Where is this stricken Stag?" "In Coventry." Apparently a lock had failed. I said I practically never lock a car, especially other people's. But it wasn't that kind of lock which had gone wrong; apparently either you could have ignition but couldn't steer, or could steer but not run the engine. . . .

I felt they had left it rather late to discover this, the test having been agreed some three weeks beforehand; eventually I left London at 10.30 p.m., arriving home in the early hours of the Saturday morning. It's all part of this road-testing job, of course, but I wonder how often the BLMC PROs hang about of an evening waiting for motor cars which haven't arrived? If Lord Stokes thinks my September editorial was spiteful, he should now think again! As I waited I contemplated testing a Saab 99 instead, which would at least be driving behind *half* a Stag power unit, and I wondered how this Coventry Alfa/Mercedes-eater would impress one who has named the Stuttgart products as the best-engineered cars in the World and who enthuses over those from Milan, although I have yet to drive an Alfa Romeo with power steering.

The late arrival of this long-postponed Stag at least enabled me to have a good preliminary thrash in it over empty night roads. But taking over an unfamiliar car in the dark isn't the most pleasant of motoring experiences, especially when you discover that, because of changed arrangements, you haven't sufficient money to fill the tank. However, I contrived to get the petrol tank about ¾-full, sank gratefully into a reasonably comfortable but spongy driving seat, was faced with a generously-stocked facia and a multiplicity of controls, found the

wiper blades worthless, cleaned the screen with my handkerchief, and set off. Very powerful Lucas dual quartz-halogen headlamps were obviously going to make light of my darkness but the cut-off was an alarming contrast. I found substantial left and right stalks, the right one doubling up for turn-indicators, flashers, a muted, shrill horn, and the lamps-dipping, which I don't altogether approve of on expensive cars, and which on this one had a long reach when in the full-beam setting. The 16-in. dia. steering wheel had a laced-on leather-covered rim, a gale of air blew at me from universally-adjustable central fresh-air grilles, until I later discovered how to turn them off, when the dribble of air from the additional gimballed vents at the facia extremities was quite inadequate. Interior trim is in non-dazzle black and the screen sill has a tray for picnic cups.

I thrust my way out of night-life London and was skirting Oxford satisfactorily soon, without seeming to have driven very hard. The Stag's V8 2,997-c.c. five-bearing engine is very reasonably smooth, but without the exhaust "wuffle" characteristic of its type. No need to discuss why a two-plane instead of a single-plane crankshaft is used, but worth noting that it is drastically over-square, at 86 × 64½ mm., and that it develops 145 (net) b.h.p. at its peak speed of 5,500 r.p.m., which is only 8 b.h.p. more than Ford's push-rod o.h.v. V6 gives at 750 fewer r.p.m. There is a chain-driven overhead camshaft above each cylinder bank, and twin sidedraught Zenith-Stromberg 1.75CD carburetters. An 8.8-to-1 c.r. provides for 4-star fuel.

Fumbling about in the dark I found a slide on the gear-lever knob which brings in, or cuts out, overdrive in 3rd and top gears (it is an optional extra). Later I found that this slide uncovers legends "In" and "Out", which would be appropriate to some cars which have to be rowed along on the gear-lever but not to the Stag which, although it is outpaced by almost all the comparable GT cars such as the Ford Capri 3000 GT, Reliant Scimitar GTE, Alfa Romeo 1750 GTV, Porsche 911T and Mercedes-Benz 280SL, somehow feels faster than it is and decently disposes, smoothly and quickly, of the slower-moving traffic. In fact, a 0-to-60-m.p.h. time of 10.7 sec. and a s.s.

¼-mile time of just under 18 sec. is not impressive for a 1970 3-litre car; the product of 145 b.h.p. and a kerb weight of 26 cwt. puts the Stag in the fast-tourer rather than the GT category.

As the traffic thinned out I was able to go faster and it was then that I added a third dislike to two others already evident, and one which makes the Triumph Stag quite unacceptable to me in its present form. The too-notchy gear-change controlled by a tall central lever I had put down as typically Triumph, and an irritating rattle coming from my left was traced later to free movement on the part of the empty passenger's front seat. This third misfortune relates to the power steering. The action is fairly light and smooth but with no feeling whatsoever of what the front wheels are doing, to the extent that they feel all the time to be about to break away their road grip. This prompts the driver to apply correction, and as the steering then becomes lighter, to over-correct, with the result that cornering becomes a very ragged business, feeling as uncomfortable as it looks. I tried to get used to this inconsistent power-assisted steering and in sober driving it is just about acceptable and reasonably geared (2¾ turns, lock-to-lock), with a very small turning circle. But as soon as I drove fast, or anticipated having to chuck the car about, the lack of positive feel to the steering became horrid. The sensation of the front wheels tucking in as lock is applied to combat what is actually understeer, is most disconcerting. Another peculiarity is that when the throttle is backed off after accelerating hard, as when changing gear, the back-end of the car gives a slight lurch.

Power-assisted rack-and-pinion steering is standard on the Stag—there is no option. This was presumably deemed necessary to woo the American customers but it is a great pity that something better isn't used. I would put up with strong-arm parking for precise manual control. I recall how the 4-litre Rolls-Royce-engined Vanden Plas Princess was ruined by light/heavy/light power-assisted steering, and the unfortunate Stag seems to have inherited something of the same sort. There is no excuse, for ZF and others have shown how effective good power steering can be. The column is adjustable to suit different requirements.

The Triumph Stag sent for appraisal was Pre-Production Car No. 7, but I assume catalogue cars are identical. The horrid steering apart, what can one make of it ? It is a convertible after the style of a Mercedes-Benz 280SL, with very nice Michelotti styling reminiscent of a Fiat Dino, and it was conceived for release in 1968. It is nicely finished, is a reasonable 2+2 coupé for which an optional hard-top is available to cover the fully-disappearing hood which has the protection of an elaborate (detachable) roll-over frame and would have been an excellent substitute for the late-lamented Sunbeam Alpine which Chrysler UK had to put down because of its 4.2-litre Ford V8 engine. It is, alas, not an easy hood to stow or erect—see pictures.

The interior of the Stag is essentially Triumph, which I have never much liked. The simulated dull-wood facia has a rather restricted lockable cubby on the left, complete with map-cum-interior lamp, dials to the right. The latter comprise expensive-looking, closely-calibrated Smiths speedometer and tachometer, a clock, fuel gauge, water thermometer and voltmeter but, curiously, no oil-pressure gauge. As my early-morning journey progressed I had not only a blue full-beam light shining in my weary eyes but an intermittent flashing from Triumph's all-services-light-up cluster, saying "Fuel". As petrol supplies scarcely exist in darkest Radnorshire after dusk and I was skint anyway, this ruined what pleasure I was trying to obtain from flinging the Stag in untidy lines round the corners. I need not have worried, for this unwelcome warning starts when the tank has nearly seven gallons in it—and is consequently ignored, dazzle apart, until the thing runs dry with the fuel gauge indicating just above empty! The lamps are switched on by a rotary switch to the right of the steering column, as if Spencer King or one of his team had been driving a Hillman Avenger, although the Stag's rotator is smaller and provides for parking lights as well as side- and headlamps. The controls are well endowed with symbols or lettering; there was one small knob with a mysterious inscription on it, which turned out to be the rheostat panel-lighting control. Down on the console you get

"The office" of the Stag, showing leather-rimmed, padded steering wheel, layout of instruments, long central gear-lever and press-buttons on the console for the electric window-lifts and interior lights.

the tumbler switches for the window lifts, rather close set to the central hand-brake, with its man-sized grip, when this is off; a similar switch which brings in two forward-facing interior lamps, and on the upright wall of the console there is a Triumph press-button radio, some well-contrived heater controls, and a pull-out ash-tray. The facia has a brake-failure warning light directly before the driver and other lights, in the cluster, for brake on, choke in use, high temperature, etc. Big knobs on the console wall pull out to bring in the choke, effective Triplex rear-window heater (with tiny in-use light), cigar-lighter and 2-speed fan. The l.h. stalk control provides for two-speed screen wiping and washing. The electric window lifts are excellent in themselves but as they require the ignition to be on before they will function, and the ignition key is needed to open the fuel-filler flap, the uncomfortable situation arises where the driver has to wait in driving rain with his window down while the car is refuelled.

British Leyland have really tried to make this the Mercedes-Benz/Alfa Romeo slaughterer they claim it to be. The doors have convenient grabs-cum-arm-rests, locks actuated by convenient small handles below the door-opening handles, knob-operated quarter-lights, and red warning lights on the doors. Opening them brings in courtesy lights, one each side of the transmission tunnel. The external lift-up handles, however, were stiff and not pleasant to use. The rear compartment side windows have good toggles for opening them as vents. The three-spoke steering wheel has a padded boss. The front-seat passenger has an under-facia shelf and there are elastic-edged pockets on the doors and backs of the seat squabs. The front seats have conveniently-placed release knobs for tilting them to give access to the deeply-welled back seat, and they can be raised by winding a crank handle, as well as possessing squabs easily adjustable by operating the long, plated side levers. These seats are of decidedly generous size but the upholstery is, alas, p.v.c. The doors have effective "keeps". The very deep, if shallow, 9 cu. ft. boot has a clear floor and is illuminated. Reversing lamps, and the heated back window if a hard-top is ordered, are fitted as standard. There is a wobbly anti-dazzle mirror. The wheels have stainless-steel trims and are shod with 14-in. Michelin XAS tyres. There are two keys, the long one for ignition/steering lock, the fuel-filler flap and the doors, the smaller one for the boot and glove box. An instruction book covered in imitation leather has been prepared for the Stag. The shape of the back seat well suited the Motoring "Dog", who, after having had her first puppies at the age of nine is now travelling again, and far more comfortably in the Stag than on the bouncy shelf of a Triumph TR sports car in which she once rode. But for adult humans head-room is somewhat restricted.

The conservative iron-block, alloy-heads, cross-flow engine of the Stag fires up promptly if given a little choke and runs smoothly from idling speed to its peak of 5,500 r.p.m. It will pull away from 500 r.p.m., or 10 m.p.h., in the 3.1-to-1 top gear and, trying to conserve fuel on my initial nocturnal acquaintance with it, I used top and o/d top gear satisfactorily for most of the journey. At 2,500 r.p.m. in top gear the accurate (at this speed) speedometer shows 50 m.p.h., or 60 if o/d is engaged, and at the legal maximum permitted on British roads the engine is running smoothly at less than 3,000 r.p.m. It will give just over 100 m.p.h. in o/d 3rd gear, and achieve 117 m.p.h. flat-out in its highest ratio. Two tail-pipe extensions which look like something from a battlefield probably assist in muffling the exhaust note so that Stag motoring is accomplished quietly, yet with a satisfyingly purposeful exhaust note when you open up. A good deal of exhaust heat is exuded under the car.

The gear-lever has weak synchromesh and a long travel, the clutch is heavy and a trifle sudden, and the servo-assisted disc/drum brakes are satisfactory without being outstanding; they tended to squeal under light pressures. The ride is soft but well-damped, roll being well controlled except when cornering very fast. Rear-wheel adhesion is apt to be lost on slippery surfaces when accelerating, but the ultimate cornering ability is high.

The fuel tank holds 14 gallons and in fairly gentle motoring I obtained a m.p.g. figure of 22.5 m.p.g. Driving harder, this decreased to 20.8 m.p.g., an overall figure of 21.4 m.p.g. The petrol gauge is unhappily optimistic.

The exterior *decor* of this handsome, compact fast-tourer is confined to the name "Triumph" in the rubber-tipped bumpers and a stag badge, although zoologists will not recognise the depicted animal as a stag, with its unnatural, overweight antlers. The name "Stag" is engraved on the steering-wheel boss, and name and motif appear on the sides of the tail. A clever feature is angling of the o/s wiper blade so that it sweeps right to the edge of the screen.

Turning to details, some of the luxury aspect of the Stag rubs off when gear-lever rattle between about 3,300 and 4,300 r.p.m. in 3rd gear (I thought this was a thing of the past, but met it again recently on the smallest Opel and now on the Stag) and wind noise, even with the hard-top in use, intrude. The pedals, labelled "L" for Leyland, are large but off-set to the o/s, which brings the brake pedal rather close to the accelerator, but gives room for a clutch-foot rest. There is not a lot of elbow-room and the padded anti-dazzle vizors are rather tricky to clip back; the driver's carries the five-stage sequence for lowering and raising the soft-top. The self-propping forward-hinged bonnet is light and lifts to reveal the V8 engine with its Stanpart

**Continued on page 106**

## THE TRIUMPH STAG CONVERTIBLE

*Engine :* Eight cylinders in a 90° vee, 86×64.5 mm. (2,997 c.c.). Overhead valves operated by single overhead camshafts. 8.8 to 1 compression ratio. 145 b.h.p. (net) at 5,500 r.p.m.

*Gear ratios :* 1st, 11.08 to 1; 2nd, 7.77 to 1; 3rd, 5.13 to 1; o/d 3rd, 4.2 to 1; top, 3.7 to 1; o/d top, 3.04 to 1.

*Tyres :* 815×14 Michelin XAS, on bolt-on steel wide-rim wheels.

*Weight :* 26 cwt. 0 qtr. 0 lb. (Empty, but ready for the road with hard-top on and approx. half-a-gallon of petrol.)

*Steering ratio :* 2¾ turns, lock-to-lock (power-assisted).

*Fuel capacity :* 14 gallons. (Range approx. 300 miles.)

*Wheelbase :* 8 ft. 4 in.

*Track :* Front, 4 ft. 4½ in.; rear, 4 ft. 4⅞ in.

*Dimensions :* 14 ft. 5¾ in. × 5 ft. 3½ in. × 4 ft. 1½ in. (high—hood up).

*Price :* £1,602, plus £491 15s. 10d. purchase tax. Total, as tested, £2,173 1s. 5d.

*Makers :* Standard-Triumph International Ltd., Canley, Coventry, England.

### Performance Data

*Acceleration :*

| | | | | |
|---|---|---|---|---|
| 0-30 m.p.h. | .. | 3.8 sec. | 0-70 m.p.h. .. | 14.2 sec. |
| 0-40 ,, | .. | 5.5 ,, | 0-80 ,, .. | 18.2 ,, |
| 0-50 ,, | .. | 7.5 ,, | 0-90 ,, .. | 23.4 ,, |
| 0-60 ,, | .. | 10.7 ,, | 0-100 ,, .. | 33.9 ,, |

*Speeds in the gears :*

| | | | |
|---|---|---|---|
| 1st .. | 41 m.p.h. | O/d 3rd .. | 101 m.p.h. |
| 2nd .. | 60 ,, | Top .. | 114 ,, |
| 3rd .. | 92 ,, | O/d top .. | 117 ,, |

*Overall fuel consumption :* 21.4 m.p.g.

*Views of the Stag's V8 power unit and,* right, *its luggage boot.*

# Attractive 2 plus 2 with V8 engine

Before writing a road test report of the Stag, it is necessary to analyse exactly what sort of a car it is. It is certainly a GT, whatever that may mean, but probably not a sports car in the traditional sense of the term. It is far from being a sort of junior E-type Jaguar, for no attempt has been made to place an undue emphasis on performance, and it is altogether more luxurious than a TR6 or an MGB.

The Stag is propelled by a very over-square engine of extremely modern design which bears no relationship to any previous Triumph power unit. Of moderate overall length, this V8 packs 3-litres under the bonnet that is fashionably short and permits a body with really useful rear seats to be carried on a wheelbase of only 8 ft 4 ins. The result is a car that is just the size everybody wants and in appearance it is an absolute winner—I parked it and watched the reaction of passers-by. It is only sold with an open body, with hood or very attractive hard-top, and the problem of rigidity has been ingeniously solved with an upholstered roll-over bar that also extends forward to the centre of the windscreen. The independent rear suspension permits the rear seats to be carried low, which allows the roof line to sweep down attractively in a way which is usually denied to four-seaters.

The Stag, then, is a compact car of delightful appearance which is rather more than a 2 plus 2. It has a big enough engine to have an effortless performance but, although this is of advanced overhead-camshaft design, it is in relatively "soft" tune. Very great trouble has been taken to insulate the suspension from the body, with the object of

reducing road noise to a minimum, and a bridge-pipe between the two exhausts systems avoids "V8 beat," while a viscous fan drive cuts down noise from this source.

The car is of solid construction and not particularly light, which one can sense from the way the doors close. The steering can be instantly adjusted both for rake and length of column, without stopping the car if desired. The seats are very comfortable and there is ample leg room for tall drivers, the minor controls all being sensibly positioned, as in the other Triumph models. The engine starts instantly from cold, but the test car sometimes had to be spun for a few moments when hot.

Very quiet and smooth, the V8 engine seems equally happy at any speed in its considerable range up to the red line on the rev-counter dial at 6500 rpm. Though an overdrive is available, it is impossible to over-rev on the direct top gear of the manual transmission or of the Borg-Warner automatic. The latter option was fitted to the test car and suited the characteristics of the engine very well. In manual form the Stag is capable of about 118 mph, but in automatic form I timed it at 115.4 mph, while one loses about half a second from a standstill to 60 mph. The speedometer registered 124 mph at maximum speed. Obviously the choice of transmission is a personal matter, the automatic appealing more particularly when there is a lot of town and traffic work.

The Stag has power-assisted steering as standard, and this is in keeping with its luxurious rather than sporting character. The lightness of the steering may be a little disconcerting at first but one soon takes it

for granted, and I often forgot about it. The car is very stable at high speeds, required absolutely no "holding," even in gusty winds. This exceptional stability is bought at the expense of a fairly marked understeering characteristic which can be reduced by the application of power but not entirely cancelled out. The tyres stick to the road very well, never bouncing over bumps, and in general

one does not feel that the car has a relatively short wheelbase.

A very comfortable ride is achieved without any wallowing, and there is also less roll on corners than would be expected. Perhaps the most impressive feature is the insulation of road noise ; in spite of using Michelin XAS steel-braced tyres, there is literally not a sound when one deliberately drives down a line of catseyes. Like all the other controls, the brakes are light to use and stand up to hard driving.

With the hood down and stowed beneath its decking and the windows raised—at the touch of a switch—the car is remarkably silent and most of the draughts are subdued. This is ideal motoring when the weather permits it. With the hood raised it is a different story, for the wind roar at speed makes conversation impossible, though the fabric does not actually flap. The hood is very neat but by no means easy to raise or lower—it would be best to fit the hard-top when winter comes. The hard-top looks very much a part of the car and at a casual inspection the Stag could be mistaken for a fixed-head coupé.

Much of the charm of this Triumph resides in the effortless performance of its engine. It has no preferred cruising speed, and there is a reserve of power that makes this an easy car to drive under all conditions and a splendid companion for long journeys. The four quartz-halogen headlamps let the miles or kilometres pass easily at night, and there is copious adjustable fresh air ventilation to stop the interior from becoming stuffy. The detail refinements are too numerous to mention, but this is a very well-equipped car.

At the start of this report, I said that it was necessary to analyse exactly what sort of a car the Stag is. Perhaps it is a revival of the " open touring car " of ancient times, or maybe it is something entirely new. It is certainly a very pleasant vehicle to use for getting around, and nobody minds being seen in really good looking cars. Above all, it is more than reasonably priced as such things are reckoned today, and the demand is bound to be enormous.

**SPECIFICATION AND PERFORMANCE DATA**
**Car tested :** Triumph Stag open 2 plus 2. price £1996 including tax. Extra : Automatic transmission, £105.
**Engine :** Eight cylinders, 86 mm x 64.5 mm, 2997 cc. Single chain-driven overhead camshaft per bank Compression ratio 8.8 to 1. 145 bhp (net) at 5500 rpm. Twin horizontal Stromberg carburetters.
**Transmission :** Borg-Warner Type 35 automatic transmission, ratio 1.0, 1.45 and 2.39 to 1, multiplied by torque converter reduction of 1.0-2.3 to 1 Hypoid final drive, ratio 3.7 to 1.
**Chassis :** Combined steel body and chassis. Independent front suspension by MacPherson struts with lower wishbones and anti-roll bar power assisted rack and pinion steering. Independent rear suspension by semi-trailing arms and coil springs with telescopic dampers. Servo-assisted disc front and drum rear brakes. Bolt-on steel disc wheels fitted 185-14 radial ply tyres.
**Equipment :** 12-volt lighting and starting with alternator. Speedometer. Rev counter. Voltmeter. Water temperature and fuel gauges Clock. Heating, demisting and ventilation system, 2-speed windscreen wipers and washers. Flashing direction indicators. Cigar lighter. Reversing lights. Electric window actuation. Radio (extra).
**Dimensions :** Wheelbase 8 ft 4 in. Track (front) 4 ft 4½ in; (rear) 4 ft 4¾ in Overall length 14 ft 5¾ in. Width 5 ft 3½ in. Weight 1 ton 5 cwt.
**Performance :** Maximum speed 115 mph. Standing quarter-mile 16.9 s. Acceleration : 0-30 mph. 4.0 s. 0-50 mph, 7.2 s. 0-60 mph, 10.0 s. 0-80 mph, 18.8 s. 0-100 mph, 29.8 s.
**Fuel consumption :** 18 to 24 mpg.

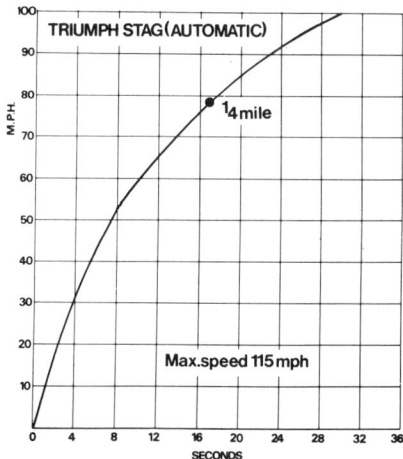

The Stag has clean lines and is attractive as either an open car or with the hard top.

The 3-litre V8 engine is very over square, being of extremely modern design (above). The interior is very comfortable and the controls sensibly positioned (below).

TRIUMPH STAG (AUTOMATIC)

¼ mile

Max. speed 115 mph

M.P.H.

SECONDS

# another angle on

WHEN the STI admen got to work on Stag a few months back and warned Mercedes and Alfa to watch out because their traditional sports limousine market was in danger, I thought they'd gone too far. I was wrong. Stag is the challenge to Merc. and Alfa that Standard-Triumph claim it to be.

With strictly few reservations after 1500-miles I can find little to change my original impression that the new V8 is going to be a winner of Jaguar XJ-6 proportions. And it deserves to be.

If STI can get it out at a better rate it could be an even bigger seller. Certainly the 15,000 a year intended from the new Speke factory is not going to be sufficient to meet demand for years.

I placed an order for a Stag after my brief encounter with it during pre-release testing in May. The promise then was that I might see one in September. Since then this date has been stretched to October, then November (due to strikes at component manufacturers) and now I think I will be lucky to see it before January!

Overseas sales aren't expected to start for 12 months after the release in Britain, though a few might find their way abroad before then.

Let's hope BLMC aren't going to be caught up in another Jaguar XJ-6 deal, where supply has never ever looked like catching up with demand.

The Stag is a good-looking two plus two convertible (with room for real people in the back seats), two doors, and a lightly stressed, torquey, all-aluminium three litre V-8 producing 145 bhp and using chain driven sohc.

Underside it borrows Triumph 2000 and TR-6 suspension, with independent semi-trailing arms and coil springs at the rear and struts, coil springs and dampers up front.

It comes with folding roof drophead which can be tucked away in 280SL style under a neat hood lid. For extra money you can have a snug-fitting hardtop with electrically heated rear window.

Other extras include overdrive on third and top gear of the standard four speed all-synchro box, automatic transmission and air-conditioning. Apart from a radio I don't really think there's much else anyone might want.

The unusual T-piece roll-over bar between the window pillars and the front screen adds to the torsional rigidity of the car. The bar can be taken off by removing three bolts. While I didn't try this on test I can imagine some owners might decide to do so when the hardtop is in position.

## Faults

My test car was pre-production prototype number 5. It had covered 5500 miles including a lot of testing and the pre-release drives back in May. By now any inherent faults should have started to show up. They had.

There'd been trouble with the steering lock ignition keys so that number 5 now had a key for ignition, a separate one for door and fuel tank cap and another for the glove box and boot, which caused no end of confusion.

When I first took over the test car I was disappointed that it had come without the hardtop. With 1500 miles

# the triumph stag

ahead of me through France, **Belgium and Germany** I had visions of a draughty and noisy journey.

As it turned out I had the top up only on the odd occasion. The noise level and buffeting even at cruising speeds up to an indicated 128 was less than any open car I have tested in almost 20 years.

I even left the hood down (and the heater at full blast) during an 80-mile drive in dense foggy conditions through Belgium and France. With just a jacket on I couldn't have been more comfortable.

Thanks to adjustable steering column and seats, the driving position is excellent.

The steering wheel of 15¾ in. diameter is leather-bound and nice to the feel even in very hot weather. The standard Adwest power-assisted rack-and-pinion steering which is used now on Jaguar, Rover and Triumph has four turns lock-to-lock.

I would prefer the wheel to be a little smaller and perhaps have a turn less, but in fact one quickly becomes accustomed to it and combined with

*(Continued on page 99)*

**Triumph's new-concept Stag will be a winner in the XJ-6 mould, reports Harold Dvoretsky after a 1500-mile test over some of Europe's toughest terrain**

Ian Fraser takes the

# *Stag* express to *Scotland*

YOU MAY SAY THAT WE WERE QUITE MAD. AND certainly that we cheated as well. After all, who but the deranged bent on deception would take a drophead to John o' Groats for a winter weekend *and* break the rules by not starting from Land's End? Brainstorms notwithstanding, we decided that we owed the Triumph Stag a healthy run within its homeland and to do this utilised a spare weekend by converting it into a leisurely 1700mile journey between 9.30pm on a Friday and 1.30am on the Monday.

Although we did not start from Land's End, since we were not indulging in anything as vulgar as a record attempt, our journey began and finished at Portsmouth; in a sense the south-north-south aspect was achieved but possibly not to the entire satisfaction of those people who have in the past set out with great determination to go from one end of the country to the other and back again in the shortest possible time. Our journey was somewhat different since it did not matter two hoots if we got to John o' Groats or not—which is probably the reason we made it without any difficulty at all.

Primarily we proved some things about the Stag which shorter jaunts would have failed to reveal. The car itself turned out to be a soft-top with a manual transmission and overdrive, although we had vaguely hoped that it might have been equipped with the optional hardtop. In retrospect it was to our benefit that our friends at Triumph provided a drophead, despite its shortcomings. The Stag is not a sports car and does not claim to be,

although its lack of clear definition is one of its sales advantages. For the sort of journey we did it would have been hard to find a better car and certainly difficult to locate one as suitable among the ranks of sports cars of comparable price. But please do not think that we are defending the Stag. Far from it; in fact we were able to turn up some basic weaknesses which would have made sports cars puff out their power bulges with pride.

As L J K Setright said recently, the only true sports car feature of the Stag is the size of the luggage compartment, which was brought home to us as we started to pack the vehicle. We were able to fit a couple of reasonable bags and odds and ends into the boot but the leftovers ended up on the back seat (we had hoped that the Stag would prove to be a family two-plus-two—and it would be for gentle journeys requiring not much more than a picnic lunch to be carried in the stern—but it would be out of the question as a holiday transport for two adults and, say, two children). Perhaps the best thing about the Stag's luggage compartment is its clear and unobstructed shape, and, being fully lined, it would be possible at a pinch to dispense with bags altogether and turn the boot into a wardrobe. Some of the weather we encountered on the way north was so bad that had our bags come floating out of the boot like barrels over Niagara Falls we would not have been all that surprised, but the compartment proved to be completely waterproof. The disadvantage of carrying tempting things like camera bags on the back seat is that the light-fingered fraternity would have very little trouble helping themselves.

Our voyage started in teeming rain at 9.30pm when we set ▶

sail for London, our first refuelling stop. Mistakenly, we decided to press on for a couple of hours instead of having dinner right there and then in London. When we did stop an hour or so later we must have selected the worst motorway eatery known to man. For that reason we will not alarm you with the sordid details of what must be the epitomy of poor eating. By 3am we were well clear of the M1 and making reasonable progress on the M6 despite violent crosswinds, heavy rain showers and gallons of dirty water being thrown up by lorries. Rather than continue the unequal struggle in a tired condition we found ourselves a quiet corner in a motorway-services car park, reclined the Stag's seats and slept for two hours before continuing northwards. At the moderate speeds dictated by the conditions and not having any desire to see if we could outrun the law in their XJ6s, the Stag proved to be remarkably economical, allowing us to reach Carlisle before refuelling—some 305 miles on 12.5 gallons, or 24.4mpg. The fuel-tank capacity of 15 gallons therefore provides a moderate-cruising range of well over 300 miles with reserves, thus making reasonably high point-to-point averages possible without really trying.

Our aim was to make our headquarters in Perth before setting course to John o' Groats. We arrived in the tidy, prosperous town just on 9.30am as the sun was making its attempt to break through. Our AA guide indicated a good hostelry called the Station Hotel, so we checked into that large and extremely clean establishment, obviously built during the great Victorian train era but having declined little in subsequent years, especially the service. The staff cheerfully prepared a late breakfast while we bathed and changed clothes prior to continuing north. We figured that if we were to get to John o' Groats before sunset we would need to be there by 3.30 or 4pm at the latest, so the Stag would have to work quite hard on the narrow but virtually traffic-free roads.

The overnight run to Perth

had revealed several important things about the Stag, not the least of which was its comfort. Neither of us had suffered any discomfort from the seats which, admittedly, were pushed back far enough to make it impossible for even children to be carried in the rear, thus leaving ample legroom in the front. While the Stag's seats really are good, we were less impressed by the hood, which flapped very badly at speeds around 85 to 90mph whereas the racket at 100mph must surely be akin to going through the sound barrier in a Vickers Vimy—unbearable enough to make 85mph just about the maximum sustained cruising speed. We were also driven to distraction by the totally inadequate illumination for map reading. There are a number of lights in the cockpit but none of them are bright enough or sufficiently well placed to make navigation anything less than a hideous task of torment.

Despite the rush from Perth to the Inverness fuel stop, the Stag managed a highly creditable 25.6 mpg overall from Carlisle at an average speed for the 269 miles in the mid-60s. The remaining 139 miles to the top of Scotland were a matter of some desperation to try to get there while there was still daylight. This was helped by the discovery that turning off the A9 at Bonar Bridge and rejoining it near Mound Station represented a 12 mile saving as well as providing some truly spectacular scenery, again with very little traffic apart from the occasional transient fisherman or hunter, lurching dog-laden Renault 4Ls along at a healthy rate.

Inevitably, John o' Groats was mildly grotty and bleak, although the air was clear enough for the houses on Stroma to be visible, with South Ronaldsay merely a shape in the distance. Our calculations revealed that we had averaged 62mph from Inverness and had, indeed, arrived pretty much as planned. Again the Stag proved itself as a comfortable tourer with very safe and predictable roadholding, although we became disenchanted with the standard-equipment power steering and its inherent lack of feel as we rushed

up the coast. Maybe the steering would have been better had the excessively large wheel been of smaller dimensions. It felt too big and awkward and there was just not enough feel to get the understeer set-up nicely in corners. On a similar tack, the brakes—disc front, drum rear you will recall—were not totally to our liking, the main problem being lack of feel once more, due perhaps to the use of split hydraulics. Anyway, that does not alter the fact that they worked well. Only after some really violent use did they show any sign of fade and even then recovery was fast enough and complete. The balance was good, too. Once or twice we got onto the brakes very hard indeed, but the Stag pulled up straight without the rear drums locking-up prematurely, as we half expected they might, and there was little enough dive to keep the headlamps effective.

Although not as sophisticated as it could be, the rear suspension worked well enough for us. As with the TR6 there is quite a lot of squat under hard acceleration, the disadvantages of which are

largely cancelled out by the well-controlled and comfortable ride. On the sometimes choppy and difficult roads in the Highlands the Stag's suspension smoothed out the worst of it quite fusslessly and also gave the impression of being much more rigidly constructed than is the case with the majority of open vehicles. Of course, what success the Stag has in this direction is helped enormously by that controversial roll-cage arrangement which, we suspect, is intended more to tie the body together than to overcome problems of gravel-rash on the scalp. Some of the hood noise that assailed our ears at high speeds was due to the fabric belting out a tattoo on the cage.

The three-litre V8 engine with its single overhead camshafts, behaved faultlessly, providing good power and torque up to and beyond the red line on the tachometer. However, its standing start acceleration was somewhat handicapped by the very high first gear which gives better than 40mph, but leaves second too low with its maximum of just on 60, while third runs to 91. On a number of occas-

ions we found ourselves floundering about between second and third and then having to make do with third. It would have been rather nice if the overdrive also worked on second to bridge the gap. As it stands an indicated 80mph in direct third equals 5500rpm or 4600 in o/d third, while it represents 4100 in top and 3250 in o/d. We thought the change lever was a bit notchy but having the overdrive engagement switch on the knob simplified things quite a lot. We also thought the instrument display should have included an oil pressure gauge while the legibility could have been better, especially at night. On a similar tack, the dipper arm on the steering column went too far away on the high-beam position to be readily retrieved in a hurry without taking one's right hand off the wheel.

Our return from John o' Groats (so named after Dutchman John de Groot who is reputed to have run the Orkney ferry in the late fifteenth century; eight of his descendants became joint owners of the estate and to obviate problems of precedence they built an octa-

gonal house with eight doors leading to an octagonal table; the house has long since vanished, although the local hotel, built in 1875, recalls it with an octagonal tower) to Perth was more leisurely and the refuelling stop at Iverness revealed that the 278 miles from there to the top and back again had consumed 12.2gallons which was rather better than we had anticipated. In sudden fear of missing our only real meal of the day, we pressed on back to Perth on mostly wet roads, the slop thrown up from them seriously damaging our capacity to illuminate the way ahead. And slipping along the floor of the valley between the Monadhliath Mountains and the Cairngorms we were attacked by an enormously strong wind and rainstorm which tried very hard to push the Stag off the road or, at the very least, carry off the hood. Such an acid test proved convincingly enough that the car is very stable directionally and that the weather equipment is made of stouter stuff than anything the Scots can blow at it. We were less fortunate in resisting the onslaught of a particularly

well-driven and extremely fast Lotus-Cortina in the hands of a local who was obviously practising for something important.

The obliging people at the Station Hotel, back in Perth, were quick to organise us a pleasant meal even though we had arrived back right on the death knock at 9.30pm. We were gone again by 10am Sunday, the Stag making its first cold start since some time in the forenoon on Friday. In a rare moment of bravado we decided to take advantage of crisp sunshine by lowering the top, the instructions for which are placarded on the driver's sun visor. Best as a two-man task, the hood nevertheless folds quickly and disappears from sight under the rear-hinged metal tonneau. Having released the tonneau it then refused to lock down again on the driver's side so we had to tolerate a few rattles and booms on our gentle journey over the black ice en route to Crieff via Dunkeld—a notably attractive little town on the banks of the Tay, with a recorded history of more than 1000 years and a medieval cathedral, to say nothing of the fact that this is exactly

where the Jacobite Highlanders routed William III's army in 1689. Besides providing us with a warming lunch, Crieff is also the tourist gateway to the Highlands, a role it fulfills more in the warmer months than during our visit. The insanity of open-air motoring having lapsed after our aimless wanderings through the highly diverse and staggeringly beautiful scenery, we turned the Stag's nose south again, towards the Forth Bridge, Edinburgh, the A1 and then the M1 to London. Cruising gently in the -70 to 80mph region in overdrive top (traffic density, accident rate and police activity making anything higher a risky procedure) the Stag returned 29.5mpg. We were back in Portsmouth, opening the front door, by 1.30am Monday, regretting only the limited time we had been able to spend in the Highlands. The Stag performed faultlessly, having found few rattles and lost none of its tune, but just as we left the petrol station in London, the Stag's speedometer groaned as the needle whipped around the dial before dropping back to the zero stop. Stone dead. ●

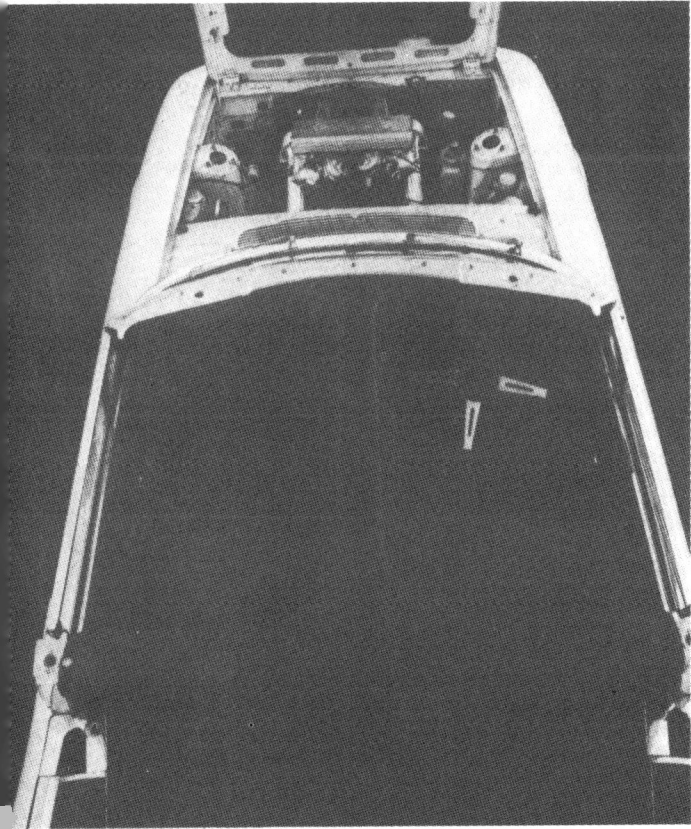

**The Stag's hood (far left) proved to be weatherproof but too noisy at speed, although it was quick to raise and lower, while U-shaped panel is metal tonneau to cover hood well. Controversial roll cage helped keep the body nice and tight (centre) over the uneven roads we encountered in the Highlands. We braved the freezing temperatures (top) to lower the top and take advantage of the few moments of feeble winter sunshine that did nothing to disperse some evil patches of black ice**

# AUTOTEST

## Triumph Stag Automatic (2,997 c.c.)

**AT-A-GLANCE: British Leyland's sporting 2+2 with automatic transmission. Performance still very brisk, but some worsening in fuel consumption. Noise level not much affected. Very comfortable and well equipped, with good handling but power steering lacking in feel for fast drivers.**

---

### MANUFACTURER
Standard-Triumph International Ltd., Canley, Coventry.

### PRICES
| | |
|---|---|
| Basic | £1,825.00 |
| Purchase Tax | £559.93 |
| Seat belts (approx.) | £14.50 |
| Total (in G.B.) | £2,399.43 |

### EXTRAS (inc. P.T.)
| | |
|---|---|
| Automatic Transmission* | £104.45 |

*Fitted to test car

**PRICE AS TESTED** .................. £2,503.88

### PERFORMANCE SUMMARY
| | |
|---|---|
| Mean maximum speed | 112 mph |
| Standing start ¼-mile | 17.9 sec |
| 0-60 mph | 10.4 sec |
| 30-70 mph through gears | 10.1 sec |
| Typical fuel consumption | 20 mpg |
| Miles per tankful | 300 |

---

IT is now just a year since the Triumph Stag was first announced. For some people, it has been a frustrating year of waiting for their Stag to materialize, because the moment the car arrived, demand shot well ahead of supply and stayed there. As we said in our Stag road test (30 July 1970), we were hoping to add an example to our long-term test fleet. Even though we were high in the queue, we had to wait some considerable time before we were able to take delivery of the Stag (with automatic transmission) which forms the subject of this test and will of course also be reported on in the usual way when it has done 10,000 miles.

It may be remembered that our original, manual transmission test car (one of the pre-production batch) returned rather disappointing performance figures. We therefore took the opportunity of trying another manual car while we were completing the test on the automatic, with the double object of updating our performance figures and comparing the two versions.

The automatic car is the result of a simple substitution of the Borg-Warner 35 transmission for the original four-speed manual gearbox and overdrive; nothing else is changed. Since the tyre size and the final drive ratio are unchanged, the overall gearing (giving 19.8 mph per 1,000 rpm, assuming no torque converter slip) remains the same. The transmission is controlled by a lever on the central console, moving in a conventional gate marked with Park-R-N-D-2-1 positions.

### Performance and consumption

The Borg-Warner transmission on our test car was set to give fairly relaxed performance, with automatic up-changes coming at 37 and 63 mph, even on full throttle. These change points did not take the engine to the 5,500 rpm power peak, and contrasted with the 54 and 89 mph which represented the 6,500 rpm limit in the lower ratios. The kick-down points were similarly low; 34 mph for the change down from intermediate to low, and 53 mph from high to intermediate.

Clearly, it should be possible to improve on the times for a fully automatic acceleration run by holding on to the lower ratios. It is a tribute to the stepless nature of the Stag engine's torque curve that, while this proved to be the case, the improvements were far from spectacular. In particular, there appeared to be little point in going beyond the power peak, and it must be said that the present settings are a good deal more intelligent than a keen driver might think at first. In absolute terms, the automatic Stag is respectably fast. By taking the engine up to the 2,100 rpm torque converter stall point against the brakes, a standing start can be made with a little cheep of protest from the tyres. The time to 30 mph was 4.1 sec; this was the only time which did not represent an improvement on the original manual-gearbox test car. For example, 60 mph took only marginally more than the magic 10 sec, and the quarter-mile was covered in under 18 sec.

The manual gearbox car tried this time, however, proved a good deal faster. In spite of having to carry two of our heavier testers, it matched the manufacturer's claims very closely, and actually improved on the claimed 0-60 mph time. Our table compares the performance of this car with that of the first test car; it can be seen that while there is little change in the bottom-end figures, those in the mid-range and at the top end are much improved. Perhaps strangely in view of this, we were unable to obtain much improvement on the original maximum speed, the car stabilizing at 116 mph (123 mph indicated) on a very long, level run. This is well below the power peak in overdrive top gear, whereas the maximum in direct top takes the engine to the far side of the peak. The automatic car achieved a mean of 112 mph, coming very close to the maximum of a non-overdrive manual car.

Despite the improvement in the mid-range performance of the manual car, it returned a marginally better fuel consumption than before. The automatic, driven in much the same way during the test period, managed only 17.2 mpg, a deficiency of almost exactly 20 per cent, and rather more than one would normally expect. It could well be that the looseness of the manual car (which had covered almost three times the mileage) accounts for a good deal of this difference. The steady-speed figures for the automatic show that it is using no more fuel at 40 mph than at 30, and scarcely any more at 50 mph; so it looks as though low-speed commuting is hardly the most economical way of using the automatic.

### Good cruising

Both cars proved very relaxed when cruising at 70 mph, with very little difference in noise level, even though the automatic's engine was turning over a good deal faster. As speed was increased, it was wind rather than mechanical noise which became intrusive, so that there was still little to choose between the two. Only at really high cruising speeds – 100 mph or more – did the automatic start to suffer from an

The Stag's air of interior quality has hardly suffered in this test car, despite the fact that it has covered over 15,000 miles. The seats are very comfortable, the controls good, and the instruments clear – although some drivers missed an oil pressure gauge

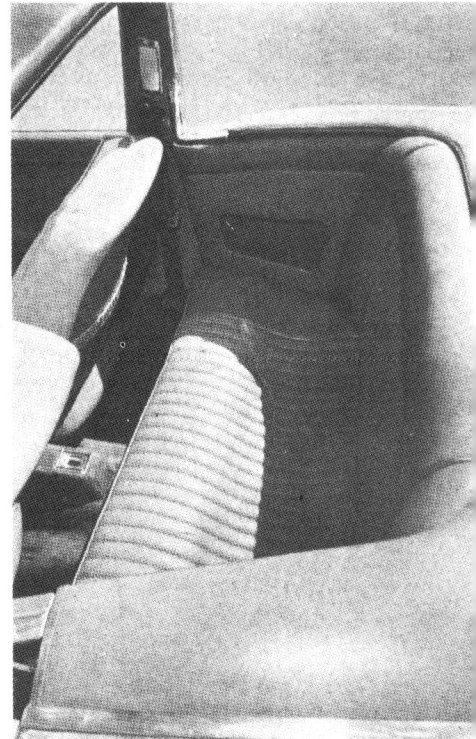

*Above: the Stag's boot is large in area but rather shallow, because of the spare wheel and large fuel tank lodged beneath the floor. It is carpeted throughout, but the wooden base panels are not easy to remove and replace*

*Above left: because of the room taken up by the hood stowage, the back seat is rather small. If its occupants are not too large it is quite comfortable, but the lack of legroom is a problem*

*Left: under-bonnet accessibility in the Stag is quite good (by British standards), and the engine looks relatively small. The dipstick, however, is difficult to get at, and very awkward to replace in the dark*

# PERFORMANCE CHECK

## Maximum speeds

| Gear | mph | | kph | | rpm | |
|---|---|---|---|---|---|---|
| | R/T1 | R/T2 | R/T1 | R/T2 | R/T1 | R/T2 |
| O.D. Top (mean) | 115 | 116 | 185 | 187 | 4,770 | 4,810 |
| (best) | 117 | 116 | 188 | 187 | 4,850 | 4,810 |
| Top (mean) | 113 | 113 | 182 | 182 | 5,710 | 5,710 |
| (best) | — | 113 | — | 182 | — | 5,710 |
| O.D. 3rd | 100 | 105 | 161 | 169 | 5,790 | 6,080 |
| 3rd | 92 | 92 | 148 | 148 | 6,500 | 6,500 |
| 2nd | 61 | 61 | 98 | 61 | 6,500 | 6,500 |
| 1st | 42 | 42 | 68 | 42 | 6,500 | 6,500 |

**Standing ¼-mile, R/T1:** 18.2 sec 75 mph **Standing kilometre, R/T1:** 33.4 sec 98 mph
**R/T2:** 17.1 sec 82 mph **R/T2:** 31.3 sec 102 mph

| Acceleration, R/T1: | | 3.9 | 5.8 | 8.1 | 11.6 | 15.1 | 19.6 | 25.7 | 36.9 |
|---|---|---|---|---|---|---|---|---|---|
| R/T2: | | 3.5 | 5.1 | 7.1 | 9.3 | 12.7 | 16.5 | 21.8 | 29.2 |
| Time in seconds | 0 | | | | | | | | |
| True speed mph | | 30 | 40 | 50 | 60 | 70 | 80 | 90 | 100 |
| Indicated speed MPH, R/T1: | | 30 | 40 | 50 | 61 | 73 | 85 | 97 | 108 |
| Indicated speed MPH, R/T2: | | 31 | 41 | 52 | 63 | 73 | 83 | 94 | 105 |

### Speed range, Gear Ratios and Time in seconds

| Mph | O.D. Top (3.04) | | Top (3.70) | | O.D. 3rd (4.20) | | 3rd (5.13) | | 2nd (7.77) | | 1st (11.08) | |
|---|---|---|---|---|---|---|---|---|---|---|---|---|
| | R/T1 | R/T2 | R/T1 | R/T2 | R/T1 | R/T2 | R/T1 | R/T2 | R/T1 | R/T2 | R/T1 | R/T2 |
| 10–30 | — | — | 9.1 | 9.0 | 8.2 | 8.0 | 6.4 | 6.3 | 4.1 | 3.9 | 3.0 | 3.0 |
| 20–40 | 10.8 | 11.0 | 8.2 | 8.2 | 7.1 | 7.1 | 5.5 | 5.6 | 3.5 | 3.4 | 3.1 | 2.8 |
| 30–50 | 11.1 | 10.4 | 8.4 | 7.8 | 7.0 | 6.6 | 5.4 | 5.3 | 3.8 | 3.7 | — | — |
| 40–60 | 11.3 | 10.5 | 8.4 | 7.8 | 7.1 | 6.8 | 5.6 | 5.4 | 4.9 | 4.3 | — | — |
| 50–70 | 11.6 | 11.6 | 8.8 | 8.3 | 7.8 | 7.6 | 6.7 | 6.0 | — | — | — | — |
| 60–80 | 13.2 | 13.1 | 9.9 | 9.2 | 9.1 | 8.7 | 8.5 | 7.4 | — | — | — | — |
| 70–90 | 16.1 | 14.4 | 12.6 | 11.2 | 11.9 | 10.2 | 13.7 | 9.8 | — | — | — | — |
| 80–100 | 21.9 | 17.1 | 18.0 | 13.7 | 17.9 | 13.3 | — | — | — | — | — | — |

### Fuel Consumption
Overall mpg, **R/T1:** 20.6 mpg (13.7 litres/100 km)
**R/T2:** 20.7 mpg (13.6 litres/100 km)

NOTE: "R/T1" denotes performance figures for Triumph Stag tested in *AUTOCAR* of 30 July 1970.

# ACCELERATION

SECONDS

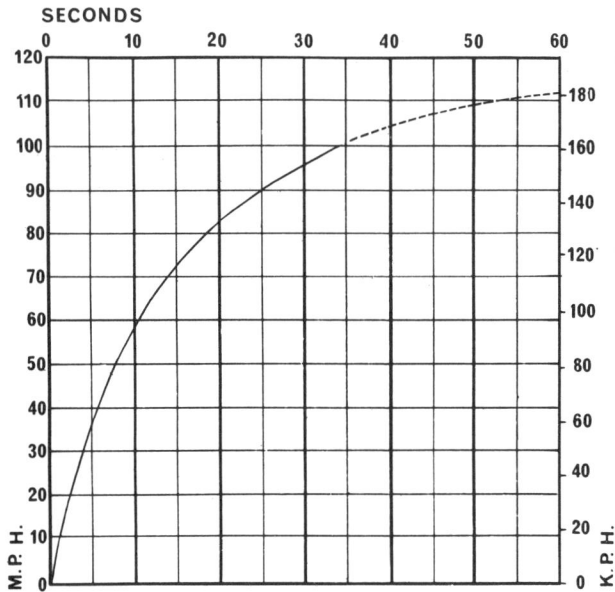

## SPEED RANGE, GEAR RATIOS AND TIME IN SECONDS

| mph | Top (3.70-8.50) | Inter (5.37-12.37) | Low (8.85-20.40) |
|---|---|---|---|
| 10-30 | — | — | 3.0 |
| 20-40 | — | — | 3.0 |
| 30-50 | — | — | 3.7 |
| 40-60 | — | 5.3 | — |
| 50-70 | — | 5.9 | — |
| 60-80 | 8.6 | 7.4 | — |
| 70-90 | 9.9 | — | — |
| 80-100 | 14.8 | — | — |

| SPEED MPH TRUE INDICATED | TIME IN SECS |
|---|---|
| **30** | 4.1 |
| 29 | |
| **40** | 5.8 |
| 39 | |
| **50** | 7.9 |
| 50 | |
| **60** | 10.4 |
| 61 | |
| **70** | 14.2 |
| 72 | |
| **80** | 18.6 |
| 84 | |
| **90** | 24.9 |
| 95 | |
| **100** | 34.5 |
| 107 | |

**Standing ¼-mile**
17.9 sec 78 mph

**Standing kilometre**
32.6 sec 98 mph
Test distance
1,050 miles
Mileage recorder
0.7 per cent over-
reading

# PERFORMANCE
## MAXIMUM SPEEDS

| Gear | mph | kph | rpm |
|---|---|---|---|
| Top (mean) | 112 | 180 | 5,660 |
| (best) | 113 | 182 | 5,710 |
| Inter | 89 | 143 | 6,500 |
| Low | 54 | 87 | 6,500 |

# BRAKES

**(from 70 mph in neutral)**
**Pedal load for 0.5g stops in lb**

| | | | |
|---|---|---|---|
| 1 | 32 | 6 | 35 |
| 2 | 28 | 7 | 35 |
| 3 | 30 | 8 | 40 |
| 4 | 30 | 9 | 50 |
| 5 | 35 | 10 | 45 |

## RESPONSE (from 30 mph in neutral)

| Load | g | Distance |
|---|---|---|
| 20lb | 0.36 | 84ft |
| 40lb | 0.69 | 44ft |
| 50lb | 0.82 | 37ft |
| 60lb | 0.95 | 32ft |
| 70lb | 1.0 | 30.1ft |
| Handbrake | 0.25 | 120ft |
| Max. Gradient 1 in 4 | | |

## MOTORWAY CRUISING

Indicated speed at 70 mph . . . . . . . 72 mph
Engine (rpm at 70 mph) 3,540 . . 1,495 rpm
(mean piston speed) . . . . . . 1,495 ft/min.
Fuel (mpg at 70 mph) . . . . . . . . . . 23.0 mpg
Passing (50-70 mph) . . . . . . . . . 5.9 sec

# COMPARISONS

## MAXIMUM SPEED MPH
Lotus Elan Plus 2S 130 (manual) 121 (£2,616)
Mercedes 280SL (automatic) . . 121 (£4,791)
BMW 2800 (automatic) . . . . 120 (£3,719)
Alfa Romeo 1750GTV
(manual) . . . . . . . . . . . . . 116 (£2,450)
**Triumph Stag (automatic) 112 (£2,489)**

## 0-60 MPH, SEC
Lotus Elan Plus 2S 130 . . . . . . . 7.4
Mercedes 280SL . . . . . . . . . . 9.3
BMW 2800 . . . . . . . . . . . . 10.1
**Triumph Stag .. .. .. .. ..10.4**
Alfa Romeo 1750GTV . . . . . . . 11.2

## STANDING ¼-MILE, SEC
Lotus Elan Plus 2S 130 . . . . . . 15.4
Mercedes 280SL . . . . . . . . . 17.0
BMW 2800 . . . . . . . . . . . . 17.2
**Triumph Stag .. .. .. .. ..17.9**
Alfa Romeo 1750GTV . . . . . . . 18.0

## OVERALL MPG
Alfa Romeo 1750GTV . . . . . . . 23.9
Lotus Plus 2S 130 . . . . . . . . 23.3
BMW 2800 . . . . . . . . . . . 19.9
Mercedes 280SL . . . . . . . . 19.0
**Triumph Stag .. .. .. .. ..17.2**

## GEARING (with 185-14in. tyres)
Top .. .. . . . 19.8 mph per 1,000 rpm
Inter .. .. . . . 13.7 mph per 1,000 rpm
Low .. .. . . . 8.3 mph per 1,000 rpm

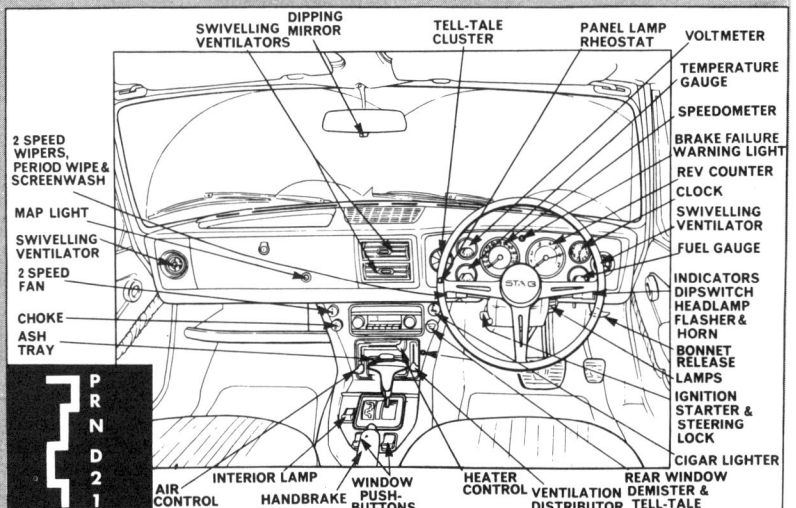

# CONSUMPTION

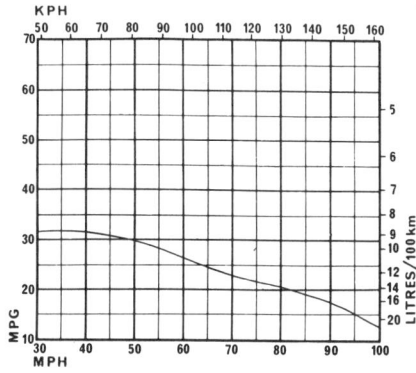

## FUEL
**(At constant speeds—mpg)**

| | |
|---|---|
| 30 mph | 31.5 |
| 40 mph | 31.5 |
| 50 mph | 29.8 |
| 60 mph | 26.3 |
| 70 mph | 23.0 |
| 80 mph | 20.6 |
| 90 mph | 17.1 |
| 100 mph | 12.3 |

| | |
|---|---|
| **Typical mpg** | 20 (14.1 litres/100km) |
| Calculated (DIN) mpg | 20.9 (13.5 litres/100km) |
| Overall mpg | 17.2 (16.4 litres/100km) |
| Grade of fuel | Super, 5-star (min. 100 RM) |

## OIL
Miles per pint (SAE 10W/40) . . 1,000

# SPECIFICATION FRONT ENGINE, REAR-WHEEL DRIVE

### ENGINE
| | |
|---|---|
| Cylinders | 8, in 90 deg. vee |
| Main bearings | 5 |
| Cooling system | Water, pump, fan and thermostat |
| Bore | 86.0mm (3.39in.) |
| Stroke | 64.5mm (2.54in.) |
| Displacement | 2,997 c.c. (182.9 cu. in.) |
| Valve gear | Single overhead camshaft per bank |
| Compression ratio | 8.8-to-1 Min. octane rating: 97 RM |
| Carburettors | 2 Zenith-Stromberg 1.75 CD |
| Fuel pump | SU electric |
| Oil filter | Full flow, replaceable element |
| Max. power | 145 bhp (net) at 5,500 rpm |
| Max. torque | 170 lb. ft. (net) at 3,500 rpm |

### TRANSMISSION
| | |
|---|---|
| Gearbox | Borg-Warner 35, 3-speed automatic |
| Gear ratios | Top 1.0-2.30 |
| | Inter 1.45-3.34 |
| | Low 2.39-5.50 |
| | Reverse 2.09-4.81 |
| Final drive | Hypoid bevel, ratio 3.70-to-1 |

### CHASSIS and BODY
| | |
|---|---|
| Construction | Integral with steel body |

### SUSPENSION
| | |
|---|---|
| Front | Independent, MacPherson struts, lower links, coil springs, telescopic dampers, anti-roll bar |
| Rear | Independent, semi-trailing arms, coil springs, telescopic dampers |

### STEERING
| | |
|---|---|
| Type | Power-assisted rack and pinion |
| Wheel dia. | 15¾in. |

### BRAKES
| | |
|---|---|
| Make and type | Lockheed disc front, drum rear |
| Servo | Lockheed vacuum |
| Dimensions | F 10.6in. dia. R 9.0in. dia. 2.25in. wide shoes |
| Swept area | F 220 sq. in., R 127 sq. in. Total 347 sq. in. (182 sq. in./ton laden) |

### WHEELS
| | |
|---|---|
| Type | Pressed steel disc, 4-stud fixing, 5.5in. wide rim. |
| Tyres—make | Michelin |
| —type | XAS radial ply, tubeless |
| —size | 185-14in. |

### EQUIPMENT
| | |
|---|---|
| Battery | 12 Volt 56 Ah |
| Alternator | Lucas 11AC, 45 amp a.c. |
| Headlamps | Lucas 4-lamp tungsten-halogen, 110/220 Watt (total) |
| Reversing lamp | Standard |
| Electric fuses | 8 |
| Screen wipers | Two-speed |
| Screen washer | Standard, electric |
| Interior heater | Standard, air-blending type |
| Heated backlight | Standard with hard-top |
| Safety belts | Extra, mounting points standard |
| Interior trim | PVC seats and headlining |
| Floor covering | Carpet |
| Jack | Screw scissor type |
| Jacking points | 2 each side under body |
| Windscreen | Toughened |
| Underbody protection | Phosphate treatment prior to painting |

### MAINTENANCE
| | |
|---|---|
| Fuel tank | 14 Imp. gallons (64 litres) |
| Cooling system | 18.5 pints (including heater) |
| Engine sump | 8 pints (4.5 litres) SAE 10W/40. Change oil every 6,000 miles. Change filter element every 12,000 miles. |
| Transmission | 11.5 pints ATF. Check fluid every 6,000 miles. |
| Final drive | 2 pints SAE 90EP. Change oil every 6,000 miles. |
| Grease | No points |
| Tyre pressure | F 26; R 26 psi (normal driving) F 26; R 30 psi (full load) |
| Max. payload | 728lb (330kg) |

### PERFORMANCE DATA
| | |
|---|---|
| Top gear mph per 1,000 rpm | 19.8 |
| Mean piston speed at max. power | 2,300 ft/min |
| Bhp per ton laden | 99 |

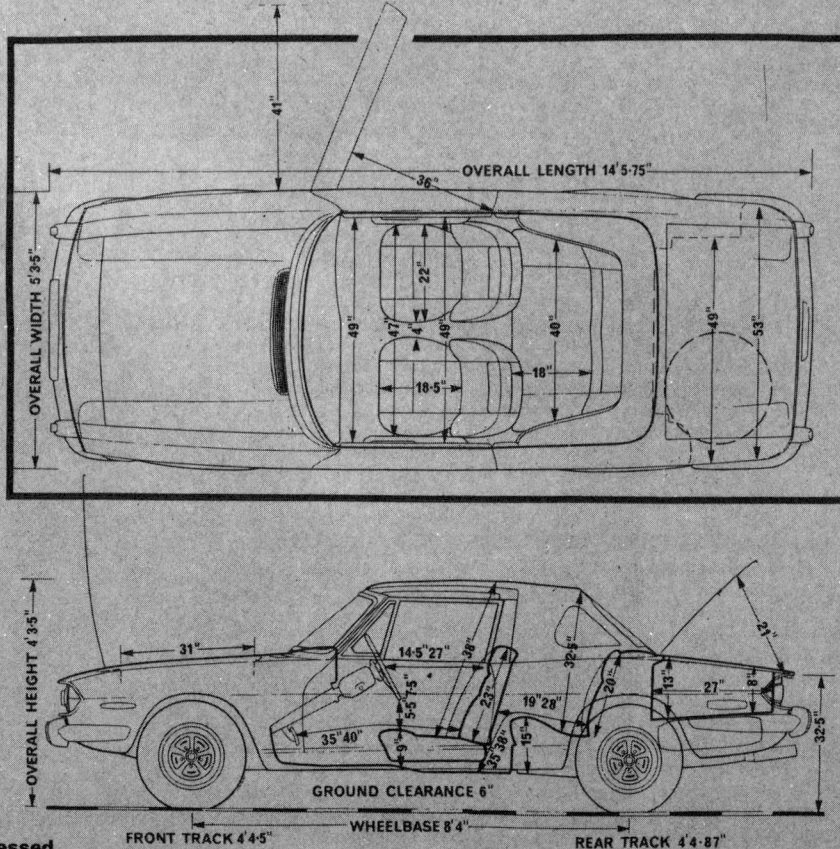

**STANDARD GARAGE 16ft x 8ft 6in.**

OVERALL LENGTH 14'5·75"

OVERALL WIDTH 5'35"

OVERALL HEIGHT 4'3.5"

GROUND CLEARANCE 6"

FRONT TRACK 4'4·5"    WHEELBASE 8'4"    REAR TRACK 4'4·87"

**SCALE 0.3in. to 1ft**
**Cushions uncompressed**

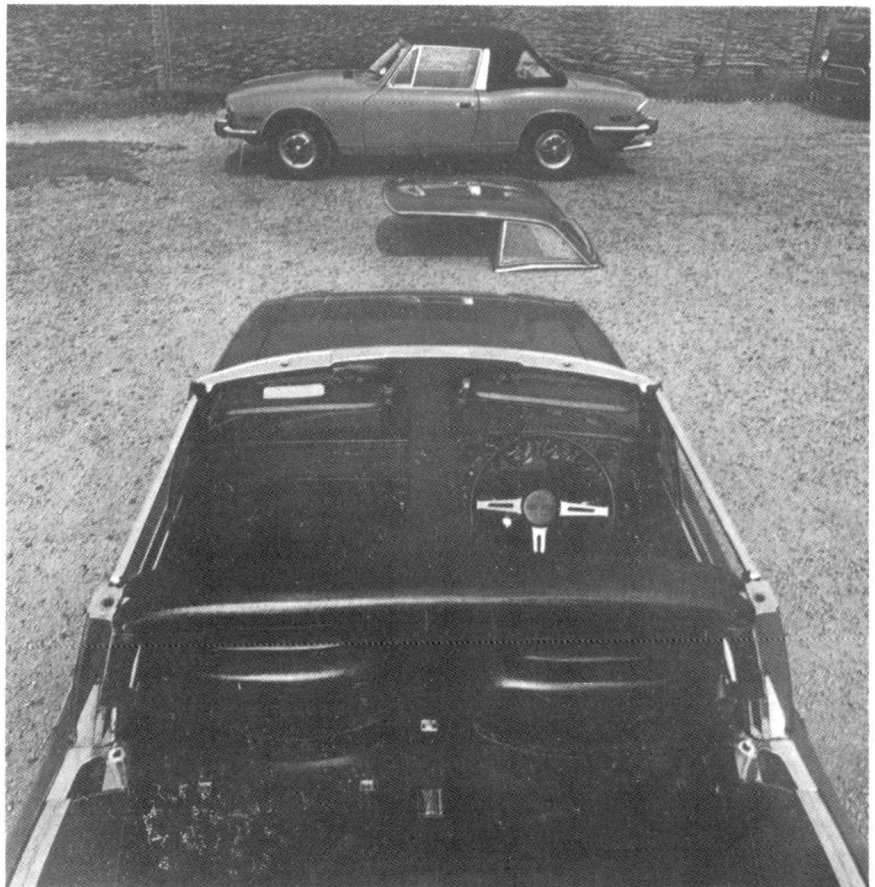

Stags at rest, rather than at bay. Both test cars were equipped with soft and hard tops; the hard top (which contains a considerable weight of glass) looks small enough here, but is a job for one man to lift and needs two to remove it from or replace it on the car. Nothing is changed in the cockpit of the automatic transmission car except for the transmission lever itself. The massive fuel tank, greatly appreciated during long journeys, is shaped to accept the spare wheel beside it; but beyond the usual wheel-changing equipment, there is little in the way of a tool kit

annoying low-frequency throb from the transmission. This noise, once started, could persist until the car was slowed below 70 mph.

Gear changing in the automatic car was usually smooth and quiet, even if the driver held the intermediate ratios to high speeds. The only thing to mar the performance of the transmission was the sometimes-violent 'clunk' as it changed down to first gear, or when engaging 'D' from neutral. At times, the kickdown was rather slow to operate.

In our original road test, we criticized the clutch action and gearchange of the manual car, while expressing the hope that it was non-typical in this respect. We can now establish that this was so; on this manual car the clutch was smooth in operation and required a medium operating effort, while the gearchange, though still having rather long movements, was light and precise. It was notable that the rushed change across the gate from second to third (which causes us difficulty during acceleration runs with some cars) could be made cleanly and consistently.

Comparing the brakes of the automatic with those of last year's car, their characteristics appear to have changed somewhat. A best stop, still of 1g, can now be made with only 70lb pedal effort instead of 120lb. There is a marked tendency for the rear wheels to lock first. On the other hand, the brakes no longer stand up quite so well to our fade test, showing a consistent rise in the pedal effort required and starting to smell after a few stops; it would seem that our automatic must have softer linings. The handbrake did not prove very effective on either car, whether on the test hill or in an emergency stop on the level.

### Handling and comfort

The Stag's power steering (which is standard equipment, of course) has a sort of Jekyll and Hyde character which takes quite a lot of getting used to. There is a certain amount of load built into the system which, for small wheel movements or at low speeds, almost disguises the fact that there is any power assistance at all. When one really comes to throw the car around, however, one finds that the load does not increase beyond a certain point, so that the motorway impression of a system with lots of 'feel' disappears and the driver tends to over-control and has to adjust his line in mid-corner.

Once the driver is used to this, the Stag handles remarkably well for a fairly nose-heavy car. Most of the time it understeers, tending to almost neutral behaviour before the track and camber changes caused by the semi-trailing rear suspension take charge and bring the tail round. Opening the throttles wide in mid-corner on a dry road can delay the onset of the final oversteer; there is a comforting feeling of added stability as the tail tucks itself down and the car spurts on along the chosen line. In the wet, however, such a technique results only in an abrupt loss of adhesion at the back end.

Straight line stability is normally good, but is rather upset by gusty sidewinds. The Stag particularly objects to being taken through the wind shadow of large lorries in these conditions.

For long distance driving, the Stag really scores with its good ride and excellent seating. If anything, the ride feels a little firm at first, but

there is plenty of spring travel to soak up mediocre surfaces and the damping is very good indeed. The seats are well shaped and adjustable (along with the steering column) to suit almost any size and shape of driver.

In the rapidly changing weather conditions of our test period, we really appreciated the excellent heating and ventilation system in the Stag, with its easily adjusted air-blending heater. Our only criticism was that the output of fresh air from the butterfly-type vents at the ends of the facia was not very great (in either car). The heater controls are not quite as easy to reach as most of the other controls, which are mostly in the form of column-mounted stalks or knobs.

### Needed in numbers

There is no doubt that the Stag is a very attractive car on many counts. We are now happy that the manual car goes better than our original figures suggested, while the automatic transmission suits the Stag very well, taking away the strain of city driving while leaving a very respectable performance. The car's handling is good, and its comfort and equipment excellent. Since we first tested it, it has risen in price, but this seems to be almost inevitable these days. More expensive or not, the queue for it is still long, and we hope that as much effort as possible will be put into satisfying the market.

# TRIUMPH STAG

## . . . V-8 Grand Tourer targeted to sell for just under $5,000 in the U.S.

Offering an unusual mixture of characteristics from luxury cars and from open sports jobs, the Triumph Stag is difficult to assess. As a luxury car it is small, a 2+2 (with moderate space in the trunk) which is 175 inches long and 63.5 inches wide. As a 3-liter sports job it takes almost 10 seconds from zero to 60 mph but goes on to 120 mph or more, handling easily but not inviting you to throw it around. As something between sports and luxury, it is an open job with excellent roll-over protection which can be converted quickly into a hardtop. It is tremendously comfortable for two people and practical for four, running very quietly indeed when hard-topped. It is very effortless to drive at normal-to-brisk speeds on any sort of road. I enjoyed my week with a Stag

tremendously, and thought it excellent value for money at a factory price about 20% below a Jaguar XJ6.

Triumph developed this Stag out of their 2000-2500 series of six-cylinder sedans. There is pretty much the same underframe, shortened by six inches to give a 100-inch wheelbase. Four-wheel independent springing is much the same, with revised settings and with larger-diameter wheels to accommodate bigger brakes. Instead of an old and highly-developed in-line six, however, there is a new V-8 based on the SOHC slant-four engine which Triumph have been building for the SAAB 99.

Fewer linear inches and more cubic inches make the Stag quite fast. In overdrive 4th, 120 mph is well within the "per-

**With a target price in the U.S. of about $5,000 for a bare bones car, the Stag is in a sense a replacement for the 2000-2500 sedans, not the TR6, which will be continued.**

mitted continuous" sector of the tachometer, and may be an underestimate of true maximum speed across flat desert. At about 2,900 pounds curb weight with the optional hardtop, however, the Stag's luxury comes rather heavy, and the docile motor does not give spectacular acceleration. Reaching 80 mph in a 17.4 second quarter-mile, the Stag is clearly towards the "roomy and luxurious" rather than the "sporting" side of Porsche's 911 range. It is also likely to sell in the U.S. for about $5,000, less than the least exciting Porsche 911T and with equipment of a very high standard.

I tested the Stag in all the three different

The standard soft top looks as ungainly when erected as those of any previous Triumph design. It also is noisy at speed. However, one man can fold it neatly under a steel lid.

Full instrumentation in a solid walnut panel is augmented by back-up warning lights at far left. Tach reads to 7,000 rpm and the outlet at the far right is for ventilation.

roles which one example can play. As an open car with the top folded completely out of sight, as an open car with the top raised, and as a closed car made by fixing a hardtop above the folded-away soft top. Two men can lift the optional hardtop on or off with reasonable ease, and the car may be purchased with hardtop only, with soft top only or with both, as on the test job. Emphatically, the Stag is more pleasing open or with the hardtop in use than with the soft top raised.

Open, the Stag does not expose the driver and front seat passenger to as much buffeting and back-draft as do many sports two-seaters, in spite of having an open rear passenger compartment. During some of the less-cold weather of England's winter, I was able to enjoy driving the Stag as an open car without needing more than an ordinary topcoat, the heater keeping my feet comfortably warm.

There is a reassuring impression of strength when this car is open, derived from the broad roll-over hoop and its sturdy link to the windshield frame.

Whereas many convertibles shake and rattle on bad roads, this one feels stronger than many steel-topped sedans.

Some open Triumphs of the past were notable for soft tops which could be raised or lowered very quickly and easily. Times change, though, and while the Stag s soft top can be furled completely out of sight below a hinged metal cover (and I did manage to raise it and stow it again without anyone to help me), neither process was quick. With the soft top up the car was truly weatherproof, and rear quarter panels as well as the zip-out rear window gave plenty of all-around vision. Unhappily, by the time you were doing 60 mph under the soft top there was too much wind noise inside the car. This became louder as the speed increased, so long

Optional overdrive, with which our test car was equipped, has a handy selector on the shift knob which eliminates the time-consuming lift-off on the accelerator of past designs.

76

journeys under the soft top were unexpectedly tiring.

Drop the hardtop over the folded soft top with a friend's help, move four accessible levers to lock it in position, and the Stag is an extremely refined car once again. Automatically, as you lowered the hardtop into place, you coupled up the electric circuit to a heated anti-mist and anti-ice rear window. Except that covered-over space behind and at each side of the rear seat, in which the soft top is stowed, reduces rear seat roominess, nothing hints that this is not an ordinary two-door hardtop. Nothing except acceleration-reducing weight which has been added to make a really strong open car, with its soft top, and with a hundred pounds of hardtop incorporating three more glass windows.

I'm afraid that perhaps I'm being *too* rude about the Stag's performance. It certainly isn't a car which has problems keeping up with its own shadow, but it could so easily handle more horsepower. Acceleration in overdrive is moderate, and feels downright slow because it is so quiet. In 4th you pick up more quickly and in 3rd better than your ears suggest, but you must downshift to 2nd for real lift-off sensations. Once down there you are in a really useful ratio which jumps you to 50 mph at 5,500 rpm where a yellow sector on the tachometer ("not recommended for continuous use") starts, and on to about 60 mph before you reach a red-marked 6,500 rpm limit.

Probably if the Stag had a less tractable engine, one would be in the acceleration-providing ratios more often. The way it is, you can crawl smoothly along at 10 mph in 4th with 500 rpm on the tachometer, and whether you press gently on the gas pedal or kick it down to the floor, the Stag will accelerate away unhesitatingly. In the snarled traffic of central London, the V-8 idled and moved me a few yards and idled and moved me a few yards in the smoothest possible manner, never vibrating or stalling. Up towards the top of the speed range, there was more of a tremor on the shift lever than I had expected, but except when you touch the shift lever this is a silkily smooth V-8, and a pretty quiet one also.

Three kinds of transmission can be ordered in a Stag. I've spent a couple of hours with an automatic transmission example, and really it was altogether too smooth and lacking in sparkle. There simply wasn't quick enough response to match the car's rather sporting appearance. Four-on-the-floor is much to be preferred if your wife knows how to use it. The full maximum speed of 120 mph is available without over-revving, and you can do 110 mph within the rpm range

approved for continuous driving.

A good way to spend more money is, however, to add the overdrive option to the four-speed box. Typically British, the overdrive gives a "hot shift" up or down when you move a switch, which, on the Stag, is recessed into the top of the shift lever. In 4th overdrive you could exceed 130 mph downhill without infringing the "continuous" rpm limit, or if you knew a sufficiently steep and straight hill, you could do 155 mph without actually over-speeding the engine! More to the point, 100 mph is a quiet and easy cruising gait when the law permits it, at which you have barely more than 4,000 rpm on the tachometer and are burning fuel at a rate no faster than 16 mpg. Overdrive also gives you an extra-versatile 3rd, the easy 5,500 rpm accelerates you to nearly 80 mph and then a flick of the switch without lifting off the gas produces a half-ratio shift, taking the speed on up to 95 mph at the same 5,500 rpm.

Gas mileage does not sell this kind of car, but the 16.5 gallon tank will usually take an overdrive-equipped Stag more than 300 miles. You are a hard driver if you don't get nearly 20 mpg, and it is quite easy to do better than that figure. Without the overdrive you would probably be about 2 mpg worse off on the open road, but no worse off in city traffic. An automatic transmission (overdrive cannot be added to this) might make 16 mpg look pretty normal.

Seats provided for British buyers do not have the headrests demanded in the U.S., So I can only hope that the jobs sent over the Atlantic are as fabulously comfortable as those kept at home. You can waste a lot of time fooling with non-powered adjustments for seat height, reach and backrest angle plus moving the steering wheel up-and-down and to-and-fro, but wherever the five adjustments are when

*Visibility is marred only by the excessively thick "B" posts, but these, of course, are necessitated by the integral roll bar. Rear quarter windows are fixed.*

*Someday stylists will realize that a functional wheel with, perhaps, a little more polish and chromed lugs is more attractive than any cover, particularly when the phoney lugs don't add up.*

you get bored, you'll probably find that you are very comfortable and that you stay that way for hours and hours. Side windows on the two big doors do have power lifts, one of which was finding its job a struggle on the pre-production car which had already suffered a lot of mileage in the hands of factory testers. Air conditioning will be available in cars delivered to the U.S., whereas in more temperate Britain we are content with plenty of fresh air vents plus a heater.

Quite a lot has been spent on furnishing and equipping the Stag well, for function as well as for appearance. Minor controls and switches are all easy to find when you are wearing the safety harness. A sports car's set of clear circular-dial instruments is supplemented by a group of eight warning lamps on the polished wood panel.

Without its four-wheel independent suspension feeling exaggeratedly flexible, the rigidly-built Stag rides very, very well indeed over most any kind of surface. Bulkier sedans from which the chassis derives, after all did take 2nd place in both the London-to-Australia Marathon and the London-South America-Mexico World Cup Rallye. You never wallow, yet you never seem to hear or feel the bumps

very much. Spencer King, who moved from Rover to complete the Stag's development, shares with Colin Chapman of Lotus a belief in soft springs controlled very firmly by powerful shock absorbers. This does allow some body roll when you corner fast, but attitude changes are slow enough not to be much noticed.

Powered rack-and-pinion steering is standardized on this car, plus high-speed radial tires, which, on the test example, were Michelins. Results are good, but with no need for compromises such as limited self-centering to suit cars without power steering, I would have hoped for something even better. Ordinary driving is extremely pleasant, the steering (geared

*A roll bar reminiscent of that in the last of the Shelby Cobras is standard equipment and fits under both the standard soft and optional hard tops.*

*Trunk holds 9 cubic feet of luggage but the prominent hinges mean that not all of this is usable unless soft duffle is wrapped around them. Note the full carpeting.*

2.75 turns from lock to lock) is quick in response and you soon forget that its lightness results from hydraulic power. Ordinary driving includes staying in your lane effortlessly through curves when cruising at 100 mph. Sadly, however, there is no real ''feel'' to warn you when you approach the limit of tire grip, so this isn't a sports job to be thrown around for the fun of controlling slides.

Yes, it will power slide. Traction in a straight line is quite good, but when accelerating hard out of a side street you lift a lot of weight off the inner rear wheel. No spin-limiting differential is offered. Other than when using 1st or 2nd out of a tight turn, road grip seems to be good in either wet or dry weather.

At the risk of re-starting the Anglo-American War of 1812, I'll describe the Stag brakes as somewhat American in their behavior. Discs up front and drums down aft have eager power operation, so a gentle push on the pedal produces quite sharp retardation. Measuring the best stopping distance from 60 mph, 125 feet on a good dry road which represents almost 1g deceleration throughout, was a matter of judging pedal pressure rather nicely to prevent the four wheels locking. A bit too much push, or a length of blacktop with less broken stone rolled into it, would see the Stag sliding along without any wheels rotating—and a locked radial does not give anywhere near its best grip. Also, too frequent stops without any cooling time would lead, very temporarily, to uneven braking with one or two wheels locking up earlier than the others. I can be unreservedly polite, though, about the handbrake, a sturdy pull-up lever between the seats which feels good and works as

powerfully as a handbrake on the wrong (i.e., rear) end of a front-engined car possibly can.

Although this is essentially a 2 + 2 rather than a full sedan, it is entirely possible to carry two men in the rear seat, and children are really comfortable behind even the tallest driver. Luggage capacity in a fully carpeted trunk is not intended for four people, especially with concealed hinges intruding upon space for rigid cases. As a two-seater there is, of course, plenty of capacity split between the trunk and the +2 rear seat. Against what standard should one judge the Stag when tabulating a rating?

It's utterly unreasonable to complain about a car being different, yet it does make the Stag difficult to assess. Emphatically, it is not a ball-of-fire sporting job as some TR owners may have hoped, but rather, it's a luxury 2 + 2 which runs quite fast over really long distances and has a gently sporting flavor. Very few teen-age males are likely to want Stags, whereas quite a lot of girls and mature men would seem likely to think it an extremely pleasing car at an entirely sensible price.

Joe Lowrey

*V-8 adapted from two SAAB 99 blocks*
*Twin Solex carburetion*

*Eight-blade fan*
*MacPherson strut tower*
*Brake servo*
*Distributor*

*Unlike Rover, which borrowed its V-8 from ex-Buick tooling, the SOHC Triumph consists essentially of a pair of fours as sold by them to SAAB on a common crank.*

# TRIUMPH STAG 2 + 2 OPEN/CLOSED COUPE

## PERFORMANCE AND MAINTENANCE

Acceleration:                                    Gears:

| | |
|---|---|
| 0-30 mph | 3.6 secs.— I |
| 0-45 mph | 6.6 secs.— I, II |
| 0-60 mph | 10.0 secs.— I, II |
| 0-75 mph | 15.5 secs.—I, II, III |
| 0-1/4 mile | 17.4 secs. @ 80 mph |

| | |
|---|---|
| Ideal cruise | 100 mph |
| Top speed(est) | 120 mph |
| Stop from 60 mph | 125 ft. |
| Average economy (city) | 16 mpg |
| Average economy (country) | 21 mpg |
| Fuel required | Premium |
| Oil change (mos./miles) | —/6000 |
| Lubrication (mos./miles) | —/6000 |
| Warranty (mos./miles) | 12/12,000 |
| Type tools required | SAE |
| U.S. dealers | 500 total |

## SPECIFICATIONS AS TESTED

| | |
|---|---|
| Engine | 182.9 cu. in., SOHC 2V V-8 |
| Bore & stroke | 3.39 x 2.33 ins. |
| Compression ratio | 8.8 to one |
| Horsepower | 145 (SAE net) @ 5500 rpms |
| Torque | 170 lbs.-ft. @ 3500 rpms |
| Transmission | 4-speed, manual w/overdrive |
| Steering* | 2.75 turns, lock to lock 32.0 ft., curb to curb |
| Brakes* | disc front, drum rear |
| Suspension | coil front, coil rear |
| Tires | 185 HR 14, Michelin XAS radial |

Dimensions (ins.):

| | | | |
|---|---|---|---|
| Wheelbase | 100.0 | Front track | 52.5 |
| Length | 173.8 | Rear track | 52.9 |
| Width | 63.5 | Ground clearance | 4.5 |
| Height | 49.5 | Weight | 2900 lbs. |

Capacities: Fuel ...16.5 gals.    Oil ...4.8 qts.
       Coolant ....11.0 qts.   Trunk ....9.0 cu. ft.

*Power assisted as tested

## RATING

| | Excellent (91-100) | Good (81-90) | Fair (71-80) | Poor (60-70) |
|---|---|---|---|---|
| Brakes | | 88 | | |
| Comfort | 96 | | | |
| Cornering | | 85 | | |
| Details | 92 | | | |
| Finish | 92** | | | |
| Instruments | 94 | | | |
| Luggage | | 88 | | |
| Performance | | 84 | | |
| Quietness | 92 | | | |
| Ride | 96 | | | |
| Room | | 88 | | |
| Steering | | 84 | | |
| Visibility | 92 | | | |
| Overall | | 90 | | |

**Assessed on pre-production example

## BASE PRICE OF CAR

(Excludes state and local taxes, license, dealer preparation and domestic transportation): $5,000 at P.O.E. East Coast (estimated).

Plus desirable options:
$ n/a Hardtop
$ n/a 3-speed automatic
$ n/a Air-conditioning
$ n/a AM/FM radio
$ n/a Chrome wire wheels
$ n/a Luggage rack
$ n/a TOTAL
$ n/a Per lb. (base price).

## ANTICIPATED DEPRECIATION

(Based on current Kelley Blue Book, previous equivalent model): $ n/a 1st yr. + $ n/a 2nd yr.

n/a — not available

# TRIUMPH STAG

*A luxury convertible coupe with no direct competitor*

PHOTOS BY CHUCK BOONE

THE VERY APPEARANCE of a new open car on the market these days is heartening; with the various influences of heavy traffic, smog, high speeds and overzealous legislators (who may legislate open cars out of existence to protect us from ourselves), open cars are fast becoming a rarity and in the few new ones that are appearing there is a definite trend to rollbars included in the bodywork.

The Triumph Stag is indeed a new open car—actually an open/closed car. In effect it's a brand-new car, even if it's basically a variation on the Triumph 6-cyl sedan series. This line began with the 2000 (2-liter six) several years ago, and Triumph sold it in the U.S. before Rover—and the Rover 2000—became part of British Leyland Motors. They didn't

# TRIUMPH STAG

want two 2-liter sedans to sell here, and as the Rover was already established they dropped the Triumph 2000 from the U.S. market. At home in England the 2000 has been updated to a Mark 2 version and expanded to include the 2.5PI (for petrol injection) with the TR-6's 2.5-liter engine. Michelotti, the Italian designer who often designs or restyles bodies for Triumph, was called in to style the Stag some five years ago, using this sedan series as a basis; this they did by shortening the sedan platform 6 in. and the external sheet metal an almost corresponding 8.5 in., meanwhile designing new external panels. In turn, this new styling treatment was grafted onto the sedans to give their Mk 2 series a new look.

So the Stag is a sedan transformed into a convertible, just as the typical American convertible is. Unusual, however, is the integral rollbar structure that arcs over the central body section and goes forward in the center to meet the windshield frame. There is a folding soft top and an optional hardtop. The suspension systems front and rear are those of the sedans but with minor modifications: MacPherson struts with coaxial coil springs and tube shocks plus semi-leading lower A-arms and an anti-roll bar that is not found on the sedans; semi-trailing arms, coil springs and tube shocks at the rear. The brakes are disc front, drum rear with the discs larger than those of the sedans.

The important mechanical difference between the Stag and the sedans is its new V-8 engine. Many readers will remember that Triumph developed the single-overhead-cam inline four for the Saab 99 and continues to build it for Saab. The Stag's V-8 is essentially a pair of these fours (which in the Saab are mounted at a 45° angle) with the stroke reduced from 78.0 to 64.5 mm and the bore increased from 83.5 to 86.0 mm. These new cylinder dimensions mean that the V-8 isn't twice the displacement of the four: whereas the Saab displaces 1709 cc (and now a new version with 87-mm bore displaces 1850 cc), the Triumph V-8 is just 2997 cc or three liters. Its two camshafts are driven by chain, the block is cast iron and the heads are aluminum; two Stromberg emission-control carburetors and mild valve timing give it a low-key output of 145 bhp net at 5500 rpm. At British Leyland there is a trend away from gross power figures and these are not given; by the usual 15% rule-of-thumb one can estimate the gross output as about 167 bhp.

A manual 4-speed gearbox, same as that in the 2.5PI sedan or TR-6 but with a taller 1st, is standard gear for the Stag; one can specify overdrive in addition to it or the Borg-Warner Type 35 automatic in place of it. Our test car had the automatic. The final drive ratio is 3.70:1 regardless of transmission; and there are a minimum of options available over and above the extensive standard equipment, which includes electric window lifts, power steering and brakes, a tilt-telescope steering wheel, radial tires and (for the U.S. market only) knockoff chrome wire wheels.

Triumph will be advertising that this is the only imported car with a V-8 selling for less than $10,000 in the U.S., and it is, at least for the moment. One will not buy this V-8 for performance, though; it is typically (for the layout) quiet and smooth and offers a discreetly sporting exhaust note from its paired twin pipes, but its refinement is more notable than its power or torque. It has a manual choke and in times past we might have complained of that in a $6000 luxury car; but in these days of cold-start and cold-running difficulties in emission-controlled cars a manual choke can be a saving grace. Manual choke aside, the Triumph V-8 (by the way, not to be confused in any way with the Buick-Rover 3.5-liter V-8) simply blends into the Stag's mild-mannered character and disappears.

An engine like this is well suited to an automatic and we concentrated on the automatic version for this test. It could be a better automatic—the Model 35 Borg-Warner has never been our favorite one, with its somewhat reluctant upshifting and the necessity of a floored throttle to get it to downshift for extra power—but it, like the engine, blends into the general character of the car and is unobtrusive. For contrast we sampled a Stag with the manual gearbox and found it both more and less satisfactory: more in that it can knock more than a second off the 0-60 time and improve fuel economy, less in that its shift linkage is anything but pleasant to operate. We think most Stag customers will specify the automatic.

The standard power steering, operating on a rack-and-pinion steering gear, was something we expected to be good; but it isn't. Road feel is sadly lacking, there seems to be a lot of friction in the system when one is making gentle maneuvers near the center of steering-wheel travel, and the hissing sounds it makes on corners turn into grating noises in parking. Surely with all the years of power steering experience behind the industry Triumph could have done better

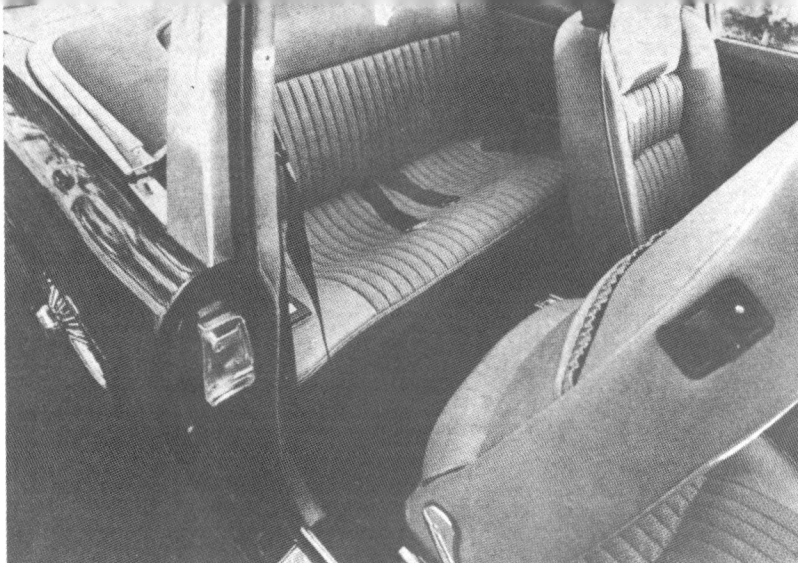

# TRIUMPH STAG

on this, and it's one of those details that can spoil a car.

Once past the steering deficiency, one finds generally good handling. The Stag is noseheavy in an amount normal for front-engine cars but handles close to neutral, and the 185-14 tires give it plenty of cornering power; being Michelin XAS they also do very well in the rain. But there's an odd quirk; often but not every time, when one lifts the throttle foot after accelerating hard, the car does a sidestep as if the whole rear suspension subframe had shifted on its rubber mounts. This may well be what is happening. Another theory is that the axle halfshafts bind on their splines under high torque, forcing the rear suspension into an unintended attitude that makes the semi-trailing arms' bushings give, and letting them return to normal when the torque is removed and the shafts can again slide on their splines. This is, we have found, a common characteristic of both Stags and 2.5PI sedans. We never noticed it in the 2000, but it doesn't have so much torque.

As do many cars with fully independent suspension and adequate wheel travel, the Stag combines a good ride with its good roadholding; the suspension is supple, neither too firm nor too soft. The body structure is adequately rigid—impressively so in view of its moderate weight—and generally rattlefree, so that rough-road driving is not a traumatic experience.

In normal, moderate everyday use the brakes feel good under-foot and behave in a reassuring way; they also are highly resistant to fade even if there aren't discs all around. In a panic stop, though, they don't behave themselves very well—at least with a lightly loaded car—as the rear wheels lock up and slide all too easily, making directional control a bit skittish. And the handbrake doesn't begin to hold on a steep grade.

The all-vinyl interior is simple and tasteful: comfortable and pleasing to the eye if not sumptuous. The seats are good, with reclining backrests and even a firmness control for the seat cushion, and one is restrained in a crash by inertia-reel 3-points belts that don't restrain one from reaching the controls. The controls are excellent. First, there's a padded-rim steering wheel that adjusts in-and-out and up-and-down via a single release lever. Then, on the steering column, there

are two stalks: one does wiper-washer functions and includes the possibility of one-stroke wiping by pulling it toward the wheel rim; the other works the directionals, horn, high beam-low beam and headlight flasher.

The instrument layout comes from the sedans, and it is also excellent with good-looking, legible round dials. Especially nice is an intriguing, effective round warning-light cluster consisting of radial segments in various bright colors for generator, oil pressure, handbrake, choke, fuel, high beam and directionals. There's also good ventilation through the dash air outlets which also handle cooled air when air conditioning is included. A lot of thought went into the Stag cockpit.

The two tops are, except for the fact that they cover four seats instead of two, a direct copy of the Mercedes SL equipment, right down to the method of folding the soft top and the levers and clamps for attaching either. There's a difference, though; on the Stag, nothing works quite right. It was only with great difficulty that we could get the cover panel secured down over the folded ragtop, and when lifting the hard-top off we found it easy to knock off the stainless trim strips along its rear lower sides. Finally, when either top was up the wind noise was anything but minimum, and there were rain leaks from the soft top and elsewhere. There were other evidences of poor assembly and shoddy design, but to enumerate them all would be to belabor the point that true quality of assembly is not one of the Stag's outstanding features.

In most of our office talks we compared the Stag to the SL. There's a big difference in price and quality as well as seating capacity, but that's the only car we know of that comes close to the Stag in concept: an open/closed GT of refinement with modest styling and performance, good roadability and extensive comfort-convenience equipment. Perhaps the Stag will find its greatest potential market among those who want an SL but can't afford it. With upgraded assembly quality, improved power steering and a little recalibration work on the brakes the Stag could fill that niche nicely, but as it stands it has too many detracting irritations to be a really satisfying car even after its basic character is accepted.

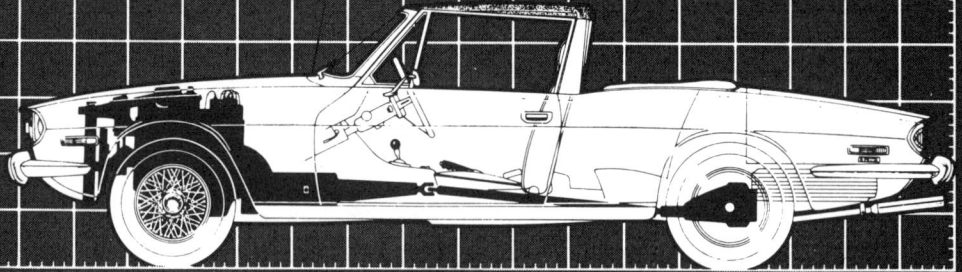

SCALE: 10" DIVISIONS

## PRICE

List price, all POE......... $5525
Price as tested............ $6622
Price as tested includes std equip (radial tires, power steering & brakes, electric windows, tilt-telescope steering wheel, chrome wire wheels), hardtop ($248), automatic transmission ($219), air conditioning ($495), AM/FM radio ($135)

## IMPORTER

British Leyland Motors Inc., 600 Willow Tree Rd., Leonia, N.J. 07605

## ENGINE

Type...................sohc V-8
Bore x stroke, mm...86.0 x 64.5
  Equivalent in.......3.38 x 2.54
Displacement, cc/cu in...2997/183
Compression ratio.........8.8:1
Bhp @ rpm....145 (net) @ 5500
  Equivalent mph.............101
Torque @ rpm, lb-ft.....170 (net) @ 3500
  Equivalent mph.............69
Carburetion: 2 Stromberg 175 CDSE
Type fuel required.premium, 97-oct
Emission control.....engine mods

## DRIVE TRAIN

Transmission....automatic; torque converter with 3-speed planetary gearbox
Gear ratios: 3rd (1.00)......3.70:1
  2nd (1.45)..............5.37:1
  1st (2.39)..............8.83:1
  1st (2.39 x 2.0).......17.66:1
Final drive ratio..........3.70:1

## CHASSIS & BODY

Layout.....front engine/rear drive
Body/frame..............unit steel
Brake type....10.6-in. disc front, 9.0-in. drum rear; vacuum assisted
  Swept area, sq in..........347
Wheels....chrome wire, 14 x 5½
Tires...... Michelin XAS 185-14
Steering type......rack & pinion, power assisted
  Turns, lock-to-lock.........4.0
  Turning circle, ft.........33.0
Front suspension: MacPherson struts, lower A-arms, coil springs, tube shocks, anti-roll bar
Rear suspension: semi-trailing arms, coil springs, tube shocks

## ACCOMMODATION

Seating capacity, persons...2 + 2
Seat width, front/rear.2 x 22.0/40.0
Head room, front/rear...38.0/32.5
Seat back adjustment, degrees..30

## INSTRUMENTATION

Instruments: 140-mph speedo, 7000-rpm tach, 99,999 odo, 999.9 trip odo, coolant temp, voltmeter, fuel level, clock
Warning lights: oil pressure, generator, brake system, handbrake, choke, fuel level, high beam, directionals

## MAINTENANCE

Service intervals, mi:
  Oil change.............6000
  Filter change.........12,000
  Chassis lube...........6000
  Minor tuneup..........6000
  Major tuneup.........12,000
  Warranty, mo/mi....12/12,000

## GENERAL

Curb weight, lb.............2795
Test weight.................3190
Weight distribution (with driver), front/rear, %...55/45
Wheelbase, in..............100.0
Track, front/rear.......52.5/52.9
Overall length.............173.8
  Width...................63.5
  Height (soft top)........49.5
Ground clearance............6.2
Overhang, front/rear...31.8/42.0
Usable trunk space, cu ft....8.2
Fuel tank capacity, U.S. gal..16.8

## CALCULATED DATA

Lb/bhp (test weight)........20.5
Mph/1000 rpm (3rd gear)....20.1
Engine revs/mi (60 mph)....2990
Piston travel, ft/mi.......1270
R & T steering index.......1.32
Brake swept area sq in/ton...218

## RELIABILITY

From R&T Owner Surveys the average number of trouble areas for all models surveyed is 11. As owners of earlier models of Triumph reported 7 trouble areas, we expect the reliability of the Triumph Stag to be better than average.

# ROAD TEST RESULTS

## ACCELERATION

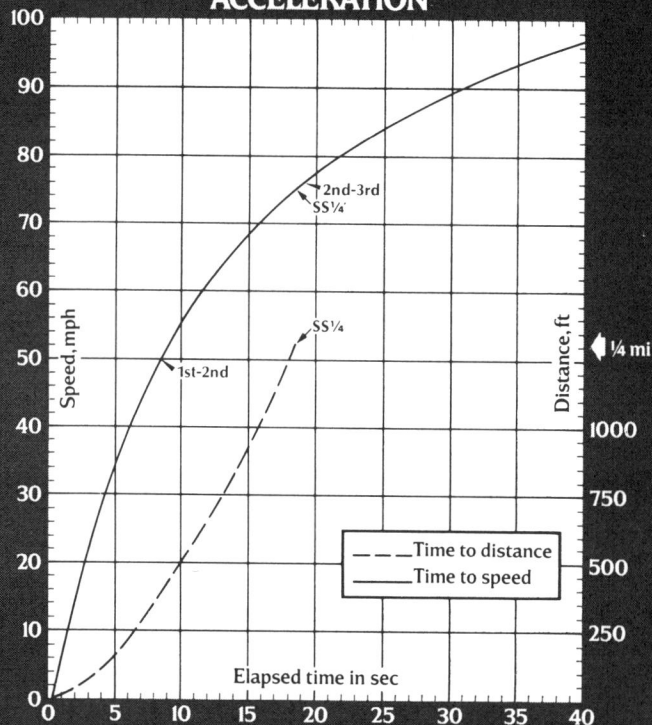

Time to distance, sec:
  0-100 ft.................4.0
  0-250 ft.................6.6
  0-500 ft................10.0
  0-750 ft................13.2
  0-1000 ft...............15.9
  0-1320 ft (¼ mi).......18.5
Speed at end of ¼ mi, mph....75
Time to speed, sec:
  0-30 mph................4.4
  0-40 mph................6.1
  0-50 mph................8.5
  0-60 mph...............11.5
  0-70 mph...............15.9
  0-80 mph...............21.9
  0-90 mph...............30.9
Passing exposure time, sec:
  To pass car going 50 mph...7.9

## FUEL CONSUMPTION

Normal driving, mpg.......17.9
Cruising range, mi.........300

## SPEEDS IN GEARS

3rd gear (6200 rpm).........112
2nd (6500).................82
1st (6500).................54

## BRAKES

Panic stop from 80 mph:
  Max. deceleration rate, % g..78
  Stopping distance, ft.......340
  Control..................fair
Pedal effort for 50%-g stop, lb..35
Fade test: percent increase in pedal effort to maintain 50%-g deceleration rate in 6 stops from 60 mph...................0
Parking: Hold 30% grade?......no
Overall brake rating.........fair

## HANDLING

Speed on 100-ft radius, mph..32.7
Lateral acceleration, g......0.714

## SPEEDOMETER ERROR

30 mph indicated is actually..30.0
40 mph....................41.0
50 mph....................51.0
60 mph....................61.0
70 mph....................70.0
80 mph....................79.0
Odometer, 10.0 mi.........10.0

# ACCELERATION

2nd-3rd SS¼
SS¼
1st-2nd
¼ mi

- - - - Time to distance
——— Time to speed

Speed, mph / Distance, ft

Elapsed time in sec

# TRIUMPH STAG

*Is it possible to say "Gran Tourismo" with a Coventry accent?*

● "Ha! Porsche-Romeo, or whatever you call that little thing you're sitting in over there, prepare yourself for the worst. After all those years of seeing cars like you braaaa-aaa-aaaping away from stoplights and scooting around corners at speeds that had my Plushmobile (I only bought it for my wife, you know) leaning over almost on its side . . . after all those years I'm going to get back at you.

"I see you staring at me. Noticing the wire wheels, no doubt, envious of all these dandy gauges and . . . uh, indicator *things* . . . that are mounted in front of me on a genuine wood dashboard. Don't sit there trying to be nonchalant. I can *feel* you listening to my V-8 motor. Yes it's a *V-8* and maybe I don't know what's in that little thing of yours, probably it's just a Four, but even if it's a Six, no matter. This car has an overhead cam V-8 and they told me only four other production cars in the world have anything like it so I know you've got to be jealous.

"Ah, there goes the light. Now's your Moment of Truth. Ooops, well you got me at the start but (braaaaaaaaaaaaaaa) look at me go by you now . . . and I'm still in first gear. *I* have three more gears, plus overdrive . . . *two* overdrives, actually. What a mistake that I didn't buy a car like this before. . . . it really is smashing."

And the answer to that question of why he didn't buy a car like the Stag before is that there simply never has *been* a car like the Triumph Stag. It finds its definition more by what it is not than by what it is. It, first of all, is not a sports car despite the fact that it does, in fact, have a Triumph-designed overhead cam V-8 engine and all-independent suspension. And it really isn't a *grand* touring car either, at least not in the classic sense of, say, a Mercedes 350 SL or even a BMW 3.0 CS. For, despite all the accouterments of luxury, such as power windows, leathery interior, fully carpeted trunk, it is very definitely a Triumph. A new type of Triumph, but a Triumph nevertheless, with the same feeling of rough-hewn capability that has been a trademark of all Triumph cars sold in this country.

And it's possible that it will be the last Triumph car to have this readily identifiable trait, now that Triumph has been merged into the British-Leyland conglomerate which also includes such traditional Triumph competitors as MG, Austin, and Jaguar. The development of the Stag was nearing completion when the merger came into effect and rather than cancel the car and the work that had gone into it, BLMC decided to proceed with putting it into production. But with certain modifications. When Triumph existed as a separate entity, the car was slated to be an out-and-out sports car, a vehicle that would compete with Porsche, Alfas, Austin-Healeys and even Jaguars in the marketplace. But BLMC wisely decided that it would not do to have the Stag selling directly against some of its other divisions' models and altered the concept of the Stag in order to create an entirely new market.

Perhaps the idea came from looking back on the success of the MG Midgets and Austin-Healey Sprites. Those cars were strictly intended for people who wanted to get their feet wet in the sports car market. Cars priced low enough to be attractive to someone who had never owned a sports car but at the same time promised to deliver a real sports car feel. Much in the same way the Stag is intended for someone who has never owned a GT car. It is priced well below the current Mercedes and BMW, Jaguar and even Porsche GT cars, yet it has the cachet of offering much the same feel and image as those cars: an image of luxurious, yet unostentatious, sportiness that is not presently available from any Detroit manufacturer and still costs thousands of dollars more in the import market. The only other car that approaches this concept is the Volvo 1800E—but even then the Triumph stands apart simply because it is so undeniably *British*.

The styling, which like most recent Triumphs, was done by Giovanni Michelotti, a designer who manages to create cars that look like evolutionary descendents of the strictly functional pre-war English cars despite the fact that his studio is located in Turin. Therefore the Stag disdains the currently popular trend toward swooping thin profiles in favor of a bulky high silhouette. And, in anticipation of future roll-over protection standards, the Stag incorporates a massive Corvette-like three-pronged roll bar. Recently BLMC decided to make the hardtop version standard with

# STAG

a soft top version available as an option. The hardtop hides the roll bar if you find its appearance that objectionable, but it also weighs 80 lbs. and requires two people to sweat it off (or on) should you ever decide to damn the aesthetics and get that wind in the face feeling.

The interior of the Stag holds two massive tucked and rolled, pleated bucket seats plus a rear bench seat which really is comfortable enough to accommodate two children or a single adult for relatively long trips. Everything is covered in leather-like vinyl or carpeted, with enough map pockets, parcel shelves, gloveboxes and armrests to make you think that a non-stop trip to and from Boston to Ketchikan would not be out of the question. In addition, the seats have adjustment levers and cranks sticking out everywhere so that you can personally tailor seat back angle, seat height, and fore-and-aft position. You also will find a flip-lever on the steering column that will adjust the wheel for height and reach.

Immediately behind the steering wheel are fingertip lever controls for the lights and windshield washers so that you never have to take your hands off the wheel when you are driving. On a wide center console are the window controls and interior light rocker switches and the gear shift lever (with a thumb switch mounted in the knob for the third and fourth gear overdrive—which is optional on manual transmission models).

Everything looks right, just the way you would expect of a British GT car—a place for everything and everything in

its place. Even the seat belts are kept neatly stowed away thanks to an inertia reel system (which also allows the wearer complete freedom to lean over to reach the glovebox or whatever, but clinches him tight under hard braking or in a crash, a truly excellent system that should be made universal). And you climb into the seats and begin making the proper adjustments to tailor the car to you, noting as you go along that visibility is excellent with the exception of directly overhead where the low windshield line will block off overhead streetlights. But for some reason as you continue to make adjustments you find that you can't *really* get comfortable—at least not in the way that you can in an Alfa or a Porsche. The problem is that even with the adjustable steering column you can't find a proper compromise between your legs and arms. The enormous steering wheel is canted so that it is impossible for many drivers to find a position—even with the fore and aft and vertical adjustment of the wheel column—that allows comfortable reach both to the top and bottom quadrants of the wheel and most drivers end up sitting closer to the wheel than they would like. The solution would be to make the steering wheel itself adjustable in the manner many Detroit manufacturers offer as an option. Then anyone could find a perfect seating position to go along with the comfort of the seats.

Turn the ignition key, adjust the manual choke, and

*Text continued on page* 89

87

## ACCELERATION standing ¼ mile, seconds

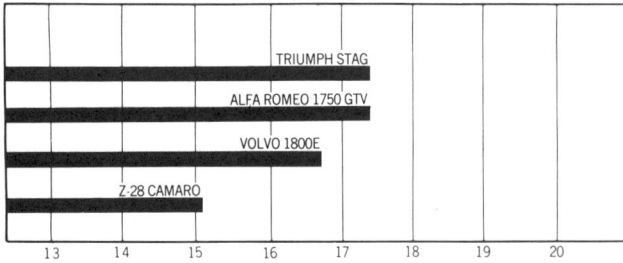

| | 13 | 14 | 15 | 16 | 17 | 18 | 19 | 20 |
|---|---|---|---|---|---|---|---|---|
| TRIUMPH STAG | | | | | | | | |
| ALFA ROMEO 1750 GTV | | | | | | | | |
| VOLVO 1800E | | | | | | | | |
| Z-28 CAMARO | | | | | | | | |

## BRAKING 80-0 mph panic stop, feet

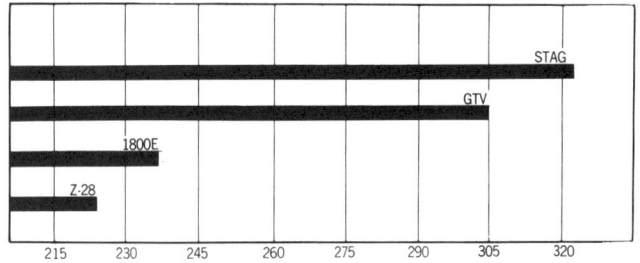

| | 215 | 230 | 245 | 260 | 275 | 290 | 305 | 320 |
|---|---|---|---|---|---|---|---|---|
| STAG | | | | | | | | |
| GTV | | | | | | | | |
| 1800E | | | | | | | | |
| Z-28 | | | | | | | | |

## FUEL ECONOMY RANGE mpg

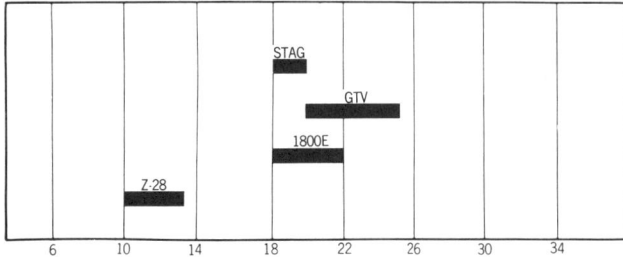

| | 6 | 10 | 14 | 18 | 22 | 26 | 30 | 34 |
|---|---|---|---|---|---|---|---|---|
| STAG | | | | | | | | |
| GTV | | | | | | | | |
| 1800E | | | | | | | | |
| Z-28 | | | | | | | | |

## PRICE AS TESTED dollars x 1000

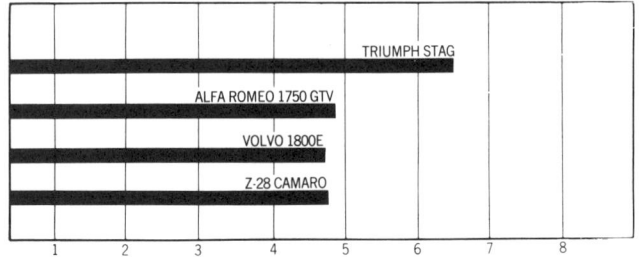

| | 1 | 2 | 3 | 4 | 5 | 6 | 7 | 8 |
|---|---|---|---|---|---|---|---|---|
| TRIUMPH STAG | | | | | | | | |
| ALFA ROMEO 1750 GTV | | | | | | | | |
| VOLVO 1800E | | | | | | | | |
| Z-28 CAMARO | | | | | | | | |

## TRIUMPH STAG

**Importer:** British Leyland Motors, Inc.
600 Willow Tree Road
Leonia, New Jersey 07605

**Vehicle type:** Front engine, rear-wheel-drive, 2 + 2-passenger convertible

**Price as tested:** $6465.00
(Manufacturer's suggested retail price, including all options listed below, Federal excise tax, dealer preparation and delivery charges, does not include state and local taxes, license or freight charges)

**Options on test car:** Base car, $5,650.00; Air conditioning. $495.00, Overdrive, $175.00, AM/FM Radio, $145.00

### ENGINE
Type: V-8, water-cooled, cast iron block and aluminum heads, 5 main bearings
Bore x stroke . . . . . . . . . . . 3.38 x 2.54 in, 85.7 x 64.5 mm
Displacement . . . . . . . . . . . . . . . . . . 182.9 cu in, 2997 cc
Compression ratio . . . . . . . . . . . . . . . . . . . . . . . 8.8 to one
Carburetion . . . . . . . . . . . . . 2 x 1-bbl Stromberg 175 CD-2
Valve gear . . . . . . . . . . . . . . . . . . . . . Single overhead cam
Power (SAE) . . . . . . . . . . . . . . . . . 145 bhp @ 5500 rpm
Torque (SAE) . . . . . . . . . . . . . . . 170 lbs/ft @ 3500 rpm
Specific power output . . . . . . . 0.79 bhp/cu in, 48.4 bhp/liter
Max recommended engine speed . . . . . . . . . . . . . . 6500 rpm

### DRIVE TRAIN
Transmission . 4-speed, all-synchro overdrive on 3rd and 4th
Final drive ratio . . . . . . . . . . . . . . . . . . . . . . . 3.70 to one

| Gear | Ratio | Mph/1000 rpm | Max. test speed |
|---|---|---|---|
| I | 2.99 | 6.7 | 44 mph (6500 rpm) |
| II | 2.10 | 9.6 | 62 mph (6500 rpm) |
| III | 1.38 | 14.5 | 94 mph (6500 rpm) |
| IV | 1.00 | 19.9 | 100 mph (5020 rpm) |
| IV/O.D. | 0.82 | 24.3 | 100 mph (4100 rpm) |

### DIMENSIONS AND CAPACITIES
Wheelbase . . . . . . . . . . . . . . . . . . . . . . . . . . . . . 100.0 in
Track, F/R . . . . . . . . . . . . . . . . . . . . . . . . . 52.5/52.9 in
Length . . . . . . . . . . . . . . . . . . . . . . . . . . . . . . . . 173.8 in
Width . . . . . . . . . . . . . . . . . . . . . . . . . . . . . . . . . 63.5 in
Height . . . . . . . . . . . . . . . . . . . . . . . . . . . . . . . . . 49.5 in
Ground clearance . . . . . . . . . . . . . . . . . . . . . . . . . 4.0 in
Curb weight . . . . . . . . . . . . . . . . . . . . . . . . . . . 2945 lbs
Weight distribution, F/R . . . . . . . . . . . . . . . . . 55.9/44.1%
Battery capacity . . . . . . . . . . . . . . . . . 12 volts, 56 amp/hr
Alternator capacity . . . . . . . . . . . . . . . . . . . . 540 watts
Fuel capacity . . . . . . . . . . . . . . . . . . . . . . . . . . 16.8 gal
Oil capacity . . . . . . . . . . . . . . . . . . . . . . . . . . . . 4.8 qts
Water capacity . . . . . . . . . . . . . . . . . . . . . . . . . 11.0 qts

### SUSPENSION
F: Ind., MacPherson strut, coil spring, Anti-sway bar
R: Ind., semi-trailing arm, coil spring

### STEERING
Type . . . . . . . . . . . . . . . . . . . . Rack and pinion, power assist
Turns lock-to-lock . . . . . . . . . . . . . . . . . . . . . . . . . . 2.25
Turning circle curb-to-curb . . . . . . . . . . . . . . . . . . . 34.1 ft

### BRAKES
F: . . . . . . . . . . . . . . . . . . . . . . . 10.6-in disc, power assist
R: . . . . . . . . . . . . . . . 9.0 x 2.2-in, cast iron drum, power assist

### WHEELS AND TIRES
Wheel size . . . . . . . . . . . . . . . . . . . . . . . . . . 5.0 x 14-in
Wheel type . . . . . . . . . . . . . . . . . . . Wire spoke, center lock
Tire make and size . . . . . . . . . . . . . Michelin 185 HR 14 XAS
Tire type . . . . . . . . . . . . . . . . . . . . Radial ply, tube type
Test inflation pressures, F/R . . . . . . . . . . . . . . 26/30 psi
Tire load rating . . . . . . . . . . . . . . 1450 lbs per tire @ 36 psi

### PERFORMANCE
| Zero to | Seconds |
|---|---|
| 30 mph | 3.3 |
| 40 mph | 4.9 |
| 50 mph | 7.1 |
| 60 mph | 9.5 |
| 70 mph | 13.4 |
| 80 mph | 17.8 |
| 90 mph | 24.0 |

Standing 1/4-mile . . . . . . . . . . . . . . 17.4 sec @ 79.3 mph
Top speed (estimated) . . . . . . . . . . . . . . . . . . . . 114 mph
80-0 mph . . . . . . . . . . . . . . . . . . . . . . . . 323 ft (0.66 G)
Fuel mileage . . . . . . . . . . . . . . . 18-20 mpg on premium fuel
Cruising range . . . . . . . . . . . . . . . . . . . . . . . 302-336 mi

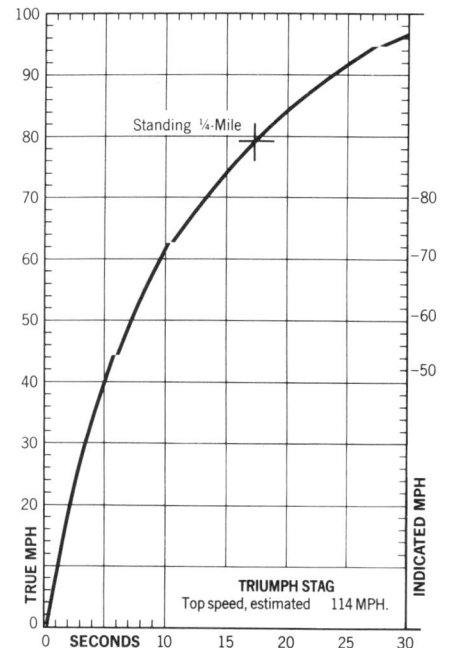

TRIUMPH STAG
Top speed, estimated 114 MPH.

*(Continued from page 87)*

there is a gratifying rumble from the twin exhaust pipes. Not the sharp rasp you might have expected from a Triumph V-8 but a definite sound of power nonetheless. The engine is a clever engineering utilization of the Triumph-designed overhead-cam Four (that powers Saab 99s and 99Es) displacing just under 3-liters and rated at 145 hp. The cast iron V-8, with its 90° Vee and aluminum heads, has a redline of 6500 rpm but is definitely more notable for its torque than high horsepower output. Rather than matching throttle inputs on a one-to-one basis there is a gradual build-up of power as the twin sidedraft carburetors open—a gentlemanly acceleration.

Because of our traditional dislike of the Borg-Warner Type 35 automatic transmission, we requested that our test car be equipped with the standard Triumph 4-speed manual transmission with a Laycock de Normanville overdrive operating on the top two gears. The Stag's all-synchro standard transmission is very notchy and will not tolerate lazy straightline shifts—"tolerate" may be the wrong word as the transmission actually gave evidence of being nearly impervious to abuse, perhaps "intolerant" would be better . . . you either line up the shift linkage properly or you won't get into another gear—Triumphs make it as straightforward as that.

The transmission itself is very versatile. First gear is a tall ratio, taking advantage of the V-8's torque, and is useful well into speed ranges where most imports would require a shift into second. It is also enjoyable to have the option of using overdrive, not only for interstate highway cruising in fourth but on secondary highways where third gear overdrive turns out to be an ideal ratio for moderate speed traffic.

The Stag's ride and handling characteristics are too much of a compromise, however. The ride quality is undoubtedly the smoothest of any Triumph ever sold in the U.S. It is almost sedan-like in cushioning you from bumps and irregularities—as well it should considering the MacPherson strut front suspension and semi-trailing arm independent rear suspension were originally designed for the Triumph 2000 sedan. It is evident that a lot of soft rubber has been put into the Stag's suspension systems to damp out that ox-cart harshness that has typified Triumph sports cars. Unfortunately, this new-found softness does absolutely nothing good for the handling characteristics of the car and in the long run makes it more difficult to drive than its sportier brethren.

The car weighs nearly 3000 lbs. with 56% of that weight located on the front wheels. As expected, the Stag understeers. And despite a front anti-sway bar and independent rear suspension, there is a lot of body roll in anything but gently sweeping curves taken at moderate speeds. Still the car feels capable on the highway, but extend the Stag and you'll find that Triumph is serious when it says that the Stag is not intended to be a sports car.

The power steering transmits very little road feel and, because of the fact that only two-and-a-quarter turns are required to go from lock to lock, initially you'll find yourself cranking in way too much lock on tight corners. Once you become more accustomed to the lack of sensitivity in the steering system you'll discover that what you do really doesn't make that much difference. The front pair of Michelin radial ply XAS (asymetrical) tires on our test car

had to do almost all of the cornering work—much of it in the form of simply scrubbing off speed—as the Stag would impolitely lift its inside hind wheel in hard cornering. It all looks very dramatic, what with the lack of camber compensation in the front causing the tires to roll under but soon you discover that no matter how you drive the Stag—even forcing the rear end to slide out and attempting to drift a corner—it will not generate much cornering speed. Fortunately it is at the same time very predictable in letting you know what it intends to do and no matter how radically you may be experimenting there is never an occasion when the car turns vicious.

The same predictability was found in braking. In moderate-use situations they are confidence-inspiring and easy to modulate, but once again widen the focus to include an extreme driving situation, like a panic stop, and performance is not what you hope for. On our test car the front wheels (equipped with power-assisted disc brakes) had to do almost all the work as the proportioning of effort back to the rear drum brakes was hardly a proportioning at all—the rears locked almost instantly. Because of this the Stag recorded a dismal 0.66 average G stopping force.

Like we said earlier, the Stag in every respect is a personification of the traits Americans tend to identify as being British. It is not an exciting car to the cognoscenti, but neither is it excitable in the hands of a novice. As a GT car in the traditional bucks-up sense there are other cars on the market that can offer more performance, or more well thought out amenities, and certainly sexier styling—but they don't sell for a base price of $5600.  •

# THE TRIUMPHANT MONARCH OF THE GLEN

THE BRITISH INDUSTRY HAS BEGUN
TO FIGHT BACK, AND A
WORTHWHILE EFFORT IS THE
NEW TRIUMPH STAG

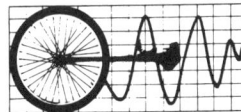

**SCG ROAD TEST**

FOR SOME TIME NOW BRITISH LEYLAND MOTORS—formerly British Leyland Motor Holdings, nee British Motor Corporation, Jaguar, Standard-Triumph, *et al* — has been telling us they were experiencing growing pains with their new conglomerate, but we felt it was just so much gas. Wait, they said. Wait and see what we have coming. We waited, and we saw . . . Austin America, MGC. A few dramamine tablets settled our queasy stomachs. Is that all there is to British Leyland? Is that all there is?

We watched as Fiat 124 sales climbed. Then Opel GT captured esthetic minds. And Datsun 240Z showed the country that Britain had lost its touch for designing sports cars. Then, after we had given up on the British industry as nothing more than a fond memory, along came the Jaguar XJ6, followed by an XKE V-12, and now a Triumph Stag, conservatively styled by Michelotti. We're told more replacements are on the way for the intrepid America and MG.

Through its family tree, British Leyland was once the *only* constructor of sports cars in the minds of carefree Americans. Oh, there was the occasional Mercedes-Benz 300SL or Ferrari America, but it was the MG, Austin-Healey and Triumph that people were buying, not to mention the venerable Jag XK. In other words, British makers had never been pressured into building anything different from what they had . . . until recently. The MG, for example, is as anachronistic as the Gatling gun. But we're not here to talk about how British Leyland used to be, rather how they're progressing in their new outlook.

The Triumph Stag is one of the new arrivals. It's not really a sports car, but a 2+2 touring vehicle that is quite unlike Triumph in approach. Other than a noisy ragtop, the car is quite solid by Triumph standards. It operates smoothly and quietly in the best American non-car tradition, which should make it an immediate hit. Only 2000 Stags are projected for sales in the first year, and we think that the figure might be pessimistic. The Stag fits into a price range ($5500 to $6500) unoccupied by any car of its type. The design isn't attuned to the young, enthusiastic driver . . . instead, it's aimed at the over-30 group which wants a modicum of performance and ease of operation, which the Stag fits nicely. And that rear seat is actually usable, able to accommodate two passengers with some comfort.

Although British Leyland officials in the U.S. will readily admit this is not really a sports car, they definitely would not turn off the older generations by denying it has sports car tendencies.

What seems a little irrational, if not tasteless humor, is the name: Stag. Quote — "Stag is rightly named after that noble

animal, noted for its speed and grace, but fierce devotion. The Monarch of the Glen is a monarch of the road." Let's just hope it makes some doe for the company.

We don't mean to sound anti-Stag, because as it turns out, it's not bad at all. Oh, there are a few things, and let's just get them out of our system before we continue:

The price tag is $1000 higher than we might expect from a car of this type, but then, so is everything else. The fuel light tends to flash ominously at the half-tank level. The engine overheats too readily under hot-weather, load conditions. When equipped with automatic transmission, the car suddenly becomes a slug . . . sluggish to the point of irritability. And that's really about all.

British Leyland seems to have exhibited foresight in its design of the Stag. First, the British sports car industry is totally dependent on the American market, since most (better than 90 percent) of these vehicles are sold in the U.S. Maybe it's time foresight became important. Anyway, British Leyland was the first to come up with a sales trend for sports cars . . . at least from what we've seen. It indicates that 1970, when some 130,000 units were sold, was the highest in history. And 1971 is expected to reach 170,000. But these aren't necessarily sports cars in the sense we once knew — the two-seater roadsters with the washboard ride. Instead, more and more are appearing with an occasional rear seat, and are really touring cars not sports cars. But then, that just goes to show the American point of view. Why, would you believe there are some who consider the Cadillac El Dorado to be a sports car?

Since almost every British sports car made is sold in the U.S., Englishmen aren't about to let Datsun and safety/smog regulations do it to them. There are some outstanding examples of the thought that went into Stag. The first is the overhead-camshaft, three-liter V-8. Although it has a modest 145 bhp, the engine is quiet, relatively torquey and meets federal pollu-

tion regulations easily because of its size, without forsaking performance. Another example of foresight is the roll-over bar. It's not just a styling gimmick, as many would believe. Instead it's an integral feature of the automobile in anticipation of federal legislation which may soon make it mandatory for manufacturers to incorporate such hoops on all convertibles and hardtops.

In addition to the convertible, Stag is also available with a removable hardtop. It's a garish-looking item that appears to have collapsed in the middle . . . but it works. It adds a silence and security, when in place, that the ragtop could not hope to offer. And with the growing interest in air-conditioning, the convertible market is on a rapid decline. The idea of a hardtop, if wanted, is a good idea, but its execution on the Stag could use some improvement. For instance, it requires two people to remove the burdensome hardtop . . . which is all right, assuming there are two people available. And the chrome strip at the base of the hardtop, where it meets the body, is easily pulled out of place when lifting or lowering the top.

Like most Triumphs, the Stag has a tasteful interior, with wood-grained facia, vinyl and pleated seats, and a gearshift lever and steering wheel that reek of performance. Except for the absence of an oil-pressure gauge — it was replaced by, of all things, a clock — the instruments are legible and all switches within easy reach of the driver. The steering wheel is adjustable fore-and-aft and up-and-down with the flip of a lever at the left of the steering column. The shift lever, as they say in Newport Beach, "falls readily to hand." Seating is excellent, with adjustments for height, reclinability and short- or long-leggedness.

Before we forget about it, the headlights are excellent. It's one of the few cars we've tested that actually illuminates the roadway.

The first thing you notice about Stag, other than its unexcit-

(Text continued on page 97)

# TRIUMPH STAG

## PRICE
Base . . . . . . . . . . . . $5575 (POE West Coast)
As tested . . . . . . . . . . . . . . . . . . . . $6205
With options . . .AM/FM radio, air-conditioning

## ENGINE
Type . . . . . . .90 degree V-8; water-cooled, chrome-iron block, aluminum-alloy head
Displacement . . . . . . . . .182.9 cu. in. (2997 cc)
Horsepower . . . . . . . . . .145 hp @ 5500 rpm
Torque . . . . . . . . . .175 lbs-ft @ 3500 rpm
Bore & stroke . . . . . . . . . .3.38 in. x 2.54 in.
(86 mm x 65 mm)
Compression ratio . . . . . . . . . . . . . .8.8 to 1
Valve actuation . . . . . . .Single overhead cam
Induction system . . . . . . . .Dual Stromberg IV
Exhaust system . . .Cast-iron headers, 4 into 1
Electrical system . . . . . . . .12-volt alternator, point distributor
Fuel . . . . . . . . . . . . . . . . . . . . . .Premium
Recommended redline . . . . . . . . . . . . . .6500

## DRIVE TRAIN
Clutch . . . . . . . . . . . . . . . . . .Single dry plate

| Transmission | Gear Ratio | Overall Ratio |
|---|---|---|
| 1st Synchro | 2.99 | 11.08 |
| 2nd Synchro | 2.10 | 7.77 |
| 3rd Synchro | 1.38 | 5.13 |
| 4th Synchro | 1.00 | 3.70 |

Differential . . . . . . . . . . . .Hypoid, 3.70 ratio

## CHASSIS
Frame . . . . . . . . . . . . . . . .Unit construction, front engine, rear drive
Front suspension . . .Independent, McPherson struts, coil springs, telescopic shock absorbers, anti-roll bar
Rear suspension . . . . . . . .Independent, semi-trailing arms, coil springs, telescopic shock absorbers
Steering . . . .Rack and pinion, power assist, 2.8 turns, turning circle 34.1 feet
Brakes . . . . . . . . . . . . . .Front disc, rear drum, power assist. dual systems, 10.6-in. dia. front, 9.0-in. dia. rear
Wheels . . . . . . . . . . .14-in. dia.; 5.5-in. wide
Tires . . . . . . . . . . . . . . .Michelin 185 HR 14, pressures F/R: 26/30 (rec.) 30/34 (test)

## BODY
Type . . . . . .Integral steel, 2-door, 4-passenger
Seats . . . . . . . . . . . .Front bucket, rear bench
Windows . . . . . . . . . . . . . .2 power, 2 vents
Luggage space . . . . . . . .Rear trunk, 7.5 cu. ft.
Instruments . . . . . . . . . . . .140 mph speedo, 7000 rpm tach
Gauges: . . . . . . . . . . . . . .Volts, temp, fuel
Lights: . . . . .Oil pressure, fuel, ign, choke

## WEIGHTS AND MEASURES
Weight . . . . . .2800 lbs (curb), 3025 lbs (test)
Weight distribution F/R . . . . . . . . . .58%/42%
Wheelbase . . . . . . . . . . . . . . . . . . .100.0 in.
Track F/R . . . . . . . . . . . .52.5 in./52.9 in.
Height . . . . . . . . . . . . . . . . . . . . . .49.5 in.
Width . . . . . . . . . . . . . . . . . . . . . .63.5 in.
Length . . . . . . . . . . . . . . . . . . . . .173.8 in.
Ground clearance . . . . . . . . . . . . . . .5.5 in.
Oil capacity . . . . . . . . . . . . . . . . . .5.5 qt.
Fuel capacity . . . . . . . . . . . . . . . . .16.1 gal.
Coolant capacity . . . . . . . . . . . . . . .11.1 qt.

## MISCELLANEOUS
Weight/power ratio
(curb/advertised) . . . . . . .19.3 lbs per hp
Advertised hp/cu. in. . . . . . . . . . . . . . .0.79
Speed per 1000 rpm (top gear) . . . .20.2 mph
Warranty . . . . . . . . . .12 months/12,000 miles

AERODYNAMICS FORCES AT 100 MPH

CORNERING CONDITIONS

## PERFORMANCE

Acceleration . . . . . . . . . . . . . . . . . . . . . . . . . . . . . . . . . . . . . . .0-30 (3.5 sec.), 0-60 (9.6 sec.), 0-quarter mile (17.2 sec., 80.0 mph)
Top speed . . . . . . . . . . . . . . . . . . . . . . . . .122 mph (est.) at 5500 rpm (hp limited)
Braking . . . . . . . . . . . . . . . . . . . . . . . . . .Distance from 60 mph: 150 ft. (0.80 g av.)
Number of stops to fade: Not attainable
Stability: Excellent
Maximum pitch angle: 2.2°
Handling . . . . . . . . . . . . . . . . . . . . . .Maximum lateral: 0.70 g right, 0.74 g left
Skidpad understeer: 6.3° right, 3.5° left
Maximum roll angle: 7.0°
Reaction to throttle, full: Understeer; off: Oversteer

| Speedometer | 30.0 | 40.0 | 50.0 | 60.0 | 70.0 | 80.0 | 90.0 | 100.0 |
|---|---|---|---|---|---|---|---|---|
| Actual mph | 30.0 | 40.0 | 50.0 | 60.0 | 69.0 | 77.5 | 84.0 | 94.0 |

Mileage . . . . . . . . . . . . . . . . . . . . . . . . . . . . . . . . . . . . . .Average: 17.1 mpg
Miles on car: 5800 to 8200
Aerodynamic forces at 100 mph:
Drag . . . . . . . . . . . . . . . . . . . . . . . . . . . . . .300 lbs (includes tire drag)
Lift F/R . . . . . . . . . . . . . . . . . . . . . . . . . . . . . . . . . . .80 lbs/175 lbs

## TEST EXPLANATIONS

Fade test is successive maximum g stops from 60 mph each minute until wheels cannot be locked. Understeer is front minus rear tire slip angle at maximum lateral on 200-ft. dia. Orange County Raceway skidpad.

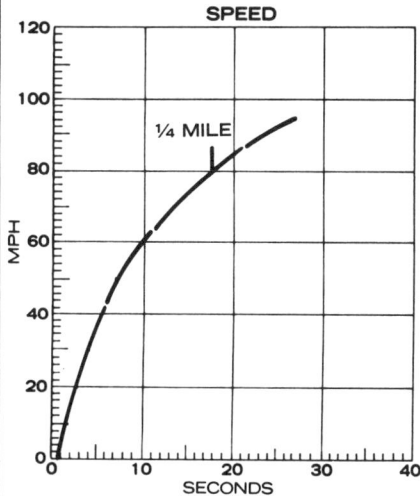

SPEED

Speed measured from standing start thru ¼ mile to maximum shown. Shift points indicated by triumph breaks.

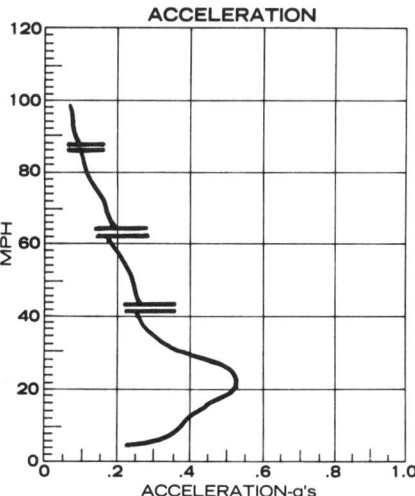

ACCELERATION

Acceleration measured in "g's" from standing start to speed shown. Shift points indicated by "spikes" on graph.

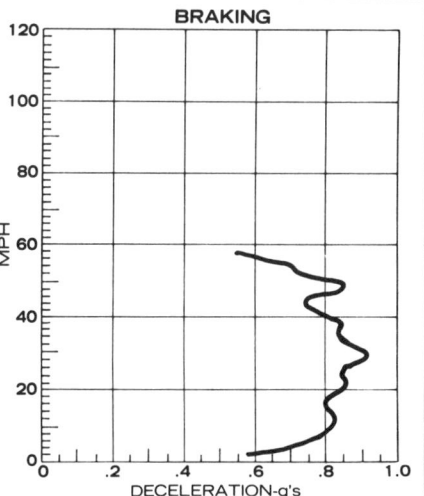

BRAKING

Brakes applied at 60 mph with maximum force, but using pedal "feathering" technique to prevent wheel lockup.

# TAKING TRIUMPH'S V8 STAG BY THE HORNS

It's not a sports car or a GT car or even a plain convertible. Peter Robinson rates BL's newest V8 as a "personal" car which doesn't make you any wiser.

ACCORDING TO one of my American friends, Triumph's V8 sporty is called the Stag because the Yankee salesmen thought the name would appeal perfectly to the masculinity and virility of their customers.

It was a good thought . . . but rather misguided. Stag comes from a Nordic word meaning not just a male reindeer, but any male animal. And there is nothing phallic about BLMC's Stag, that's for certain. It's not a red-blooded, hairy-chested beast or even a sports car in the traditional sense of the term, but a mild-mannered, gentlemanly touring machine.

*Stag styling is controversial, closely follows 2000/2.5 sedans at front and rear but middle section betrays sporting intentions.*

*It's hard to imagine any driver being unhappy with the interior layout. The steering wheel is adjustable two ways, the pedals are perfect for heel/toe changes and the switch gear and instruments ideally located. Front seats are superb.*

Neither is it a GT, although BL would like you to believe otherwise. Rather, it is a straight-forward touring drophead aimed at much the same market as was the first four-seater Thunderbird in 1958. In other words, it's designed for the man who wants to get away from the staid standard sedan to something just a little more sporty.

But it's too heavy to have the performance a keen driver would expect of a car with its specification, and its 2 plus 2 seating is considerably less than a family would require. Yet it does combine a degree of performance and comfort lacking in most sporting sedans. It is in this role the Stag reveals its true worth.

Are there any other cars of the same ilk? Perhaps the Mercedes-Benz 350SL comes closest but it is purely a two-seater, with additional padding for a dog or one or two small children. It also costs twice as much. Strangely, it is the American personal cars such as the Oldsmobile Toronado, Cadillac Eldorado and Lincoln Continental Mark

III which follow the same concept without reaching the same conclusions. The cars are poles apart but represent American and British means of achieving identical ends.

There certainly is no British equivalent unless you consider the Capri fits the bill — the closest we get in Australia is probably a Monaro 350 or Valiant 318 Hardtop but they are both symbols of the same idea rather than specifically designed personal cars.

Fiat's 124 Coupe also comes very close to providing the same kind of motoring, although at a much lower price and with greater emphasis on the sporty bit.

Well, what's the Stag doing in Australia? For starters the car we drove is the only example here. It was brought out by AMI for motor shows and has now been sold to Dick Thurston, of racing fame — a gentleman who also happens to own a prosperous Toyota/Triumph/Rambler/Lotus agency in Melbourne — for personal transportation.

95

AMI plans to bring the Stag to Australia eventually but production is now limited to supplying the home market and looking after the US — where it has been, as anticipated, a great success. I understand Lord Stokes himself has put the stoppers on any plans AMI might have had on assembling the car in Australia. It will come fully built-up or not at all.

And the anticipated price of $7000 reflects this.

The Stag obviously owes a great debt to the Triumph 2000/2.5 PI range. The styling around the front is almost identical but from there back the shape differs to suggest the car's sporting qualities although the similarity returns at the tail.

Even though the Stag was released six months after the Mk II Triumph sedans it was in fact styled, by Michelotti, before its more mundane brothers. It's rather debatable which effort succeeds best.

Most of the Stag's suspension, running gear, transmission and interior fittings come from the 2.5 PI. What IS new, is the V8 engine. Actually it is two halves of the OHC four cylinder engine Triumph has been building for Saab for the last couple of years. It is not quite as simple as that though, for the 2997 cc V8 has a bore and stroke of 86 and 64.5 mm respectively while those of the Saab engine are 83.5 and 78 mm.

Obviously Triumph hasn't just built this engine for the Swedish car — in due course we will see four cylinder examples in some of BLMCs (Triumph) range of small cars.

With Rover's 3.5 V8 and Jaguar's new V12 engine this makes three V configuration engines BLMC can call on for its future prestige cars.

For a modern 3-litre V8 it doesn't produce a great deal of power. With twin Stromberg 175 CDS carbies it puts out 145 bhp at 5500 rpm and 170 ft/lb torque at 3500 rpm. But enough for performance which all but the most lead-footed drivers will find adequate. The Saab engine now offers fuel injection as an option. If more power is ever required from the V8 this would seem to be the way Triumph would go.

There is no doubting the V8's flexibility. I had the car pulling 1500 rpm in overdrive fourth and from only 1000 rpm in direct fourth. This was achieved smoothly and quietly without transmission snatch. Above 2500 rpm it really begins to perform and develops a typical V8 burble.

During our all-too-brief impressions run we were able to test the Stag's handling to the full but a top speed run was out of the question, Melbourne weather being the way it was. Overseas tests produced a top-end speed a little on the low side of two miles a

*Completely new ohc V8 engine is virtually a doubled-up version of the Saab four which Triumph builds. Power rating of 145 bhp isn't high considering 3-litre capacity. Familiar V8 throb comes in above 3000 rpm.*

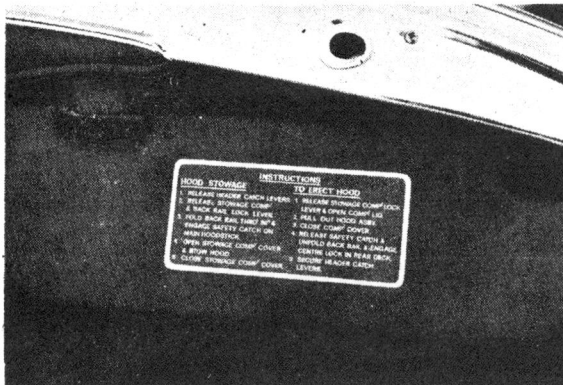

*Instructions for lowering the hood are cleverly shown on the back of the driver's sun visor but it is still a two-man job and takes at least five minutes. The hood stows away in a covered well.*

minute with the ton coming up in a whisker over 30 seconds, 60 mph in 10 seconds and the standing quarter mile in 17.2 seconds. Impressive perhaps for a touring car although there are sedans, the BMW 2800 for instance, which will out-run it with ease.

What the Stag does best is high speed touring, quietly and in complete comfort. A combination of high gearing, with overdrive (optional as is automatic transmission) on both third and fourth, and good mid-range torque, makes for a cruising speed around the ton on very little throttle. The gear ratios are identical to those in the 2.5 PI except that first gear is higher which holds back the initial acceleration. This also means second gear is a little too close to first for peak acceleration. First runs out to 40 mph, second to 60, third to 90, third overdrive to 100, fourth to 112, and fourth overdrive to 118. The overdrive switch is conveniently located on top of the gear lever.

The Stag's handling, like the engine, is a clear pointer to the car's character. The soft suspension provides a smooth and comfortable ride but does mean there is more body roll than you'd expect from a sporty car. The steering too adds to this impression for it lacks real precision. Adwest power assistance is standard and it robs the driver of road feel while being excessively low geared. The big 2.5 PI steering wheel — adjustable vertically and in reach, by the way — doesn't help.

However it is possible to come to terms with the power steering for it is not as ridiculously light as on some rival cars. Certainly the roadholding is of a very high standard. The soft suspension ensures there is plenty of adhesion on rough roads. Michelin XAS tyres, for which you pay no extra, obviously play an important role. Understeer predominates but can be overcome with the throttle and instant oversteer is there for the asking if you lift off with the right foot. In practice such traits can be used to great advantage for it means the Stag can be steered through corners on the throttle alone.

What makes the Stag unique or almost so, is its versatility. It can be a luxurious closed coupe or a topless sporty. AMI's car was the soft-top version, a detachable hardtop is optional but it's no lightweight and finding room to stow it would be a problem.

Instructions for lowering the canvas top are cunningly hidden behind the driver's sunvisor. Even when they are used to the letter, lowering the hood is a two man job and it takes time getting it down although coming the other way is quicker. As a convertible, the Stag's biggest difference is its built in roll cage. This acts as a marvellous body stiffener — something the TR6 could well invest in — and makes a fine job of reassuring passengers unhappy about travel without a roof.

The fully padded bar is made as an extension of the door pillars with a central Tee-piece linking it to the top rail of the windscreen. Both the soft top and the hardtop fit over the top of it. Its padded vinyl covering can be unzipped for cleaning and it never really interferes with visibility. You can take it as read, that this won't be the first (not forgetting the Porsche Targa) or last such roll bar we will see. Indeed if the open car is to survive at all such cages will probably be mandatory.

Inside the Stag looks like a more luxurious version of the 2.5 sedan. The steering wheel is less vertical than I like and, as mentioned previously, too large in diameter, but otherwise the controls and driving position are almost perfect.

The instrument layout is virtually identical to the 2.5 with a generous range of dials plus switches which are easy to locate in a hurry. Electric windows are standard and the Stag boasts the 2.5's comprehensive heating/ventilation system that is complimented by opening quarter vents.

Interior lighting is a special feature.

## TRIUMPHANT MONARCH
*Continued from page 92*

ing styling, is its smoothness and quietness. British builders are becoming more and more enthused about the V-8 as a powerplant. "This is the engine of the future," say Triumph engineers. Funny, we've been saying that since maybe 1932. This aluminum-head V-8, incidentally, is only one of four in the world (in production) with overhead camshafts.

The Stag cannot be equated with the entrail-jarring Corvette 454 when it comes to acceleration. Yet, there's enough there to give the driver a sense of minor hemorrhoidal distress. Acceleration is smooth and steady, to the tune of 17.2 seconds and 80 mph in the quarter-mile. The time could be considerably better, except that the Stag continually balked immediately from the standstill.

The transmission gearing is tall in both the four-speed and automatic versions, which seems a nice compromise with the low (3.70 to 1) rear-axle ratio. Once in motion, the four-speed in combination with the V-8 produce an easily driveable machine.

One thing nice about a small V-8 in street tune is that it can give decent fuel economy. Although our Stag only produced 17 miles per gallon, we're sure you can expect 20 mpg. Most of our driving was hard and fast, probably because it's that kind of car.

The Stag becomes a problem during highway driving. Not that there's any wandering or sway . . . surprisingly, even in strong winds, the car did not sway with the gusts. No, the problem was in deciphering speed. We had to keep a constant eye on the speedometer, otherwise we found ourselves inadvertently cruising at 80 or 90 mph, a spot on the dial where the Stag seemed to settle. We weren't particularly concerned about this, except that there are things such as speed laws and red flashing lights.

The Stag brakes are excellent for an over 3000-pound automobile. As a comparison, we ran times on the Stag and GT6

There is a small light let into the glove box lid so that it shines into the compartment and at the same time sends another beam downwards for map reading. Other lights illuminate the footwells and are built into the door armrests. And when the door is open the ground is lit and a red light shines rearward.

The Stag's seats are very comfortable although rearward adjustment can be difficult. If anything bigger than small children are carried in the back the front seats would have to be moved forward to provide at least some legroom.

So that is the Stag. Best described as a "personal" car it displays the great attention to detail so beloved of Triumph designers. Its appeal to upper-crust drivers, searching for something with a little more appeal than just any old four-door sedan, must be very strong. It is a car Britain can be proud of. Even at $A7000 it would seem assured of at least a couple of hundred sales a year. If we could get them, that is.
\*

*Neat rear light clusters incorporate reversing lights. Chrome-tipped twin exhaust pipes and rubber-faced overriders are standard.*

at the same time, and found that stopping distance from 100 mph after the quarter-mile was about 75 feet shorter in the Stag.

You might question handling as we did. After all, .74 g for lateral cornering force is only about average. This was due, primarily, to the understeer in the automobile. We found it difficult to stay within the 200-foot diameter circle of the Orange County Raceway skidpad . . . it wanted to plow off course. The outside front tire tucked under drastically, as you can imagine it would with 58 percent of the weight on the front. Yet, we were surprised to hear the comments of "knowledgeable" people who said how well it handled.

Power steering and power brakes come standard with the Stag. Both afford a feel for the road not often found in power-assist units. Power steering and rack-and-pinion aren't always compatible. Many cars with this combination tend to bind at full-lock positions or are bothered by pump surge which makes the steering wheel throb slightly from side to side anytime the engine is running. Yet, on the Stag, much of this was eliminated.

One more point. The top. To drop the ragtop requires a certain sinewy ingenuity and an understanding of British directions. The first couple of times you attempt it, it's a trial-and-error adventure. If nothing breaks or bends, all future efforts should be successful.

In summary, the four-speed version of the Stag is most acceptable, but we really can't think of many good things to say about the automatic. Strange how a car's characteristics alter drastically when transmissions are changed.

Other than styling, we'd say the Stag is a four-seat Datsun 240Z . . . and we mean that as a compliment, despite what the British might think. In fact, we'd be tempted to buy a Stag . . . if we could get a deal, that is. Because again, if that price could be lowered to under $5000, we find it *most* appealing.

British Leyland is the largest manufacturer of automobiles in Britain, and it's only right that they should begin producing the best in the Isles. Maybe all they needed all along was a little prodding from the Germans and Japanese.

## FRESH FROM ENGLAND IS THIS LUXURY GRAND TOURING CAR WITH A NEW V-8 ENGINE AND AN UNUSUAL "T" ROLL BAR.

■The Triumph Stag, a new, completely equipped 2+2 luxury sports car featuring the first volume produced British-designed V-8 engine, and a unique "T-bar" safety roll bar, was announced recently, by British Leyland Motors Inc., the world's largest supplier of sports cars for the U.S. market. It was feautred at the April 3-11 International Automobile Show in New York City where it proved to be one of this year's hits.

The new Stag has a subtle family resemblance to current Triumph models. It is designed to be a long distance, high speed grand touring car offering maximum driver and passenger comfort. The new model comes with a convertible top , $5,525 (P. O. E.) or combined with detachable hard top, $5, 773 (P. O. E.)

The heavily padded "T-bar" consists of a cross piece which is an extension of the door pillars and a front-to-rear reinforcement connected to the windshield frame. It was designed specifically to add to occupant protection.

Other advanced safety features include an inertia switch that automatically shuts off the fuel supply in the event of collision and red warning lights built into the rear edges of the door-mounted arm rests. These alert oncoming drivers when the doors are open.

The Stag's single overhead camshaft, aluminum cylinder head engine is an entirely new British-designed and built V-8. It is a three litre 182.9 cu. in. displacement unit developing 145 net brake horsepower (@ 5,500 RPM) and 170 lb/ft. of torque (@ 3,500 RPM). The Stag is the only imported convertible sports model from any country with a V-8 engine, selling for less than $10,000.

The new Stag comes with a full complement of lurury features as standard equipment including power assisted rack and pinion steering, The front wheel disc and rear wheel drum brake combination is power assisted. Windows are (Continued on page 99)

**STAG** Continued from page 65

the Adwest leaves sufficient feel for spirited and accurate aiming.

The four turns lock adds to the "softness" of the car — and I mean this in the nicest possible way.

Not that it is slow. Despite its weight of more than 25 cwt. it can produce a handy turn of speed and acceleration.

STI had some initial trouble finding suitable tyres. I would think there'll be a full option of Michelin XAS, Pirelli Cinturato and Dunlop SP68 by the time the car gets into full swing.

During pre-release runs I was a little troubled by what seemed a sudden change of weight distribution under hard braking. (Front 55.7 rear 44.3 percent). In practice this is taken up so quickly it is hardly noticeable.

Cruising down at autobahn with 120 mph on the clock (about 112 mph), a drama up front required the anchors to be thrown on very quickly indeed. I needn't have worried on the Stag's account. The four discs came on squarely and despite a bit of necessary correction the car skidded along on the Michelins straight and true. No problems.

## Comparison

A friend from Mercedes Benz loaned me his factory 280SL and he took over the Stag for a run down the autobahn.

It's a bit sad to have to say it but the old 280 is dated now. The suspension is spongey, the brakes are just so-so and the "feel" and top speed, passe. My friend emerged at the end of our very fast run and commented: "It is a good car." And then, as if reading my thoughts, admitted: "It does make the 280 SL outdated."

Later, both of us were caught in a traffic jam. My friend in the 280SL was eventually forced into the side of the road with his temperature gauge "off the clock". The Stag's was just above normal which is good news for people in Australia and other hot countries.

Consumption was remarkable considering the cruising speeds of more than 100 mph that the car was subjected to for the greater part of the test. My overall was 21 mpg, around town I averaged 23. On a steady 70 mph section this went up to 24.5 mpg.

The weather during the test was great so the hood came up and down (mainly in the interests of security at night) many times. The mechanism will change slightly from this pre-production prototype and I trust some sections will be made a little stronger. As it was, a front side stay snapped at a bolt hole (which has been obviously bored too large).

Dials could be lettered more legibly. At night they are a bit hard to discern. I heartily dislike that sectionalised warning light dial. First, it contains the turn indicator (and doesn't show which way the indicator is turning). When the roof is down and sun shining, it is impossible to see the indicator light working.

I think there should be two independent warning lights for indicators. The fuel warning light is almost a waste of time. It starts blinking away annoyingly when the 14 gallon tank is about half full. It's supposed to come on full when there's just over a gallon left, but by that time the driver is usually a nervous wreck thinking he might run out of fuel or that the warning indicator isn't operating correctly.

Here's where Triumph could learn from another of the quality divisions of BLMC. They have borrowed the Adwest power steering from Jaguar. I suggest they borrow the reserve tank idea from both Jaguar and Rover — it is a far better system and much more reassuring than an inaccurate flashing light.

The eyeball vents on either side of the dashboard are next to useless in Stag. Admitted Stag has a couple of extra rectangular vents in the central dashboard, but they too are anything but efficient. With air-conditioning they may work well (I have yet to try a car so fitted) but certainly the average buyer could do with more air on his face. The gear lever could be a little shorter, say an inch, as it is inclined to be too close to the wheel in the reverse position. The overdrive switch on the top of the knob is an excellent idea.

But these are minor niggles. This is a great grand touring car.

Acceleration: 0-30 mph, 3.5 secs; 40 — 5.5; 50 — 7.0; 70 — 13; 80 — 16.5; 90 — 22.0; 100 — 28.

Top speed: 119 mph average, 122 mph best. ●

## TRIUMPH STAG

Continued from page 98

electrically operated. The fully reclining front bucket seats are adjustable for rake and height as well as for and aft position. The steering wheel can be adjusted for distance from the driver and angle. An electrical defroster is standard with the hard top. Its circuit is automatically connected or disconnected when the hard top is put in place or removed. Specially designed chrome wire wheels and radial ply tires are also standard equipment.

Factory installed air-conditioning is available for the Stag as an option.

The new Triumph is available with a four-speed, fully synchromesh manuac transmission with optional overdrive, or with a three-speed automatic

of the shift-shiftless type that allows manual selection of any of the three forward speeds or completely automatic operation.

The Stag's suspension is completely independent—front and rear—with front and rear coil springs. Liberal use is made of rubber mountings in the suspension system to insulate against noise.

Stag measurements (compared to the familiar two passenger Triumph TR6) are as follows: Length over-all—14 ft. 5 3/4 in. (TR6—12 ft. 11 in.). Wheelbase—8 ft. 4 in. (TR6—7 ft. 4 in.). Height with convertible top up—4 ft. 1 1/2 in. (TR6—4 ft. 2 in.).

The Stag was styled by Giovanni Michelotti of Turin. Features include full wrap around front and rear bumpers, bright stainless steel rocker panels, rear light clusters inset for extra protection and quad headlights in the clean, horizontally styled grille. The T-bar's uprights are finished in

bright stainless. The convertible top stows neatly out of sight under a steel deck behind the rear seat when not in use. It can be raised, or lowered, in less than a minute by one person.

The interior is completely carpeted with deep pile over felt. The deep, luxurious seats are upholstered with expanded vinyl having a basket weave pattern for maximum cooling effect. The dash is finished in walnut veneer, which attractively sets off the white on black instrumentation. A console, also finished in walnut veneer, houses the gear shift lever, switches for the electrically operated windows, ash tray, radio and heater controls. Fresh air ducts for the flow-through ventilation system are located in the middle of the dash above the console and at either side of the dash.

The Stag is a completely new addition to the Triumph line, which also consists of the TR6, GT6 Mk. 3 and Spitfire Mark IV sports models. ●

# ON TEST:

SOMETIMES one expects a car to be really good, in other cases a good car comes as a complete surprise. Such was the case with Triumph's V8-engined Stag, a car which we collected for test expecting a rather staid and placid car, but returned full of praise for the Stag's 'roadability' and advanced technical specification.

In essence a luxurious two seater with slightly more than occasional seating for two at the back, the Stag really fits into a new category of price range/appeal and we would tend to put it in a class with cars like the Reliant Scimitar GTE, BMW 2000 Touring and Volvo 1800ES, against which the Stag acquits itself very well. Priced at around £2,200, the Stag must be an ideal buy for the sort of family man to whom an Elan would appeal in his batchelor days. And, in addition to a very sound concept of design, the Stag really does reveal some advanced thinking on the part of British Leyland's designers, notably with respect to the safety features, power unit and finish.

## Michelotti design

The car tested was the convertible Stag, which to our mind must be the most attractive car to have, although a hardtop is available at extra cost. The 2000-inspired lines of the Michelotti body are pretty clean, and although they will appeal or otherwise to certain types, we liked them very much and they certainly drew admiring glances wherever we went. With the hood down, the dominating feature of the styling is the distinctive roll-over bar with rearward bracing, which does not detract from the appearance whatsoever, adds a significant safety factor and also contributes greatly towards achieving a satisfactory chassis rigidity.

Having said that the styling will either appeal or not appeal, we are sure that the interior styling and fittings will appeal to all drivers. One enters the Stag by its wide front doors, which, incidentally are fitted with red illuminating lights reminiscent of very expensive Ferraris and the like, and steps on to a plushly carpeted floor. The front seats, whilst hardly bucket-like and lacking head-restraints, are comfortable and provide good location, and can also be adjusted all ways for length and recline and padding support. With the added amenity of a movable steering column, most drivers within a large size range can make themselves feel very comfortable. Ahead of the occupants is a suitably varnished dashboard with

# RAMPANT STAG

instrumentation to the tune of speedometer, rev counter and smaller instruments for fuel and engine temperature, and a dial cluster of warning lights, one of which, the oil warning light, obviates the function of an oil pressure gauge, a feature, which like the BMWs, did not strike us as being good for such an enthusiasts driver's car. The controls of the Stag are an example to all, with the lights and wipers, with wash-wipe control and two speeds, worked off stalks together with the horn and indicators.

## Solidity

Ventilation is quite superb, with an infinite range of hot and cold adjustments, and there are nice finishing touches like the useful interior light and its well-located switch. Above all, the impression at the wheel of the Stag is of a really solid, nice vehicle. The rear seating is, naturally, somewhat short of legroom, although despite the large 15-gallon fuel tank in the back, the boot swallows up quite enough luggage for three people.

## V8 burble

On starting, the V8 makes a deep characteristic burbling sound, and it pulls away swiftly and smoothly in all gears, the 3-litres giving a delightful power range with torque galore. The four-speed gearbox has a light positive operation with the handy overdrive switch located within the gear-knob, although during our test the overdrive itself did not function, restricting top-end cruising.

The most noticeable feature when new to the Stag is the exceptionally light feeling of the power-assisted steering. Although this does not measure up to the standard of the ZF unit, it is very nice once the driver has acclimatized himself to its sensitivity. In fact, we found that the lightness of the steering added to the car's desirability as it encouraged the complete effortlessness of driving the Stag, and also after a while induced a lot of confidence when cornering rapidly.

The Stag does suffer from a certain amount of body-roll when pushed to its high limits, but with the light steering the possibilities of quick long-distance travel are great. The Stag is not really in its element when being pushed hard round narrow country lanes, comfortable and easy as it is to drive, but it is good in these circumstances and wonderful as a "GT-type" curiser over long distances. The only restricting factor to such driving is the inevitable wind noise induced at high speeds, or, in fact, anything over 60 mph, even with the hood up.

## Effective braking

Braking on the Stag never produces any worries, for the disc/front, drum/rear set-up with powerful servo-assistance, is light in operation and very effective. The V8 motor itself has power and torque enough to bowl its occupants along, wind noise notwithstanding, at a steady 100 mph, and when pushed the Stag will wind itself up to over 115 mph overdrive top.

With its lusty 140 V8 horses, fuel consumption averages out at an acceptable 18 mpg or so of the best 100+ octane fuel, giving a touring range of around 300 miles or thereabout.

Lighting, by four halogen-bulbed headlights, is superb and will not inhibit fast night driving, and as already stated, the controls of the headlamps are very well-located.

## Value for money

The Stag appealed to us so much that we hope to revisit it and explore our impressions in greater detail, but for the moment let's say that it has got to be the best car in its price range, and, oh yes, it's got lovely quick-lifting electric windows as standard. If you are looking for a car with all the right equipment *as standard,* an easy-to-erect and stow hood for the delights of open-air motoring, fast cruising and four seats, advanced engineering techniques (including a safety fuel cut-off in the event of a ruptured petrol line) and general desirability for around £2,200, then join the queue of potential Stag buyers. For the money, you couldn't do better.

# stag...
# TRIUMPH'S LONG-DISTANCE RUNNER

**A wild, high performance sports car the Stag is not, but neither is it a 2-plus-2 limousine. Waiting halfway between the two extremes, Triumph's V8 is very desirable motoring.**

IN THE DAYS BEFORE they were welded together as British Leyland, there was an outbreak of V8 engine designing at both Leyland and BMC. The first fruits appeared in the Daimler 2.5, then the Rover 3500, then half an engine appeared in the Saab 99 of all places and everyone said what a splendid unit it would be when Triumph glued both parts together and made it into an ohc V-8.

After the coming together V-8s started to be shed. The projected Jaguar one was strangled at birth, the Daimler dropped into the background and that left Rover and Triumph, Rover getting a clear lead by having its complete engine in production instead of a half for export to Sweden.

The pundits all forecast that a V-8 version of the Triumph 2000 would appear but they didn't stop to think that it would compete with the Rover 3500, made by the same firm and, in any case, they weren't aware that the petrol injection Triumph 2.5 was on the way.

Another piece of the jigsaw came to light when the design teams of Rover and Triumph were brought together and this caused everyone to forecast a Triumph 2.5 engine in the Rover 2000 body, but wrong again. Or that the Rover V-8 would be dropped. Still wrong.

*Stag interior is pleasant mixture of luxury and sportiness. General layout follows 2.5 PI sedan. Steering wheel is adjustable and combines with excellent front seats for a superb driving position.*

All right then, the Rover experimental mid-engined car would be refined and produced as a Triumph? Uh uh, but getting warmer. That Triumph would produce a 2-plus-2 V-8 based on the 2000 running gear with a removable top? Nobody guessed that one and it's not to be wondered at because the result is a car which doesn't fit into any preconceived slot and although you might just be tempted to say: "What about the Mercedes 350SL?" in truth, the only real resemblance is in the

number of seats and the fact that the lid comes off.

The Merc is unashamed luxury almost regardless of cost while the Stag is the best that can be produced within a given price limit. Both give value for money but it's a matter of whether you wish to buy your wife beaver, lamb or mink.

It's usually a good idea to study car brochures to discover how the maker sees his own vehicle, Triumph pitch in with: "Some people think that the Continentals have the reputation for really stylish grand tourers all to themselves. Well, that reputation had to run out some time. The time is now. And the car that's overtaken them in style is the Triumph Stag. The Stag's styling is sporty, but suave. Its simple low-lying lines are impeccably cool, and give it stand-out sophistication among the lumbering herd."

That's enough of that to be going on with but it is interesting that the rest of the brochure makes great play of travelling long distances on those never-ending Continental roads and, if this is what they set out to do, then they have succeeded very well.

First thoughts on driving the Stag were: "This is a car I could go a helluva distance in" and when the test was over I still felt the same about it.

*Continued on page* 105

# technical details

## TRIUMPH STAG

| | |
|---|---|
| MAKE . . . . . . . . . . . . . . . . . . . . . . . . . . . . . . . | TRIUMPH |
| MODEL . . . . . . . . . . . . . . . . . . . . . . . . . . . . . . . | STAG |
| BODY TYPE . . . . . . . . . . . . . . . . . . . . . . . . . . | 2 plus 2 |
| PRICE . . . . . . . . . . . . . . . . . . . . . . . . . . . . . . . | $8048 |
| OPTIONS . . . . . . . . . . . . . . . . . . . . . . . . . . . . . | Hardtop |
| COLOR . . . . . . . . . . . . . . . . . . . . . . . . . . . . . . | Brown |
| MILEAGE START . . . . . . . . . . . . . . . . . . . . . . . | 1570 |
| MILEAGE FINISH . . . . . . . . . . . . . . . . . . . . . . . | 1830 |
| WEIGHT . . . . . . . . . . . . . . . . . . . . . . . . . . . . . | 3000 lb |

FUEL CONSUMPTION:
| | |
|---|---|
| Overall . . . . . . . . . . . . . . . . . . . . . . . . . . . . . . | 20 mpg |
| Cruising . . . . . . . . . . . . . . . . . . . . . . . . . . . . . | 27 mpg |

TEST CONDITIONS:
| | |
|---|---|
| Weather . . . . . . . . . . . . . . . . . . . . . . . . . . | Dry, blustery |
| Surface . . . . . . . . . . . . . . . . . . . . . . . . . . . | New asphalt |
| Load . . . . . . . . . . . . . . . . . . . . . . . . . . . . . . | Two People |
| Fuel . . . . . . . . . . . . . . . . . . . . . . . . . . . . . . | Super |

SPEEDOMETER ERROR:

| Indicated mph | 30 | 40 | 50 | 60 | 70 | 80 | 90 | 100 |
|---|---|---|---|---|---|---|---|---|
| Actual mph | 30 | 40 | 50 | 59 | 68 | 78 | 88 | 97 |

### ACCELERATION (through gears):
| | |
|---|---|
| 0-30 mph . . . . . . . . . . . . . . . . . . . . . . . . . . . . . | 4.0 sec |
| 0-40 mph . . . . . . . . . . . . . . . . . . . . . . . . . . . . . | 5.7 sec |
| 0-50 mph . . . . . . . . . . . . . . . . . . . . . . . . . . . . . | 7.8 sec |
| 0-60 mph . . . . . . . . . . . . . . . . . . . . . . . . . . . . . | 10.3 sec |
| 0-70 mph . . . . . . . . . . . . . . . . . . . . . . . . . . . . . | 14.0 sec |
| 0-80 mph . . . . . . . . . . . . . . . . . . . . . . . . . . . . . | 18.4 sec |
| 0-90 mph . . . . . . . . . . . . . . . . . . . . . . . . . . . . . | 24.7 sec |
| 0-100 mph . . . . . . . . . . . . . . . . . . . . . . . . . . . . | 34.2 sec |

| | 1st gear | 2nd gear | 3rd gear |
|---|---|---|---|
| 20-40 mph | 3.0 sec | 5.1 sec | — |
| 30-50 mph | 3.7 sec | 5.2 sec | — |
| 40-60 mph | — | 5.3 sec | — |
| 50-70 mph | — | 5.9 sec | 7.7 sec |

STANDING QUARTER MILE:
| | |
|---|---|
| Fastest run . . . . . . . . . . . . . . . . . . . . . . . . . . . | 17.5 sec |
| Average all runs . . . . . . . . . . . . . . . . . . . . . . . | 17.7 sec |

BRAKING:
| | |
|---|---|
| From 30 mph to 0 . . . . . . . . . . . . . . . . . | (9.4 m) 31 ft |

## PERFORMANCE

| | |
|---|---|
| Piston speed at max bhp . . . . . . . . . | (709 m/min) 2327 ft/min |
| Top gear mph per 1000 rpm . . . . . . . . . | (31.6 kph) 19.8 mph |
| Engine rpm at max speed . . . . . . . . . . . . . . . | 5800 |
| Lbs per net bhp (power-to-weight) . . . . . . . . | 20.7 |

MAXIMUM SPEEDS:
| | |
|---|---|
| Fastest run . . . . . . . . . . . . . . . . | (186.6 kph) 116 mph |
| Average of all runs . . . . . . . . . . . . . | (183.4 kph) 114 mph |
| Speedometer indication, fastest run . . . . . . | (191.5 kph) 119 mph |

IN GEARS:
| | |
|---|---|
| 1st . . . . . . . . . . . . . | (86.8 kph) 54 mph (6500 rpm) |
| 2nd . . . . . . . . . . . . . | (143.2 kph) 89 mph (6500 rpm) |
| 3rd . . . . . . . . . . . . . | (183.4 kph) 114 mph (5800 rpm) |

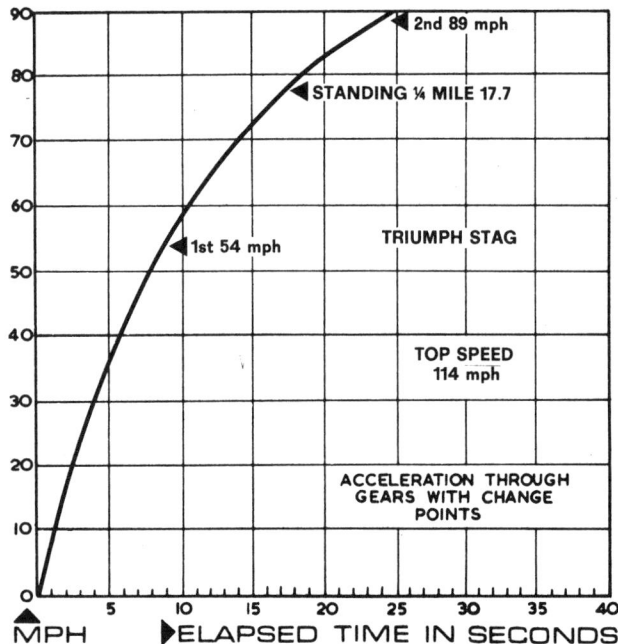

Graph — TRIUMPH STAG: MPH vs ELAPSED TIME IN SECONDS. ACCELERATION THROUGH GEARS WITH CHANGE POINTS. 1st 54 mph; 2nd 89 mph; STANDING ¼ MILE 17.7; TOP SPEED 114 mph.

## SPECIFICATIONS

ENGINE:
| | |
|---|---|
| Cylinders . . . . . . . . . . . . . . . . . . . . . . . . . . . . . | V8 |
| Bore and stroke . . . . . . | (86 mm x 64.5 mm) 3.385 in. x 2.539 in. |
| Cubic capacity . . . . . . . . . . . . . | (2997 cc) 182.9 cu in. |
| Compression ratio . . . . . . . . . . . . . . . . . . . | 8.8 to 1 |
| Valves . . . . . . . . . . . . . . . . . . . . . . . . . . . . . . | ohc |
| Carburettor . . . . . . . . . . . . . . . . . . . | Two Stromberg |
| Fuel pump . . . . . . . . . . . . . . . . . . . . . . . | Mechanical |
| Oil filter . . . . . . . . . . . . . . . . . . . . . . . . . | Full flow |
| Power at rpm . . . . . . . . . . . . . . . | 145 bhp at 5500 rpm |
| Torque at rpm . . . . . . . . . . . . . . | 170 lb/ft at 3500 rpm |

TRANSMISSION:
| | |
|---|---|
| Type . . . . . . . . . . . . . . . . . | Borg Warner automatic |
| Gear lever location . . . . . . . . . . . . . . . . . . . . . | Floor |

RATIO:

| | Direct | Overall | mph per 1000 rpm | (kph) |
|---|---|---|---|---|
| 1st . . . . . . | 2.39:1 | 8.85:1 | 8.3 | (13.2) |
| 2nd . . . . . | 1.45:1 | 5.37:1 | 13.6 | (21.7) |
| 3rd . . . . . | 1.0:1 | 3.7:1 | 19.8 | (31.6) |
| Final drive | 3.7:1 | | | |

CHASSIS AND RUNNING GEAR:
| | |
|---|---|
| Construction . . . . . . . . . . . . . . . . . . . . . . . . | Unitary |
| Suspension front . . . . . . . . . . . . . . . . | McPherson Strut |
| Suspension rear . . . . . . . . . . . . . . . | Semi-trailing arms |
| Shock absorbers . . . . . . . . . . . . . . . . | Telescopic rear |
| Steering type . . . . . . . . . . . | Adwest rack and pinion |
| Turns l to l . . . . . . . . . . . . . . . . . . . . . . . . . . . | 4 |
| Turning circle . . . . . . . . . . . . . . | (10.2 m) 33 ft 6 in. |
| Steering wheel diameter . . . . . . . . . . . . . . . . . | 16 in. |
| Brakes type . . . . . . . . . . . . . . | Disc front, drum rear |
| Dimensions . . . . . . | 10-5/8 in. dia front, 9 in. x 2¼ in. drums rear |

DIMENSIONS:
| | |
|---|---|
| Wheelbase . . . . . . . . . . . . . . . . . . . | (2540 mm) 100 in. |
| Track front . . . . . . . . . . . . . . . . . . | (1352 mm) 53¼ in. |
| Track rear . . . . . . . . . . . . . . . . . | (1362 mm) 53-5/8 in. |
| Length . . . . . . . . . . . . . . . . . | (4420 mm) 14 ft 5¾ in. |
| Width . . . . . . . . . . . . . . . . . . | (1612 mm) 5 ft 3½ in. |
| Height . . . . . . . . . . . . . . . . . | (1258 mm) 4 ft 1½ in. |
| Fuel tank capacity . . . . . . . . . . . . | (63.5 litres) 14 gal |

TYRES:
| | |
|---|---|
| Size . . . . . . . . . . . . . . . . . . . | 185 HR 14 Radial |
| Pressures . . . . . . . . . . . | 26 psi front/30 psi rear |
| Make on test car . . . . . . . . . . . . . . . | Michelin XAS |

GROUND CLEARANCE:
| | |
|---|---|
| Registered . . . . . . . . . . . . . . . . . . . . | (11.4 cm) 4½ in. |

# STAG...TRIUMPH'S LONG-DISTANCE RUNNER

It has undoubtedly been designed with long-range straight-road motoring as the chief consideration. The seats are very good, giving support right up to shoulder level and a fair bit of wrap around as well. The padding is relatively firm, as it should be.

The engine has a moderate output at 145 bhp (5500 rpm) for a modern three-litre ohc V-8. The compression ratio is a mere 8.8:1 but I suppose we'll have to get used to these lower values now that emissions have to be controlled — and rightly so because, let's face it, we're all pedestrians when we leave our cars and any bronchitis sufferer who has had a lungful of exhaust fumes will require no convincing on the subject. The torque is good at 170 lb/ft (oddly, Triumph give it in lb/inches) at only 3500 rpm and the result is a rather friendly, almost fluffy motor which doesn't show any anger until you get the revs really high. The dumble-dum V-8 exhaust note seems completely out of keeping but it is nice to listen to for all that. Like deep-throated, man.

For test we had an automatic model with our old friends Mr Borg and Mr Warner doing the changing. Borg, who makes the upward changes, is an old smoothy most of the time unless you provoke him with a downward kick in the ribs (or in the bottom, going by his reaction), while Warner changes down sweetly at all times. You can stir them both up with the manual selector, which is very nicely placed but not so easy to use by fumble until you are used to it, and to find D or R it is best to read it to be quite sure. There is very little jerk on initial engagement.

A slightly odd feature of the brochure is that the automatic box merely gets a casual mention. All the performance glamor (and details of ratios) goes to the standard manual with its overdrive on third and top, yet all the interior pictures show the auto.

This written emphasis on the manual box suggests a point where Triumph and your tester do not agree. It would seem that it regards the Stag as a sporting device that should have a stick change but I don't — as we shall have explained when we get to the suspension.

I think of it as a nice lazy-man's car for covering a long distance at high speed with the minimum amount of effort. If it were meant to be a device for revving up and throwing into bends then it would have had a rortier motor for a start.

In passing, the basic price is $7657 but you have to have the laminated screen, another $49, and the automatic adds $342 to make the total $8048 and, provided that you're not under the impression that you are buying a sports car, it is good enough value by Australian standards.

The test car had the optional hardtop and I really can't see

*Rear seat is much more comfortable than usual 2-plus-2 set-ups. It's fine for children, lacks more in width than legroom.*

a lot of point in the soft top version because there is nothing more irritating at speed that hood flap and though there may be some who like motoring with the top down, they are a dying race (not yet, Cook—Ed). Bear in mind that the Stag has very good face-level cooling and that the rearward side windows can be opened on their forward hinges to act as very efficient extractors — surprisingly quiet ones, too.

There was quite a bit of wind hiss round the roof of the test car but this doesn't seem to be typical of all Stags and some slight attention to the seals would probably fix it.

The quiet engine felt a bit tight and this seems to be confirmed by the acceleration figures which, even allowing for the losses of the automatic gearbox, are somewhat down on what Triumph claim as typical. Rear wheel traction is good and that, too, lessens initial acceleration because even by holding the car on the brakes and letting the lot go all at once it isn't possible to spin the back wheels and use them as a clutch when taking off. Thus the 0-50 time of 7.8 seconds compares with Triumph's claimed time of 7.0 for the manual and against our 0-70 of 14.0 it has 12.5. Knowing, I am sure that its figures would be correct for a fully run-in manual car.

And if it is true that the engine was still tight, then the petrol consumption figures are very good for a 3-litre automatic with eight cylinders, because some fairly hectic motoring produced an all-over consumption of 20 mpg while high-speed cruising was returning 27. Fully run in and treated gently, it should do a genuine 30 mpg. If we take 25 as a conservative figure, the 14 gallon petrol tank gives a fail-safe range of about 340 miles, which is just about right.

Big tank means small boot, though, but since the rear seats are only suitable for medium-length trips if occupied by adults, the couple nipping off to Alice Springs for a suspect weekend will have ample room for all their toothbrushes and associated gear.

In fact, the rear seats aren't too bad and certainly suitable for children up to the age of about 16. They are much more comfortable than the usual 2-plus-2 rear seats and lack more in width than in legroom.

The front seats have a long range of adjustments but even when they are fully back, with a considerable angle of recline, there is still an acceptable amount of backseat space. The steering wheel helps here by being adjustably mounted. A straight-arm position comes naturally — indeed, is almost imposed upon the driver and it is a great pity that all cars aren't made thus so that people would have to sit correctly even without the application of commonsense. But even the wheel-huggers will be happy with the four inches of adjustment axially, plus two inches vertically.

Big pedals are slightly offset to the right but very easy to use and the brake is wide enough for occasional touches by the left foot when the mood takes you at traffic lights. A lot of thought has gone into the control layout and although some initial learning is required, especially when it comes to familiarisation with the minor instruments peppered around the very pleasant and tasteful wooden fascia, they've done a pretty good job.

All-round vision is good because there is none of this up-swept cow-hip or wedge nonsense to keep down the price of the glass, and the seat is fairly high, too. The halogen headlights are magnificent, giving a clear white beam which suggests really heavy copper wiring (many cars have a 1½-volt drop at the headlamps because of thin wiring).

The body feels rigid and rattle-free and although it is apparently an open car with an attached roof, there is that massive roll-over T-bar contributing a lot of strength. But it is rather difficult to understand why there is a single central leg instead of twin ones at each side which would not only make it stronger still, but give the upper edges of the doors something more substantial to shut against and probably stop the wind-hiss referred to earlier.

It is only when you come to the suspension that there is any real evidence of the car being a cousin to the Triumph 2000. There are McPherson struts at the front and effective but not over-refined semi-trailing arms at the rear. The spring settings are relatively soft so that you get considerable straight-line comfort coupled with good stability except when the wind is very gusty (even then, there is nothing really worth complaining about).

Continued from page 105

On the other hand, despite its wide track, the Stag rolls an amount which would be little for a sedan but a lot for a sports car. The effect is minimised by a good grip offered by the seats and you have to be motoring in a fairly extreme fashion to be bothered by roll. Radial-ply thump is well damped out by the suspension mountings and you can hammer the Stag fast over rough surfaces at speeds limited only by the driver's courage.

In general, the speed limit seems to be whatever the car will do in the circumstances and cruising at over 100 mph would be a safe and sensible way to proceed even though some do-gooders will undoubtedly disagree. But then, it has always been the way of the mob to decry whatever it doesn't understand. Engine and road noise are very subdued and if it weren't for the hiss, the radio volume correct for 40 mph would still be right for 100, and you never have to raise your voice to converse.

Cornering, braking and acceleration adhesion is splendid, even on damp roads, and the Stag can be hustled through bends very rapidly and safely once you have become used to two factors.

One of these is the roll already mentioned, but you can cope with that once you realise that the car cocks up a certain extent and then goes no further. The other is the standard-fitting Adwest rack-and-pinion power steering coupled with four turns lock-to-lock (a power-steering Ford Falcon has 2½).

The initial reaction to this steering is to say that it destroys the feel, and stump off after having slammed the door behind you. But it isn't really so — there is quite a lot of feel and you just have to learn to be more fairy-fingered with it in the initial stages. I have heard it described as "too insensitive" but, after sampling it for some time, I disagree and would put any lack of sensitivity down to pilot error.

Confirmation of this can be obtained quite easily by taking the car up to its surprisingly high cornering limits when the understeer slowly and predictably changes to oversteer. You can feel it happening and, more important, you can feel a front wheel that is thinking of going for a slide on a tight slippery bend almost before it happens. In short, the steering is in keeping with the general character of the car and to condemn it is to miss the whole point of the Stag.

And, praise be, it has the screen wipers mounted the correct way for driving on the left of the road so that they clean right up to the very top corner of the screen. The car undoubtedly looks at its best with the lid taken off and we leave you with a final word from the Triumph brochure: "With the hood down and rollover bar exposed, it takes on a wicked, hot-blooded look"

Well, they said it, not me . . .                                    *

*Two Saab 99/Triumph Dolomite engines make one Triumph Stag V8. Performance is subdued and further restricted by automatic transmission but smoothness is never questioned.*

*Hard-top to soft-top in 1 min. 36 sec. L. to r.: hard-top fitted; top off, hood down showing roll-over structure (42 sec.); erecting the previously hidden hood (54 sec.); soft-top erect.*

Continued from page 61

alloy manifolding, the angled plugs accessible with the right tool, as is the dip-stick, which showed that the sump had lost none of its eight pints of oil in more than 1,100 miles. Oil changes are scheduled very 6,000 miles and there are no greasing chores. Visibility forward is good but reversing is less easy.

I tried to like this Triumph Stag, to please Lord Stokes of Leyland, T.D., D.L., LL.D., D.TECH., D.SC., C.ENG., F.I.MECH.E., M.S.A.E. It needs better steering, power or otherwise (say ZF), a better gearbox, for those who do not specify the automatic transmission, and more poke— like Eartha Kitt's Englishman, the Stag takes time to get going. (The makers quote a 0-to-60-m.p.h. acceleration time of 9.5 sec. and one journalist got 9.9 sec. but mine is closer to the average figure.) Given these changes, and a more easily erected top, the Stag could be a great success, for although its engine isn't quite as silky as I anticipated, vibration being felt through the gear-lever, the Stag is a very nice-looking 2+2 coupé/convertible, reasonably priced at £2,173 (as tested) for an eight-cylinder car having (save the mark!) power-assisted steering, electric windows, a concealed hood and Italian styling.—W. B.

# 13,000-

*Mile appraisal: Long-term Triumph*

# STAG

## A likeable and lively car, unusually well equipped.

We have been putting in about 1,000 miles a month with our Stag and have now topped the 13,000 mark, so it is time to make a first "owner's" report. To the most pointed question "would you have bought a Stag in December, 1970 if you had known what you know now?", the answer, without hesitation, is "yes". To a further question, "has it turned out as expected?", the reply is, "not entirely".

**By Maurice A. Smith DFC**

THE Stag was bought more or less as a replacement for our Sunbeam Tiger II (two-and-a-bit seater), which itself followed a Tiger I and an E-type Jaguar (two-seater). It is at least a 10-year later design and had the Tiger been developed on, it would probably by now have cost much the same as the Stag, around £2,200 delivered. With hard top, hood and automatic transmission like ours, the cost is £2,350. One of the differences you really notice apart from the Stag being bigger, concerns the vee-8 engines in the two cars; the Tiger with 1.7 more litres and less weight, was always eager to go off like a rocket and it encouraged the driver to do so. The Stag is not particularly eager or rocket-like, but can readily be wound up to give a lively burst of acceleration. It will cruise continuously nearly as fast as a Tiger—say, 100 mph and is quieter, more comfortable, softer sprung and more spacious.

To live with then, the Stag is much more relaxed and relaxing in nature. Its 3-litre vee-8 engine gives a very smooth and gentle burble of power in traffic but gets surprisingly wound up and whirring if you kick down and use all it has got. From outside you can hear the vee-8 beat as a Stag accelerates away. Now the car is smooth, but originally there was considerable tremor from the front wheels and also vibration at high speed, owed to the Borg-Warner transmission according to Triumph and to the prop shaft according to B.W. This was finally corrected in April. How, we wonder, can you detect such transmission unbalance before delivery if it only starts to show up at 85 mph on its first trip abroad after running in?

That first foreign journey to Switzerland was great fun. Three of us went with plenty of luggage and the Stag was a pleasure to drive in the mountains. Light, power-assisted steering with quick response and good lock are just what

*A good layout. Ample and easily managed ventilation and heating, well positioned secondary controls, multiple steering wheel and seat adjustments. The warning lamp sectored dial (left) is obscured as usual; the speedometer has kph markings as well as mph. Unfamiliar box is Readycall radio telephone, and microphone. Tape recorder selector switch replaces ashtray in centre*

is needed. The engine has more torque for climbing than the normal acceleration would suggest, and it hangs on well at quite low rpm. The seats are sufficiently well shaped to give location and proper support. The vibration mentioned earlier spoiled the high-speed *auto-route* cruising but we were pleased with the mechanical quietness and relatively low wind noise.

If you get a really gusty day on a motorway, directional stability is not 100 per cent, as you might expect with that short rounded tail—those horrid fins of the 1960s had some uses—but the precise rack and pinion steering keeps you on course. It's not the smartest steering

## PERFORMANCE CHECK

**Maximum speeds**

| Gear | mph | | kph | | rpm | |
|---|---|---|---|---|---|---|
| | **R/T** | **Staff** | **R/T** | **Staff** | **R/T** | **Staff** |
| Top (mean) | 116* | 112 | 187 | 180 | 4,810 | 5,660 |
| (best) | 116* | 113 | 187 | 182 | 4,810 | 5,710 |
| 3rd/Inter | 92 | 89 | 148 | 143 | 6,500 | 6,500 |
| 2nd/Low | 61 | 54 | 98 | 87 | 6,500 | 6,500 |
| 1st/— | 42 | | 68 | | 6,500 | — |

| Standing ¼-mile | **R/T:** | 17.1 sec | | 82 mph |
|---|---|---|---|---|
| | **Staff:** | 17.9 sec | | 78 mph |
| Standing kilometre | **R/T:** | 31.3 sec | | 102 mph |
| | **Staff:** | 32.6 sec | | 98 mph |

| Acceleration | **R/T:** | 3.5 | 5.1 | 7.1 | 9.3 | 12.7 | 16.5 | 21.8 | 29.2 |
|---|---|---|---|---|---|---|---|---|---|
| | **Staff:** | 4.1 | 5.8 | 7.9 | 10.4 | 14.2 | 18.6 | 24.9 | 34.5 |

| Time in seconds | | | | | | | | |
|---|---|---|---|---|---|---|---|---|
| True speed mph | | 30 | 40 | 50 | 60 | 70 | 80 | 90 | 100 |
| Indicated speed mph | **R/T:** | 31 | 41 | 52 | 63 | 73 | 83 | 94 | 105 |
| Indicated speed mph | **Staff:** | 29 | 39 | 50 | 61 | 72 | 84 | 95 | 107 |

**Speed range, Gear Ratios and Time in seconds**

| Mph | Top | | 3rd Inter | | 2nd Low | |
|---|---|---|---|---|---|---|
| | **R/T** | **Staff** | **R/T** | **Staff** | **R/T** | **Staff** |
| 0-20 | | | | | | |
| 10-30 | 9.0 | — | 6.3 | — | 3.9 | 3.0 |
| 20-40 | 8.2 | — | 5.6 | — | 3.4 | 3.0 |
| 30-50 | 7.8 | — | 5.3 | — | 3.7 | 3.7 |
| 40-60 | 7.8 | — | 5.4 | 5.3 | 4.3 | — |
| 50-70 | 8.3 | — | 6.0 | 5.9 | — | — |
| 60-80 | 9.2 | 8.6 | 7.4 | 7.4 | — | — |
| 70-90 | 11.2 | 9.9 | 9.8 | — | — | — |
| 80-100 | 13.7 | 14.8 | — | — | — | — |

**Fuel Consumption**

| Overall mpg | **R/T:** | 20.7 mpg (13.7 litres/100km) |
|---|---|---|
| | **Staff:** | 19.2 mph (14.6 litres/100km) |

NOTE: "R/T" denotes performance figures for Stag manual tested in *Autocar* of 10 June 1971
*Road Test maxima in overdrive top gear

wheel we have seen and it might be smaller.

Some early talk of Stags "jacking up" and twitching on corners, owing to the half-shaft splines sticking under heavy torque and then slipping suddenly, has not been supported by this automatic car, although another manual Stag we had on test once showed signs of this in pretty extreme conditions. We followed a Stag fairly fast round Stowe corner at Silverstone during a recent test day (in a Reliant GTE) and were interested to see the busy behaviour of the back wheels and the early lift-off of the inside front one. It goes to show that the ordinary owner, even a fast driving one, gets nowhere near the limits to which cars are tested when their ultimate behaviour is being investigated.

We are reasonably satisfied with our Stag's road-holding, cornering and ride for all ordinary purposes. If we were asked what improvements might be made, the harshness felt on rippled surfaces and occasional thump back through the steering would be one of our first suggestions for attention. There are times on wet roads when back wheel adhesion is suspect. It is also very easy to spin the inside rear wheel when moving off on a curve.

This is the first time we have opted for automatic transmission on a sporting car. It is a sign of the changing times—British automatics changing for the better and traffic congestion for the worse. We do not regret the choice, but think it high time that a cruising over-drive were available with automatic. The automatic Jaguar XJ6 is another car that needs one. Motorways are here to stay.

In appearance the Stag is obviously of the Triumph family. One rival manufacturer remarked that it was so ugly you couldn't sell one on the Continent. We do not agree, nor would we hold it up as an example to all. With luck it will take the new mandatory front bumper heights in the USA without difficulty. Perhaps its looks are an acquired taste and after 13 months we have.

It is no good buying a Stag if you need limousine accommodation. This is a compact, convertible coupé, of the true plus two kind. Average grown-ups can sit comfortably in the back without much to spare. Tall people will find their heads touch the hard-top roof. Getting in and out of the back seats (front seat catches controlled by levers high in the seat backs) is easy enough. With the hood up, back passengers have as much space, but of course, feel a bit blinkered. Wind noise becomes considerable over 60 mph, though less than in most other open cars with hoods up.

*Alone and in company: above, with a manual-transmission Stag road test car, and below, with its predecessor, a much-loved Sunbeam Tiger. Below: In this side view, the Stag's short tail and limited keel surface is apparent*

Many owners will buy their Stags intending to keep the hard tops on all the year round. The folding hood, surprisingly, is an optional extra. Not so, us. Either you like an open car or you don't. There is something about that rush of air, even if Nader says it is polluted! Agreed, most people will not bother to wrap up like Eskimos to go motoring, so the days of the classic open tourer have passed. Despite this, a few of us still actually enjoy putting on our Jack Brabham rally jackets and going for a topless drive—so to speak. Of course, the weather has become much worse as a result of the Americans and Russians piercing the earth's atmospheric crust with their rockets, so we also feel the need of the hard top.

Lack of rigidity has been a criticism of many open-bodies, although we are now over the days of coach-built convertibles that shook like a railway carriage. It usually needs big boxy sills like those on the E-type Jaguar to keep an open car rigid. In the case of the Stag, which has wide doors, body stiffness is assisted by having a roll-over 'cage'. This is made up of a roll-over bar extending from the rear door pillars and joined to the top of the screen frame by a central ridge member. Padded and trimmed, this cage doesn't look untidy nor detract from the open car effect.

Included in the fairly high price of the Stag are the exceptional number of desirable features which make up the value. Power-assisted steering is one of them. This gives unusually quick, light response from straight ahead, which would be particularly valuable for emergency evasive action. It also makes the car very handy in traffic and for parking. Next, there is the two-axis adjustment on the steering column. It moves in and out and swings up or down when you release the clamping lever. This, together with the seat adjustments – back and forth, cushion tilt-up and back-rest angle, should allow anyone to get set. The low instrument panel and screen base with short, slope-away nose help to give a first class forward view. This is very much a girl's car as well as a man's.

Going on with the equipment, you get bright red lamps on the doors in the back ends of the arm rests. These are an obvious safety-feature and good courtesy lamps to help you step in and out at night—puddle lamps, Triumph call them. The hard top comes with built-in electrically heated rear window, arrangements reminiscent of those on the Mercedes 250-280 SL.

Electric operation of the windows is something we really appreciate and specially when driving solo. Opening quarter lights may not be necessary, but if you have them they should pivot open by twisting their knobs, as they do on the Stag. The hard top has extractor-type hinged side windows.

The driver's wiper arm has an articulated linkage which lines the blade up with the screen pillar, so eliminating a blind sector. Then there are the quartz halogen lamps, proper fuse panel, locking fuel filler (which like the cigarette lighter some of us would happily do without),

rheostat for instrument lighting, electric windscreen washer, boot interior lamp, map-reading lamp, twin reversing lamps, parking lamp switch, dual-intensity signal and brake lamps with signal repeaters on the sides of the wings, day-night dipping mirror, and viscous fan coupling to prevent whirr.

Some designers cut down on their bumpers. Those on the Stag are sturdy and wrap around —right round at the back—and the overriders have rubber inserts.

All the above should be standard equipment on good class cars, as they are on the Stag, yet it is surprising how few other cars have them all. We would certainly like to add the export options of fitted head-rests and hazard warning lamps. We do not like a toughened glass windscreen, Zebrazone or not. Air-conditioning can be specified but cannot be added after the car has been assembled.

Next, we turn to some troubles and shortcomings. The Stag has never left us stranded or in fact, let us down at all, so that is a main point in its favour. The nearest it came to it was a rubbed through fuel pipe, attributable to stupid detail design or assembly (see picture and check your Stag). This also accounted for several days of untraced petrol smell and re-

# 13,000

### Mile Long Term Triumph

# STAG...

duced mpg. In this context, the fuel filler is a constant source of irritation and waste. It blows back on every possible occasion and will never accept normal pump delivery.

The developers have managed to design reliability and most of the pinka-pinka out of the blinking units for the direction indicators. At each service they have been 'looked at', but have yet to work regularly or properly. The trouble is that you can be booked by the police for this. It's in the regulations.

The circle of warning lamps is carefully positioned on the panel behind the driver's left hand and forearm to make it hard to see. Triumph's service people hit on an effective way to cure a persistent wrong red warning of radiator overheating. They disconnected it. The fuel gauge continues to read ⅝ full when empty, but now we are used to this. An 8 per cent optimistic error on the speedo is too much. The odometer and trip mileage, on the other hand, are very accurate.

We are on our third Trico washer bottle. They keep splitting where the holding bracket cuts. Fortunately, it still leaves you with nearly half a bottle of liquid. We had exactly the same trouble in our Jensen FF last year. Must get around to designing a new bracket; filing the sharp edges of the one we have is not enough. The only other recurrent trouble has been with the doors and door-locks. They tend to drop, get loose, lock themselves when slammed, or catch their inside trim on the fixed rubber sealing strips.

Being a very early Stag from an unhappy production line, it is not surprising perhaps that there was a fair amount of work to be done or redone after running in and again after the Swiss journey. The biggest inconvenience was the time taken to deal with faults in body fittings, instruments and the transmission indicator which came off. The car went in for its first check on 10 February and came out on the 12th with just the mechanical bits attended to. It went in again on 16 February for "3 days" but was still waiting for parts on 26 February. Triumph's mechanical and body servicing people in London seem to be pleasant, overworked and optimistic. When planning car

*Petrol pipe clamped in such a way that it rubs through on an adjacent bolt head. Here it has been twisted round to show the puncture. Air filter box has been removed*

usage, it may be best to add a fair margin to their time estimates for completion of work.

Of the attention under warranty, the most serious trouble was a flooding carburettor, caused we were told by a punctured diaphragm. In the course of a short drive the air filter box slowly filled with petrol and made the mixture hopelessly rich. No-one likes to have hot petrol swilling around above the cylinder blocks but this was obviously an isolated fault in one of the pair of Strombergs. Otherwise the Stag engine has been very reliable and uses hardly any oil. It has been drained and refilled at 6,000 and 12,000-mile services and has otherwise taken only about 1 pint per 1,000 miles. It is a sure starter if you give it time. Both hot and cold, however, it needs several seconds of turning before it bites. It also needs half choke for two or three miles on a cold morning before it will pull and even then, it occasionally stalls. The radiator never seems to need topping up.

Fuel consumption is now consistent for given circumstances. Around cities it gives 18.5 mpg. Long, fast journeys go up to 19.5 mpg and the best leisurely journey average we recorded last summer was 23.0 mpg. The manual Stags seem to do about 1½ mpg better.

Michelin XAS handed tyres are fitted and these we changed round as recommended at 6,000 miles. Adhesion on wet roads became noticeably poor at 12,500 miles with between 1½ and 2 mm tread remaining, so at 13,000 we changed them for new ones. The spare has not been used. These new tyres have been the only sizeable expense in the first year.

Turning to practical aspects of design details, that battery mounting accessibility will have to be improved. Take the power steering pump off to lift out the battery? Inevitably there are shorts and sparks as you struggle to put a spanner on the terminal clamp bolts.

A forward-opening bonnet is right; a self-supporting one is best. If you must have a stay, we think it should be placed the same side as the dip-stick and washer-bottle.

Next, a look at the sort of conveniences and comforts which you grow to appreciate, or otherwise, when you live with a car. We have approved seat comfort and forward view and now add the following plus points: all-round look-out with hard top fitted, good ventilation and heating, handy secondary controls—dipper, flasher, wiper, washer and horn on stalks round the steering wheel. The glove-locker is rather small but in compensation there are: a shelf, two small pockets in the doors and two more bigger ones at the back. There are also little trays in the scuttle-top on the passenger side and in front of the gear selector.

The boot is, at best, of big sports car or small coupé capacity. It's shallowness and the incur-

sion of the hinges are the points of criticism. You can carry two suitcases side by side and quite a lot of small and soft things around and on top.

The folding hood was fitted as an afterthought last summer and the "folding" was a misnomer to start with since it would not go down properly. Having been persuaded to do so, over a period of half an hour, it never rose again until seen to by the body service people. We hope to be able to report faultless folding next spring.

With the exception of the battery, everything is reasonably accessible for the help-yourself man. Replacing the dip stick, when you eventually find the hole, you have to take care to push it right home. Garage-men mostly don't, so you risk overfilling.

At night the Stag is pretty good; no nasty reflections, excellent full beams and clean, not too abrupt cut off. With a full load, the tail goes down and the lamp beams up, causing hard words and embarrassment. We have not felt the need for spot lamps, particularly since the not entirely good regulations forbid the use of a single offside lamp in fog. It is not easy to read the main instruments at night, but you soon learn that 30 mph is just past 9 o'clock, 70 mph at 12 and peak rpm at 3 o'clock.

At present we have a Smiths' radio and cartridge player fitted, with a pair of speakers tucked in above the rear pockets. This equipment gives adequate two-speaker stereo; two more speakers would be better. The standard single speaker for radio (central front and temporarily disconnected) gives good, clear volume. For comparison, we also have a portable Smiths' tape recorder-cassette player wired in. This gives facilities for alternative music, or for dictation, or will record direct from the radio. Finally, we have a Readycall FM radio-telephone, which at present gives us Greater London coverage only, and goes off at 8 pm.. and weekends. You have your own call-up signal and there is private conversation both ways. Quality of reception is usually good and free from interference but there are a few areas of weak reception. No sound is heard when the set is on until you call, or are called.

As with all *Autocar's* long-term test cars, one man is in charge, but all drive the Stag from time to time. The concensus of opinion is definitely favourable. It is a nice car with a satisfying touch of elegance and spirit. Now that Triumph have been able to deliver some few thousands of them at home and abroad, we are beginning to hear from owners who agree.

Sometimes we are only too ready to sell our long-term test cars as soon as we can. We hope to keep the Stag for at least a second year. Draw your own conclusions. □

## COST and LIFE of EXPENDABLE ITEMS

| Item | Life in Miles | Cost per 10,000 Miles |
|---|---|---|
| | | £ p. |
| One gallon of 4-star fuel average cost today 35p | 19.2 | 182.00 |
| One pint of top-up oil, average cost today 17p | 1,000 | 1.70 |
| Front disc brake pads (set of 4) | 6,500 | 8.00 |
| Rear brake linings (set of 4) | 15,000 | 4.10 |
| Michelin XAS tyres (front pair) | 12,000 | 23.50 |
| Michelin XAS tyres (rear pair) | 12,000 | 23.50 |
| Service (main interval and actual costs incurred) | 6,000 | 35.44 |
| **Total** | | 278.24 |
| **Approx. standing charges per year** | | |
| Depreciation | | 150.00 |
| Insurance | | 58.45 |
| Tax | | 25.00 |
| **Total** | | **511.69** |

Appox. cost per mile = 5.1p

# Stag. A new kind of Triumph.

Stag is Triumph's first luxury car. Built with the experience obtained through decades of national and international road racing competition. Resulting in an automobile of superb handling, performance and comfort.

This sumptuous 2+2 GT is first and foremost a car to be driven. Power assisted rack & pinion steering and front disc brakes. Fully independent suspension complemented by cast aluminum alloy wheels and high-performance radial tires. And a new overhead cam 3 litre V-8 engine making Stag the fastest Triumph of all.

Once inside Stag, its luxury is all too evident. Standard equipment includes deeply cushioned, fully adjustable bucket seats. Electrically operated windows. Padded steering wheel which adjusts for height and distance. A host of controls at your fingertips. True 2+2 seating.

And then there's Stag's unique new top. Which, if you prefer, can give you a choice of three tops.

For wind in your hair driving it's a convertible with a distinct T-shaped padded roll bar.

To keep the weather out, we've given Stag a fine quality soft top which neatly encloses you, your passengers and roll bar.

Finally, there's the optional removable hard top that looks good enough to leave on all year round.

In addition, you can order automatic transmission, overdrive, AM/FM radio and air conditioning.

Come into your nearest Triumph dealer and test drive the new Stag; Triumph's first luxury car.

For the name of your nearest dealer call: 800-447-4700. In Illinois call: 800-332-4400. British Leyland Motors Inc., Leonia, N.J. 07605.

## Triumph Stag

BRITISH
LEYLAND

**We make sports cars for everybody.**

# TRIUMPH'S STAG

By PAUL WEISSLER

## THIS LUXURY CAR CAN BE TOUCHED BY A NON-PROFESSIONAL!

■ There's an old chestnut that goes something like this: working on any British car takes just four wrench sizes: 7/16, 1/2, 9/16 and 5/8, plus lots of patience.

The Triumph Stag, British Leyland's entry in the luxury touring car field, fits this description almost as well as MG, Jaguar, Rover and so many others, whether grouped under the Leyland banner or not.

Unlike so many luxury cars, the Stag requires few if any special tools. You can do anything with reasonably ordinary wrenches, screwdrivers, pullers, etc.; it just takes time.

The air filter is a simple example. The element cover is held by spring clips, which you must hold in position with a screwdriver as you flip them into position with your fingers. It's not difficult, but it does add a bit of "touch" to what is ordinarily a ten-thumbs job. The rating still is good for element replacement, but just seven points.

The oil filter is wide open--but it's underneath the car and it's that replaceable element type, instead of a modern spin-on. The replaceable element type is $2 or so cheaper, but who today really wants to save $2 that badly? The job requires jacking up the car of course, and is messy, so the rating is poor. The fact that the filter is quite accessible from underneath is a small consolation, worth no more than two points.

The fuel pump is an electric unit in the rear, again a get-out-and-get-under proposition. But our policy is to favor remote mountings of electric fuel pumps, and we discount some of the serviceability problem so long as the job isn't impossible. It isn't, and so the rating is fair and five points.

---

**SERVICE TESTING CHECKLIST**
(Max. 8 points each)

| | |
|---|---|
| 1. Spark plugs | 7. Cylinder |
| 2. Ignition | head bolts |
| breaker points | 8. Fuel pump |
| 3. Distributor lockbolt | 9. Battery cables |
| or locknut | 10. Air filter |
| 4. Generator | 11. Oil filter |
| 5. Carburetor | 12. Brake master |
| 6. Fan belt | cylinder |

13. Heavy repairs

**Bonus Points**
(Max. 2 points each)

| | |
|---|---|
| 1 Dashboard bulbs | 3. Starter motor |
| 2 Speedometer Cable | 4. Radiator Hoses |

---

There's a lot of V-8 under the hood here as well as all the piping for both air conditioning and for power steering.

Air cleaner element removal is no problem, but just be careful refitting cover spring clips after element replacement.

Distributor is at the rear but breaker point replacement is not difficult. Distributor clock is well buried, however.

Tightening head bolts is a snap because rocker covers do not have to be removed. Rating for this is good—8 points.

Spark plugs are just accessible, but when this loaded engine compartment is considered, "just" becomes quite good enough.

The side of the carburetor has to come off to get at one of the retaining nuts, but just a few screws hold the section.

The Stag's overhead camshaft V-8 engine is really packed under the hood, but happily some thought was given to component accessibility.

The distributor is at the rear of the engine, but the breaker points can be changed without any significant effort. The rating for this item is good and eight points.

The distributor lock is somewhat buried, necessitating the use of a C-shaped distributor lock wrench. Inasmuch as this tool is in such wide use today, it cannot be regarded as special. The tool is not expensive, and if you can afford a Stag, you can certainly spend the $3 for the wrench. It does drop our rating, however, to merely fair to good and six points.

The spark plugs are just barely accessible, but in all cases they are accessible and that's what counts eight points.

The carburetors are the venerable tapered-needle-in-a-jet type, and although the design is old and prone to unusual problems, it still works and often rather well. As illustrated, partial disassembly may be necessary to get at one of the retaining nuts, but removing the carburetors is only

Continued on page 115

## Road Impressions

# The Triumph Stag

WHEN I tried the Triumph Stag as a new car, way. back in 1970, I was disappointed with several aspects of the car. It has since had time for development, although it remains basically unchanged, so I thought I had bette: try it again, because it is an interesting concept – a very good-looking sporting car of two-seater size but with room for two more occupants should this be required, a good arrangement of hood when the neat hard-top is not in use, and a very refined vee-eight 3-litre overhead-camshaft power unit.

The British Leyland Press Office complied very readily with my suggestion and so, after a day which had embraced 1930 Bentley and 1973 BMW motoring, I found myself heading for home in a Pimento red Stag with chestnut vinyl upholstery which you could have fooled me was real leather. It was shod with Avon tubeless radial tyres with those smooth sidewalls that reminded me for a moment of the special track covers of the Brooklands days.

There is no need to describe this Stag in detail, because the specification has not altered since we did this fairly fully in the issue of October 1970. It is amusing that the very complete leather—bound Stag hand-book quotes nut-tightening torques but omits any reference to power output. This the catalogue gives as 146 b.h.p. at 5,700 r.p.m. and the red-area of the tachometer begins at 6,500 r.p.m., although only in some pretty brisk driving did I have to take the revs. that high.

For some reason the Stag prompts journalists to photograph it with the appropriate animal

*STAG MEETS STAG. One of our contemporaries drove a Stag into Europe to try to photograph the animal with the Triumph, so we are rather pleased that this picture was taken economically without driving far from London.*

beside it. My friend Philip Turner of a weekly contemporary once had this irresistible urge and I believe went far into Europe to try to satisfy it, but had to admit defeat. So I had been amused to learn that, before I took over the car, one of the MOTOR SPORT photographers had secured the picture which heads this article without going much more than about 20 miles from the London office. . . .

My initial impressions of the Stag were as unfortunate as before. The power-assisted rack-and-pinion steering is far too light and the weaving effect of what seemed to be too-hard independent rear suspension constitutes handling which does not appeal at all. This is a pity, because the steering is very high geared $2\frac{5}{8}$th.-turns, lock-to-lock, from a very small racing-type wheel, with a really small turning circle. On longer acquaintance with the car I got accustomed to the lack of feel, the "in-built over-steer", at the wheel but this and the dodgy running damns the Stag in my view. Admittedly the steering is consistent, which that on the 1970 test-car was not. And Rolls-Royce Silver Shadow drivers have a similar light control to put up with. So perhaps there are Stag owners who do not regard their cars as sufficiently sporting to warrent better control characteristics. There is now only a faint sound from the power assistance via its belt-driven pump, but it is the end-product which distresses me.

That apart, the Triumph Stag is a nice motor-car. It has much of the tradition of the older sports-cars about it, but with all the very smooth power delivery and excellent bottom-end torque to be expected of a multi-cylinder engine of nearly 3-litres capacity. This adds up to very real and readily delivered performance. The ride is on the hard side, but comfortable, the gears are changed by a long lever with considerable across-the-gate movement, but one which functions very nicely, reverse easily obtained by slapping the lever to the right between the 3rd and top-gear positions. Its knob contains the switch for the overdrive, labelled IN and OUT, a selection which could not be more convenient. The hand brake lies between the front seats but its hefty hand-grip needed a good pull to make it hold. In torrential rain, no water entered car or boot.

The disc-drum servo brakes are efficient. The engine has a healthy exhaust note but with the electric windows up, there is scarcely any sound, either when accelerating or cruising fast – a veritable Rolls-Royce among sporting cars.

*The V8 Triumph Stag has handsome lines.*

Instruments – m.p.h./k.p.h., trip/total 140 m.p.h., speedometer, clock, heat, volts and fuel-level gauges, and tachometer – are set on a walnut veneer dash, all on the driver's side, and have been neatly recalibrated, with new green night-time illumination. The eight warning lights are in the grouped Triumph circular cluster, there is a variety of fresh-air vents, and twin stalks control the lamps, horn, turn indicators, two-speed wipers, etc. There are elaborate arrangements for heating and ventilation, including knob-wound half-windows and openable side windows, and refinements include the essential Triplex Hotline heated rear window, height-adjustment for the front seats, adjustable steering column, red lamps in the doors, boot illumination, lockable flush-fitting fuel filler, excellent door locks, etc. Before the front passenger are a generous-size lockable glove box and stowage shelf, and elastic-topped pockets provide extra stowage space. Altogether, then, this Triumph Stag is a very well-appointed and well-contrived car, with good door locks, stainless steel body mouldings and screen surround, etc. It has anti-dazzle visors not always found on soft-top cars and for safety in open form there is the T-shaped roll-over bar. Recent mods. include the smaller steering wheel, a rest for the clutch foot, the option of tinted glass, o/d as standard, and automatic transmission for a modest extra cost.

Realising that 25 years ago I road-tested an Allard in N. Wales and that it had a V8 engine of 3.6-litres but with side valves, and that the Stag has an o.h.c. V8 power unit of all but 3-litres, I set off to cover much of the same terrain. Alas, crawling holiday traffic has increased tenfold and what with congestion in Dolgellau and flooded roads outside this busy town, I had to cut out some of the route. However, the Stag did ascend and descend Bwlch-y-Groes, which anyway is child's play to a modern car, but the track away from Lake Vyrnwy on which I had got the Allard stuck ( MOTOR SPORT, June, 1948) proved elusive.

One likely track petered out hundreds of feet above the road with a sheer drop on the right, calling for some careful reversing, and I am inclined to think that the decently-surfaced but narrow route to Bala was the one we had used previously, when as a rough track it caused us some mild adventure.

These excursions proved that the Stag has ample clearance under its twin exhaust pipes and plenty of controllable power. With judicious use of overdrive I got 20.5 m.p.g. of four-star petrol and no oil was needed after 600 miles. By the time that long day's driving was over I had almost come to accept the lack of feel in the steering and the odd running of the independent rear suspension. It is commendable that British Leyland offer multi-cylinder vee engines in Jaguar, Daimler, Rover and Triumph cars. I do not like the handling of the Stag or of the Rover 3500 V8, so I am hoping that the new MG V8, which eschews i.r.s. and a De Dion rear axle for a beam back axle, may suit me better. It is interesting to note that the price of the M.G.-B V8 GT at £2,294 compares favourably with that of the Stag, which sells for £2,615 in soft-top form, or £2,720 with soft-top and detachable hard-top; it should be remembered that this includes a well-designed hood which stows away efficiently and four Lucas halogen headlamps. I have yet to try the new M.G., but anticipate that this effectively two-seater, permanently closed coupe, will appeal to a more sporting driver than the Stag: the two V8 models should complement each other in British Leyland's extensive range. – W.B.

*Stag in stalking country.*

## STAG SERVICE

Continued from Page 113

time-consuming, not tricky. The rating is fair and five points.

Head bolts on the Stag can be tightened without even taking off the rocker covers. A universal-joint for the socket is necessary, but that's part of even a beginner's tool supply. The rating is good and eight points.

The alternator is a mess. It's under the power steering pump and the factory procedure is to remove it from underneath, after detaching the anti-roll bar. This isn't fun, and the rating is poor and zero points.

You don't have to go through the same procedure to change a drive belt, but it's still a get-under proposition, also rated poor, but worth two points.

The brake master cylinder couldn't be much easier. The rating is good and the full eight points.

Battery cables get a fair rating and five points. You can get to both ends of the ground strap and the other cable (albeit with difficulty). But the other cable goes to the firewall, where it is connected to a junction block. A second cable leaves this block for the starter, and it is this second cable that's tough to change.

The Stag picks up four points in

**Servicing the brake master cylinder couldn't be easier. Our rating: good and a full eight points.**

the bonus category, for dashboard bulbs and the speedometer cable connection. The British have always provided under-dashboard room that is just minimum adequate for normal to smallish hands, and the Stag follows this tradition. You can dock the Stag in the bonus category if your hands are oversize.

Heavy repairs are evaluated on the basis of what can be expected in a luxury car of this type, and therefore are rated fair and four points. Pulling the engine and transmission takes almost eight and one-half hours, the transmission alone five hours. A clutch job is a five and one-half hour proposition and a head gasket nearly three and one-half.

The Stag's total is 72 points, with most of the lost points based on my personal patience limits. Oh well, the car is probably about $1500 leas than anything comparable, which can either inspire patience or leave one sufficient lucre to pay a professional. ●

115

# Triumph Stag

The Stag is at its best either fully open or with the beautifully made hard-top in position. Removing or replacing the top, however, is a trial of strength

**FOR :** very well finished and equipped ; nice controls and instruments ; smooth, automatic transmission ; good ride ; enjoyable to drive

**AGAINST :** limited luggage and passenger space ; seats too flat ; hood difficult to erect ; hardtop too heavy

The predictable success of the Stag has resulted in a year's waiting list for Triumph's top-of-the-range model. This test of the latest automatic indicates that detail improvements have bettered an already good car, so the delays are unlikely to improve.

Some (subjective) improvements like the quieter and smoother engine and the tauter handling can only be explained by detail developments, perhaps better production, as there have been no design changes to account for them. Improvements that have been made include a reduction in assistance for the power steering to give the driver more "feel"; the inclusion of a divided braking system; an intermittent wipe facility; optional head restraints ; flame retardant material for the trim; and optional alloy wheels. Modifications have also been made to the folding mechanism for the hood, and the silencing system, and rationalisation has made the combustion chambers and pistons the same as those of the Dolomite. This has resulted in an insignifi-

cant increase in horsepower of 1 bhp, and a reduction in torque of 3 lb ft. Visual alterations include matt black stone-resistant paint on the sills and rear panel, and side stripes.

The Stag's alloy-headed V8—yet to be used in any other British Leyland vehicle—is based on two Dolomite engines put together, but with a shorter stroke giving a capacity of 2997 cc rather than 3708 cc. When we first tested the car we thought the engine was less smooth than it ought to be for a V8. Now we withdraw that observation as that of the latest car was very smooth indeed, and made very little clatter; the restrained burble from the exhaust is rather pleasing.

With its tiny power increase and lower overall gearing the acceleration of the automatic is almost identical to that of the manual we tested before, and there is nothing to be gained by using the manual hold and hanging on to the gears. But whereas the manual car (on which overdrive is now standard) will wind

Above: the sumptuous seats lack thigh support despite the facility of a built-in cushion tilt. Top right: a smaller diameter wheel and less assistance give more feel to the steering. Below left: the built-in roll cage not only offers protection but stiffens the framework as well. Below right: entry to the rear is awkward and headroom limited. The seat, however, is comfortable

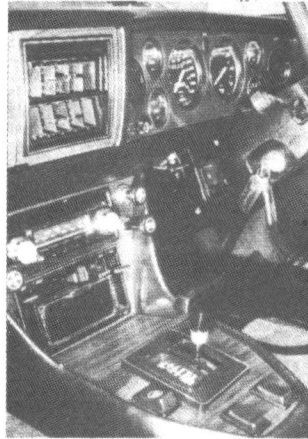

Above: the fan-boosted swivel vents in the centre are effective, unlike the ram operated side ones. Below: the stalk controls are excellent

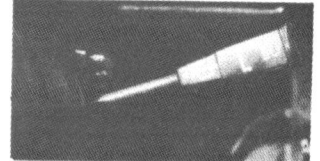

The gearchange is sited too far forward, all other controls are well placed

The comprehensive instrumentation is easy to read and well-sited. We query the fitting of a battery condition meter in preference to an oil pressure gauge

up to well over 120 mph in o/d top, the automatic managed no more than 112.4 mph for a lap at MIRA, and 116.8 mph on the most favourable flying quarter mile. Even so, the automatic would be a comfortable cruiser at more than 100 mph were it not for considerable wind noise.

Fuel consumption is also affected by the automatic transmission, the touring consumption deteriorating by as much as 4 mpg. Even so, the overall value of 18.9 mpg is not bad considering the performance and how hard we drove the car. With its 9.25:1 compression ratio, the Stag feeds on four-star fuel without pinking.

An American-style "T" shift controls the Borg-Warner 35 gearbox. Its dog-leg change pattern is precise, but the positions are unnecessarily close. Moreover, the selection guide is set for LHD models and therefore is at times masked by the lever itself. One virtue of the dog-leg pattern is that it prevents inadvertent selection of neutral, a move that's all too easy on many other automatics.

Automatic changes both up and down the gearbox are very smooth on part throttle, but can be quite jerky under full acceleration. The kickdown is also very responsive at mid-range speeds, but is reluctant to select first on

the move even at 5 mph.

Perhaps the most significant improvement is to the steering. A smaller diameter wheel and a reduction in the amount of assistance has resulted in much better response and feel. You can now use the performance with much more confidence, though some say the steering is still too light.

Understeer is the predominant handling characteristic, though such was the adhesion of the fat Avon radials on our car that the breakaway point was very high indeed. Over bumpy surfaces the car is less tidy, with traces of lurch and diagonal pitch, and it is on these occasions that still more feel to the steering would be appreciated. Under most conditions, however, the car is very well mannered and can be thrown into corners in perfect safety. A slight wriggle after powering out of a turn suggests that there is still a trace of spline lock-up in the driveshafts, though this quirk is far less pronounced than it was on earlier cars.

Another fault, perhaps peculiar to our test car, was slight sponginess in the brakes, the powerful servo-assisted disc/drum set-up needing an occasional pump to restore full efficiency.

Large catches set into the side of the seats release the backrests for access to the rear. This would be less of a scramble if the whole

# MOTOR ROAD TEST No 53/73 • TRIUMPH STAG

## PERFORMANCE

### CONDITIONS

| | |
|---|---|
| Weather | Dry, wind 4-15 mph |
| Temperature | 58-66° F |
| Barometer | 29.8 in. Hg |
| Surface | Dry tarmac |

### MAXIMUM SPEEDS

| | mph | kph |
|---|---|---|
| Banked circuit | 112.4 | 180.9 |
| Best ¼ mile | 116.8 | 187.9 |
| Terminal speeds: | | |
| at ¼ mile | 66 | 106 |
| at kilometre | 83 | 134 |
| Speed in gears (at 6500 rpm). | | |
| 1st | 55 | 89 |
| 2nd | 89 | 143 |

### ACCELERATION FROM REST

| mph | sec | kph | sec |
|---|---|---|---|
| 0-30 | 3.6 | 0-40 | 2.9 |
| 0-40 | 5.2 | 0-60 | 4.7 |
| 0-50 | 7.5 | 0-80 | 7.3 |
| 0-60 | 9.9 | 0-100 | 10.4 |
| 0-70 | 12.8 | 0-120 | 14.5 |
| 0-80 | 17.0 | 0-140 | 20.3 |
| 0-90 | 22.4 | 0-160 | 29.2 |
| 0-100 | 30.2 | | |
| Stand'g ¼ | 17.6 | Stand'g km | 32.2 |

### ACCELERATION IN KICK-DOWN

| mph | sec | kph | sec |
|---|---|---|---|
| 30-50 | 3.9 | 40-60 | 1.8 |
| 40-60 | 4.7 | 60-80 | 2.6 |
| 50-70 | 5.3 | 80-100 | 3.1 |
| 60-80 | 7.1 | 100-120 | 4.1 |
| 70-90 | 9.6 | 120-140 | 5.8 |
| 80-100 | 13.2 | 140-160 | 8.9 |

### FUEL CONSUMPTION

| | |
|---|---|
| Touring* | 21.7 mpg<br>13.0 litres/100 km |

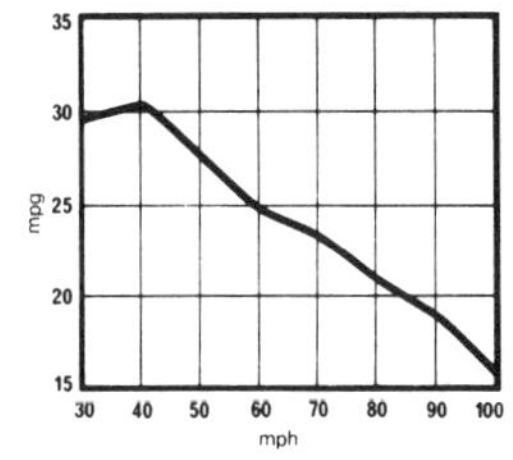


| | |
|---|---|
| Overall | 18.9 mpg<br>15.0 litres 100 km |
| Fuel grade | 98 octane (RM)<br>4 star rating |
| Tank capacity | 12.75 galls<br>57.9 litres |
| Max range | 276 miles<br>— km |
| Test distance | 2119 miles<br>3409 km |

* Consumption midway between 30 mph and maximum less 5 per for acceleration.

### SPEEDOMETER (mph)

| | | | | | | | | |
|---|---|---|---|---|---|---|---|---|
| Speedo | 30 | 40 | 50 | 60 | 70 | 80 | 90 | 100 |
| True mph | 30 | 41 | 52 | 62½ | 72 | 82½ | 93 | 104 |

Distance recorder: accurate.

### WEIGHT

| | cwt | kg |
|---|---|---|
| Unladen weight* | 24.7 | 1254.8 |
| Weight as tested | 28.4 | 1442.8 |

* With fuel for approx 50 miles.

**Performance tests carried out by Motor's staff at the Motor Industry Research Association proving ground, Lindley.**

## GENERAL SPECIFICATION

### ENGINE

| | |
|---|---|
| Cylinders | V8 (90°) |
| Capacity | 2997 cc (182.9 cu in.) |
| Bore/stroke | 86/64.5 mm (3.38/2.53 in.) |
| Cooling | Water |
| Block | Cast iron |
| Head | Aluminium alloy |
| Valves | 2-ohc |
| Valve timing | |
| inlet opens | 16° btdc |
| inlet closes | 56° abdc |
| ex opens | 56° bbdc |
| ex closes | 16° atdc |
| Compression | 9.25 : 1 |
| Carburetter | Two Stromberg CDS (E) V |
| Bearings | 5 main |
| Fuel pump | Electrical |
| Max power | 146 bhp (DIN) at 5700 rpm |
| Max torque | 167 lb ft (DIN) at 3500 rpm |

### TRANSMISSION

| | |
|---|---|
| Type | Borg Warner type 35 automatic |
| Internal ratios and mph/100 rpm | |
| Top | 1.00 : 1/20.0 |
| 2nd | 1.45 : 1/13.7 |
| 1st | 2.39 : 1/8.4 |
| Rev | 2.09 : 1 |
| Final drive | Hypoid bevel 3.70 : 1 |

### BODY/CHASSIS

| | |
|---|---|
| Construction | Unitary all steel |
| Protection | Phosphated by 7-stage process |

### SUSPENSION

| | |
|---|---|
| Front | Independent by MacPherson struts, coil springs, telescopic dampers, anti-roll bar |
| Rear | Independent by semi-trailing arms, coil springs and telescopic dampers |

### STEERING

| | |
|---|---|
| Type | Rack and pinion |
| Assistance | Yes |
| Toe-in | 1/16-1/8 in. |
| Camber | 1° positive ± 1° |
| Castor | 2° ± 1° |
| King pin | 10½° ± 1° |
| Rear toe-in | 0-1/16 in. |

### BRAKES

| | |
|---|---|
| Type | Disc/drum |
| Servo | Yes |
| Circuits | Yes |
| Rear valve | No |
| Adjustment | Self-adjusting |

### WHEELS

| | |
|---|---|
| Type | Steel 5J |
| Tyres | 185 HR14 radial |
| Pressures | 26 F; 30 R |

### ELECTRICAL

| | |
|---|---|
| Battery | 12 volt, 56 ah |
| Polarity | Negative earth |
| Generator | Alternator |
| Fuses | 12 |
| Headlights | 4 x 5¾ Quartz Halogen |

## COMPARISONS

| | Capacity cc | Price £ | Max mph | 0-60 sec | 30-50* sec | Overall mpg | Touring mpg | Length ft in | Width ft in | Weight cwt | Boot cu ft |
|---|---|---|---|---|---|---|---|---|---|---|---|
| Triumph Stag | 2997 | 2685 | 112.4 | 9.9 | 3.9 | 18.9 | 21.7 | 14 6.75 | 5 3.5 | 24.7 | 3.6 |
| Reliant Scimitar GTE | 2994 | 2517 | — | 8.7 | 7.6 | 21.7 | 28.2 | 14 2.25 | 5 5 | 22.8 | — |
| Alfa Romeo 2000 GTV | 1962 | 2849 | 115.3 | 8.9 | 10.9 | 20.8 | 26.9 | 13 4.5 | 5 2.5 | 20.2 | 6.8† |
| Datsun 240Z | 2392 | 2690 | 125.1 | 8.3 | 9.0 | 25.7 | 31.2 | 13 7 | 5 4 | 20.3 | 11.4 |
| Lotus Elan +2S 130 | 1558 | 2789 | 121.0 | 7.7 | 8.5 | 21.0 | 26.1 | 14 0.5 | 5 3.5 | 16.8 | 4.2† |
| MGB GT V8 | 3528 | 2309 | 125.3 | 7.7 | 6.2 | 19.8 | 25.7 | 12 9 | 5 0 | 21.2 | 6.6 |
| Opel Commodore Coupe GS | 2490 | 2800 | 106.8 | 12.0 | 4.7 | 17.1 | 21.3 | 15 0 | 5 7.5 | 24.3 | 11.3 |

*kickdown for Stag and Commodore †measured with boxes not cases

**Make:** Triumph
**Model:** Stag
**Makers:** British Leyland (Triumph) Ltd, Coventry England
**Price:** £2303.0 plus £191.92 car tax plus £249.49 VAT equals £2744.41. Automatic transmission £45.29 (in lieu of overdrive), Sundym glass £45.29, alloy wheels £68.52; total as tested equals £2903.51

seat tilted forward. Once you are installed in the back, headroom beneath the hood or hardtop is limited, and the backrest is a little upright. For short journeys, however, there is sufficient room for two adults.

Boot space is very limited, but there is excellent stowage for oddments, with map pockets in the front doors and the back of the seats, handy trays to the side of the rear seats, as well as a parcel shelf and glovebox in the facia and a flat area on the console ahead of the gearlever.

The driving position is very good and can be adjusted to suit all shapes and sizes. The steering column, for instance, can be altered for reach and rake and a handle on the front of the seat alters the tilt of the cushion, though we found the range of adjustment insufficient and the cushion always too flat and lacking in thigh support. The backrests will recline fully.

A large brake pedal allows for left- or right-foot braking, but some of our testers complained that both pedals were offset too far to the right. A rare and welcome refinement is a footrest for the left foot.

The traditional Triumph fingertip controls now include parking lights and electric washers, as well as the newly added intermittent wipe facility. The instruments are unchanged apart from a slightly restyled calibration.

Though Triumph say no alterations have been made to the suspension, the ride seemed better than before. It can become a little turbulent on poor secondary roads, where it feels underdamped, but unlike so many cars that are comfortable at speed the Stag has an excellent town ride.

Ram pressure through the eyeball face vents is poor but output can be supplemented by the opening quarterlights and the fan-boosted swivel vents in the centre of the facia. Overall, the heating anr ventilating system is good. our only complaint being heater tends to bleed too much air from the ventilation system.

Road noise was always subdued on this car and engine and exhaust noise have both been cut down, so it is only wind noise that breaks the peace. When the car is open or fitted with the well-sealed hardtop, wind noise is tolerable, but it is very loud with the hood up.

The Stag's equipment can only be described as lavish. Fittings include a clock, map, reversing and parking lights, heated rear window set into the hardtop.

Although the hood and hardtop have been improved, erection of both is still a trial of strength. Fixing the rear hood catch in place calls for much struggling and the hardtop, though beautifully made, requires two strong men to manhandle it on and off.

This test of the updated Stag reaffirms our initial impression that the car is not only unique in character and a highly desirable property, but that the standard of finish makes it a world-beater at the price.

# TRIUMPH STAG
## 27,000 mile report

When we wrote our first long-term test in February last year, Stags were still scarce. We had bought our red automatic in December 1970 — one of the first off the line — and it had covered its first 13,000 miles in 13 months. Now, after nearly three years, it has done more than twice that mileage and continues to be a pleasant and usually well-behaved car.

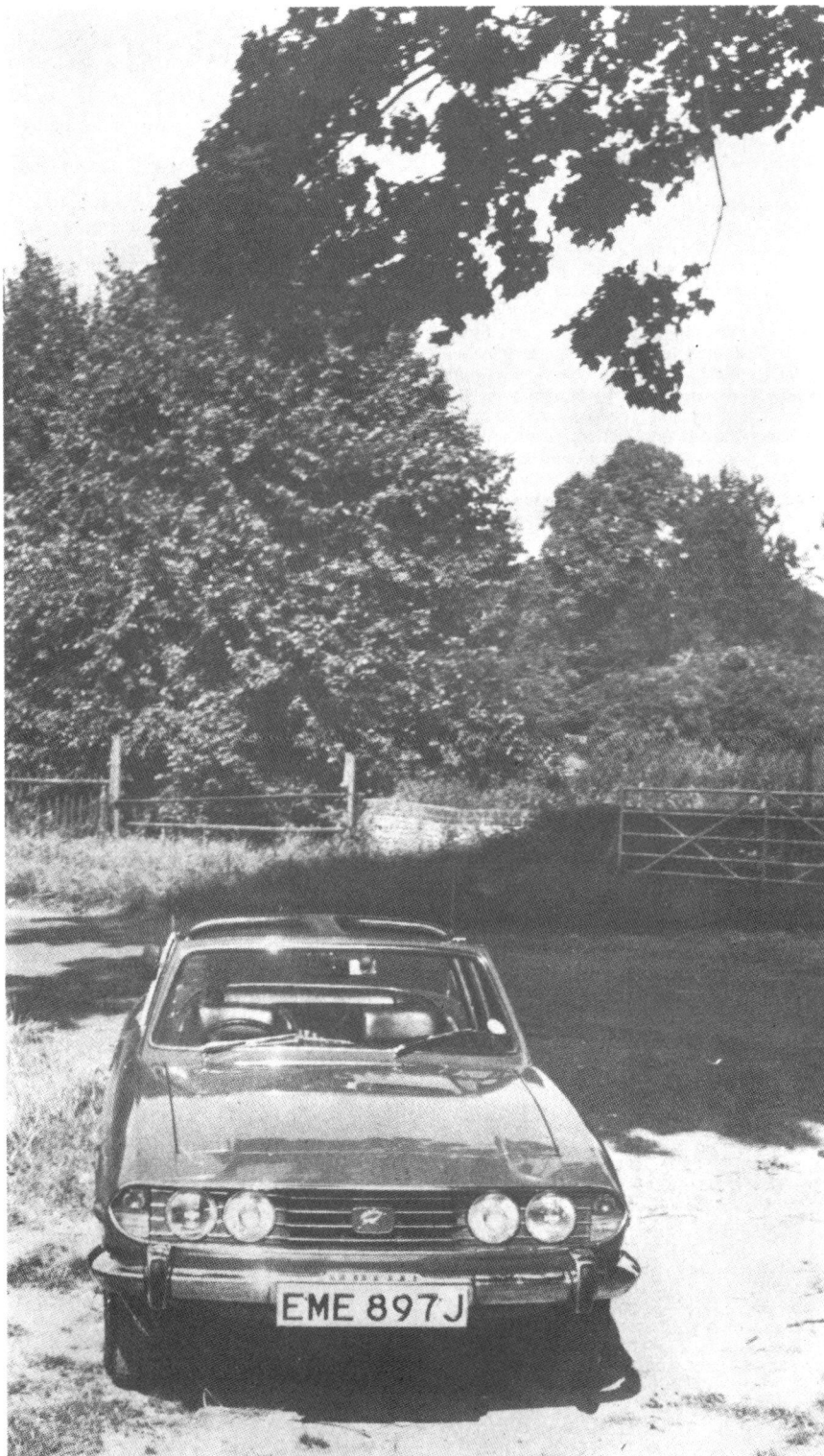

STAGS SEEM to be gregarious. Someone buys one and within a few months there are half a dozen in the immediate neighbourhood. Now, there are at least six others operating within a quarter of a mile of our Stag's garage in Surrey. Not infrequently, we are approached by complete strangers, "Oh, you've got a Stag! What do you think of it?" And there is an air of these cars being rather unknown and exciting. To the extent that they are 2 + 2 convertibles of modest overall size and cost, and with lively performance, they are still quite exclusive. There is little to compete or compare directly with them.

Once you drive V8, you are reluctant to go back to a six or a four unless it is very good and responsive — which usually means expensive too. In our previous test report, we asked for more zip from the Triumph V8 without loss of its attractive gentleness. By raising the compression ratio from 8.8 to 9.25, in common with that of the Dolomite, the makers have now taken a first step and provided a fraction more performance for those occasions when you will want to press on. Our old Stag is still good for 110 mph and will cover a standing ¼-mile in 18 seconds; and of course the manual cars are a bit quicker.

We chose automatic transmission because of the high proportion of our day-to-day driving which has to be done in built-up areas and heavy traffic. In these conditions, the overall average mpg had remained very steady at 19.3. Out on the open roads we usually manage 21 mpg.

The long warm-up period, in cold weather, and the prolonged use of choke that goes with it, seems to affect the fuel consumption more than with most cars. This may be one of the causes of the otherwise unexplained improvement in consumption for the period April to August this year for which our records show 21.6 mpg overall.

Last month we had the first signs of electric pump trouble — an occasional flurry of ticking but no fuel to the carburettors. Without doing anything about it, fuel would eventually flow again. Soon this changed to continuous, fast ticking at all times and low pressure delivery to the carburettors. To forestall trouble we fitted a replacement pump and all is well again. This is not a difficult job, the pump being accessible behind the side lining of the boot, which is "attached" by self tappers (only one of each registers with a hole in the car structure).

We have had no regrets about going automatic, in spite of the slightly higher fuel consumption. If we want to hurry, we use the central selector like a manual shift and go up into the high 5,000s of revs, which the engine willingly gives. In traffic the almost continuous use of a clutch can be very tiring and we are glad to be spared it.

After prolonged driving in traffic, the engine is sometimes hesitant and needs to clear its throat for a mile or so but it never actually misfires or "loses" plugs (except as described later) and a short spell on the open road makes it crisp again. The engine always starts, however frosty the night. In fact, like so many V8s, any reluctance to start — or re-start — comes when it is very hot, then it seems best to put the accelerator on the floor until it fires.

In our first assessment of this car we listed the large number of items of equipment which are included in the standard specification. One of these is the power-assisted steering. This takes a little time to get used to because of the lightness and very quick response, yet we now say again that it is first class and can be a real safety factor if sudden avoiding action or skid correction should be necessary. The reduction in steering wheel diameter by one inch on current production cars is a sensible change.

Another feature we continue to appreciate is the 2-way adjustment of the steering column — up and down on angle, as well as in and out. This, together with a 3-way seat adjustment, means that almost everyone can get comfortable and on long journeys alter their sitting and arm positions slightly if they begin to feel "seized up". For the over six-footers with long bodies, the head room even in front is inadequate unless they are prepared to angle the seat back and recline a little. Owing to the catch knobs on the upper outside edge of the front seats colliding with the hood-frame boxes, the recline of the front seats is limited. Someone forced one of our seats back to about a 45 deg. angle, thereby dislodging the knob from the metal-strip lever and this in turn gashed the trim material.

How has the Stag come out under the wear and tear headings? So far as the "original sin" is concerned, it has never packed up and left us stranded. It came near to doing so on one occasion and this in fact was the only serious trouble we have had. Previously, we had found that every few thousand miles, minor clunks and clonks had developed in the suspension and transmission. A check around the underneath revealed various loose attachments and sloppy bushes. Nothing dangerous; just some unwanted movement.

Having read that Stags tow well we fitted a Witter Mongoose tow-ball (20,866 miles) and shortly afterwards towed a 14cwt trailer for some 200 miles. A little later we took delivery of an Eccles Amethyst caravan from Newmarket and towed it to Farnborough and back. On both occasions the car did in fact tow very well but a few weeks later there developed more noises in the suspension. It was 40 miles later before this could be checked and on the way to our garage there were ominous signs of final-drive trouble, difficult to describe in words if you have not experienced them. Backing into our parking bay produced dental crunches which lead us to put the Stag on to a transporter to Triumph's service depot at Acton. This was at 22,213 miles. Some weeks later the car emerged in good running order but the invoice shook us somewhat. While the labour charges seemed fair enough, the £101 for two half shafts and another £69.50 for the diff. seemed pretty steep. Equivalent shafts for the Granada cost only £35.70 each with CV joint while for a Jaguar XJ12 with Hooke joints you pay only £15.61 each. Not until you get to high quality imported cars like the 280 Mercedes do you get to a figure like £62 per shaft and joint.

One day, an excited gent flagged us down and shouted "Some-one's stuffed a rag up your pipe and it's on fire!" We had in fact noticed that the stuffing was coming out of one of the end silencers which was both corroded and eroded quite badly. Two miles later another chap tapped on the window

*Our Stag in open form. The current hardtop version now includes the hood which was formerly an extra*

and said "Gorrabirdsnestinyerpipe mate!" To please him we stuffed the glass fibre back, tidied up the jagged end of the outer case and it is still in use but looking tatty, as the picture shows.

Triumph admit to early hood folding trouble. The majority of Stags seem to keep their hard tops all the year round. Some owners, if they looked in the hood box, might well be surprised to find it empty, as we originally were. Our hood, an optional extra, was fitted after purchase and we have become reasonably adept at getting it up and down, now that we know the sequence of actions must be correct and that it is really a job for two people working together. The hood and its seals are a poor fit and one result is that the wind noise is considerable above about 60 mph. Even so, it has proved to be waterproof and reasonably draught-free and has not suffered any appreciable damage or rubbing while stowed away through three winters. The car has been running open for most of four months of the year. The front-seat occupants find it cosy and free from wind buffeting.

Among the few minor troubles experienced has been the reluctance of the starter to engage for two short periods. A clean-up of the contacts and connections more or less

*Not a bird's nest up the exhaust pipe but the glass-fibre stuffing coming out of the badly corroded tailpipe silencer*

cured this problem, and a new battery fitted after 25 months also helped. The battery mounting is highly inconvenient and you almost need to "get a man in" to remove the battery from its tray.

Courtesy light switches and door latches have needed occasional attention and the outer trim strip carrying the rubber seal for the passenger side window blew out completely. Fortunately we were able to retrieve it undamaged but were not able to refit it because the clips were strained. The whole window now rattles but probably for different co-incidental reasons.

Earlier this year we ran into a spell of rough running and misfiring. The car went in for service — plugs, points and so on — and came out no better. "Like tennis elbow" the garage man said, after a second attempt, "There's nothing actually wrong, it'll work off". And later: "Give it good burst down the motorway". Being interpreted, this meant "We don't know what the trouble is and would prefer not to have any more bother with this car".

A few days later, after the burst down the motorway, the engine was positively limping and stalled at most traffic lights. It seemed to pull in bursts, if you could work the speed up. So we earmarked the next Saturday morning for personal attention at home. We went through the ignition from end to end and were satisfied that all plugs sparked. Each cylinder had proper compression and the mixture supply was apparently normal. Then we got a clue.

With the engine set to idle, as best it could on four or five cylinders, we pulled off number 1 plug lead and found it ran better. With number 2 off, it ran better still. Could we have the leads crossed? We checked this and the answer was no. One and two cylinders, at the front of each block, fire consecutively (order: 1,2,7,8,4,5,6,3).

Continuing with our elimination process, we took number 3 lead off but the engine shook once and stopped, so that one was doing some work! With 3 lead back on we tried 7 off and this had no appreciable effect on running. Now we understood. We took off the distributor, cleaned the inside very carefully and examined it in a bright light and there

# TRIUMPH STAG

they were: a tiny jagged line of hair-cracks, linking contacts 1, 2 and 7. Our spark had been jumping along these hair-cracks, sometimes firing the plug, sometimes not, and sometimes firing one of the plugs at the wrong time. The faulty distributor cap moulding was disconnected and the local Lucas agent produced a replacement off the shelf for £2.36. With the new one fitted, the engine burst instantly into smooth and healthy life — and has not looked back since then.

Stag headlamps, quartz halogen, are very good but there is one snag. If the car is

*Left: A feature of the Stag is the padded roll-over bar. In spite of its sporting looks the car is a four seater with a reasonable boot*

*Below: Persistent misfiring which professional garage attention failed to cure was finally traced to a hairline crack in the distributor cap between contacts 1, 2 and 7*

carrying four people or heavy luggage, or both, the tail goes down and the lamp beams go up — to the irritation of approaching drivers and the embarrassment of Stag owners. Except to set them so that they have only a limited range when driving solo, we have not found a solution. Some other British models have a similar problem.

So far as regular check-ups are concerned, oil consumption can be almost forgotten, still being negligible, and the power steering reservoir which originally needed frequent topping up, now remains level. The radiator also remains full.

The paint work and plate are standing up fairly well with the exception of two nasty blisters of rust adjacent to the trim shield, one on the nose, the other on the tail. We have cleared these two spots and inhibited them with Naval Jelly and after touching them up they are now just visible but no longer unsightly.

The spare wheel and tools are kept under the boot floor which is detachable, and has a loose carpet on top. The little plastic fasteners meant to hold the floor panel are brittle and some have snapped. Although still quite presentable, the carpet does not stay flat but the floor section remains in place. The size of the boot is marginal for journeys. With careful arrangement, two suitcases and other soft packages can be accommodated.

Those are some of the hard facts of Stag ownership. The list of significant trouble after 26,000 miles is shorter than average — which is a reflection on the reliability of too many modern cars. Since pride and pleasure of ownership are so important, we must record that the family enjoy the car; those who drive like driving it, all approve of its handy size and light controls and appreciate the convenience of electrically-powered windows. Next year we may well buy a replacement Stag embodying the several small improvements which are the outcome of three years' experience and development. □

## Improvements embodied in current Stags

1. Clearer instrument markings.
2. Brighter instrument lighting.
3. Provision for head-rests on front seats.
4. Smaller diameter steering wheel.
5. Intermittent wiper action.
6. Rest for driver's left foot.
7. Extra courtesy lamp on roll-over bar.
8. Under-bonnet arrangements tidied up.
9. Overdrive as a standard fitting on manual cars.
10. Stronger bumper mountings.
11. Flame resistant trim materials.
12. Tail panel and sills painted matt black.
13. Double coach line added at waist level.
14. New options.
15. Special cast alloy wheels.
16. Laminated windscreen and Sundym glass.

Stags are still in short supply and a wait of three to six months for delivery must be expected. In 1971 home market deliveries were 2,136, as compared with 1,551 for export. Last year, the figures rose to 3,262 at home but dropped to 967 abroad, many overseas orders having been lost because customers got tired of waiting and went elsewhere. Stag sales appear to have suffered considerably from a series of production hold-ups but at home the selective demand seems likely to continue, assuming that the model survives British Leyland's future rationalization plans. The proportion of women owners appears to be higher than average and there may be something to learn from this about the qualities they appreciate. □

# ESTATELY STAG

It's as quick as a Scimitar and there's room for the kitchen sink too. Gordon Bruce explains how one man put a Stag under his bonnet and gave Triumph's big estate a new lease of life. Read on and you'll see it's much more than a straight transplant. This one handles too — and all for £3000

Fat twin tail pipes, a low stance and wide boots are all clues; the V8 3 litre badge supplies at least part of the answer. Below: Snug in its new home, the Stag unit is only 70lb heavier than the old 'six.' Top: The superb Recaro seats make for comfortable long distance work and the tiny Astrali wheel adds feel to the power steering

# Estately Stag

" If you had all the money in the world, what car would you buy," she asked. Come to think of it, so have many others, though few of them were as beautiful as she. I explained that unfortunately no manufacturer made " my " car though one at least was on the right lines. She looked puzzled. Then glancing at the sea of vehicles that constituted Earls Court's ground floor she said " you mean not one of these would suit you?"

Cornered ! I had to admit that the Ferrari Boxer would keep me amused for a while and that, well yes I had always had a yen for a Porsche. But, damn it, not one of them were quite Bruce's ideal machine. " Tell me about your dream car then." I pointed out that apart from outstanding performance and handling it must be practical and spacious with comfortable reclining seats—she smiled—and a decent stereo system that the car allowed you to listen to. Then of course it must have Sundym glass, a sunshine roof and, oh yes, be able to tow the Bruce racer with ease at 70, sorry, I mean 50 mph.

Perhaps if we had then left for dinner in Del Lines' Triumph (it's all right we've got to the point at last) the evening might have continued in the same dreamy vain, but somehow jumping into the Fiat 127 had as much effect as dinner for two at the local Wimpy. Still it was a good try.

So is the Lines Triumph. In fact it does come remarkably close to " my " car. Basically it is a Triumph 2000/2.5 estate car with Stag engine, transmission and suspension and a host of goodies added to complete the effect. In reality it is one of the most pleasant multi-role vehicles I have ever driven. And thought of as a car in its own right—rather than a conversion—must be a worthy challenger to the coveted Scimitar.

It all happened by accident really. He had a Triumph 2000 with an engine that was on the verge of collapse and a redundant Stag unit. A quick survey showed that mating the two was possible and he set to work.

The result was so pleasing that friends started to order replicas. Now five have been produced and a further four are on order. The necessary body modifications are surprisingly few, namely replacing the original engine mounts with Stag ones, modifying the floor and transmission tunnel to take larger twin exhausts, moving the battery tray and changing the wiring loom. The only alteration to the exterior is where the rear arches have been " pulled " to accommodate the fat wheels and tyres.

Undoubtedly the biggest bonus of fitting the V8 powerplant is the extreme flexibility and smoothness that one enjoys, though as can been seen from our comparison chart the car is no sluggard either. In fact with the relatively slippery shape of the estate body and the transmission geared for performance rather than low rpm cruising, it will comfortably out-perform the PI and even the Stag itself come to that. Of course there is no reason, why with a little jiggery pokery, that lovely engine couldn't be made to part with a few more horses still—as you see the possibilities are endless.

Criticisms of the unit in our test car (Del's own transport as if you hadn't twigged from the registration) were of a slight vibration at high speed and relatively heavy fuel consumption, the latter being the product of the lower gearing and hard driving on our part.

At the moment Del drives through a standard Stag clutch which although unfussed about a full-blooded wheelspin start does seem to judder in normal use. Our man is aware of the problem and has persuaded AP to make some special ones for future conversions. We certainly had no other criticisms of the transmission which consisted of a competition Stag gearbox (standard ratios) and competition overdrive unit coupled to a 3.7:1 Salisbury limited-slip diff. If you're still with me, you may by now have noticed the lovely gold $7\frac{1}{2}$ in. x 13 in. Minilites, which are fitted with low profile Dunlop SPs, again lowering the gearing and, on this car at least, sending the speedometer—round to an astonishing and somewhat optimistic 130 mph.

Pleasant though the idea of the two extra cylinders is, it was the handling of the car that really won my heart over completely. The canny Lines has been underneath the beast as well you see. Stag springs and Armstrong heavy-duty dampers look after the rear and a Stag roll bar provides suitable roll stiffness at the front. Bearing in mind that the ordinary estate handles well, one cannot fail to be impressed with these alterations. Turning into a corner one is immediately aware of the extreme lack of roll and the stability afforded by taut suspension and a good helping of rubber to road contact.

Initially, understeer messages find their way through the power steering to the tip of the tiny Astrali wheel, but a touch more lock and a good helping of power will bring the tail round in a gentle, most controllable slide. There is so much feel in the car (the combination of small wheel and power steering is a lesson to Jaguar, BMW and Mercedes alike) that you are inclined to forget you have a large estate body behind you and what is well over a ton of motor car at your disposal, especially as the ride is still very acceptable. The only quirk of the handling is a slight exaggeration of the spline lock from which all big Triumphs appear to suffer.

Talking of weight brings the brakes to mind. Here a Stag dual circuit system with Stag rear drums distinguish the set-up from standard. This particular car suffered from extreme fade due to a mistake over the pad compound, but with the trusty DS11s installed there is no reason why braking should not be more than adequate.

Having built a car that performed, handled and stopped, Del then set about adding the comforts that can change such a fun machine into a Grand Tourer in the true sense of the phrase. Those Recaro seats for instance, they may cost a wopping £175 for the pair, but you'd be hard pushed to better them (Del incidentally is an agent for them). Then there are the windows—Sundym glass all round with a banded front screen and electrically operated side-windows. The final touch is a beautifully made Helandia electrically operated sun-roof fitted by Bristol Coach Builders. The Stereo Eight system is another welcome inclusion.

Take the whole package minus the LSD and you'd have to find around £3000. This must be good value when you consider the standard PI costs £2502. Of course this is assuming you buy the complete outfit new from Del. If you want your own car converted it will be a more costly exercise. Saloon versions are available as well and in both cases the specification is to your own choosing.

If you glance through *Motor's* new car price guide you will see that apart from the Scimitar the car really has no rival at home or abroad, the current Citroën Safari being about the nearest comparison. Perhaps then, while puzzling why it should take a privateer to show BLMC what sort of car they could be producing, we should express our gratitude that such cars are at least available in limited quantities. Thank you, Mr Lines.

**Gordon Bruce**

---

**Make :** Triumph.
**Model :** Stag Estate.
**Makers :** Atlantic Garage, Weston-super-Mare.
**Tel :** Weston-super-Mare (0934) 26208.
**Price :** Approximately £3000 including sun roof, Recaro rally seats and Minilite wheels.

**MAXIMUM SPEED**

| | Triumph PI saloon with o/d mph | Reliant Scimitar mph | The Lines V8 mph |
|---|---|---|---|
| Lap | 110.5 | 116.9 | 120.3 |
| Best ¼ mile | 117.6 | 120.0 | 121.9 |

**ACCELERATION**

| mph | sec | sec | sec |
|---|---|---|---|
| 0-30 | 3.3 | 3.1 | 3.0 |
| 0-40 | 5.1 | 4.5 | 4.7 |
| 0-50 | 7.1 | 6.6 | 6.6 |
| 0-60 | 9.7 | 8.7 | 8.9 |
| 0-70 | 13.5 | 11.6 | 12.2 |
| 0-80 | 17.1 | 15.5 | 15.6 |
| 0-90 | 22.5 | 20.1 | 21.4 |
| 0-100 | 31.3 | 27.8 | 29.8 |
| Standing ¼ mile | 17.0 | 16.1 | 16.8 |
| Standing Km | 31.9 | 30.3 | 31.1 |

**IN O/D TOP**

| mph | sec | sec | sec |
|---|---|---|---|
| 20-40 | 11.9 | 11.3 | 10.5 |
| 30-50 | 11.1 | 11.3 | 9.9 |
| 40-60 | 11.6 | 11.0 | 9.5 |
| 50-70 | 12.9 | 12.5 | 10.1 |
| 60-80 | 15.3 | 14.8 | 11.3 |
| 70-90 | 18.4 | 15.9 | 14.0 |

**IN THIRD**

| mph | sec | sec | sec |
|---|---|---|---|
| 10-30 | 6.1 | 5.8 | 5.6 |
| 20-40 | 5.8 | 5.7 | 5.0 |
| 30-50 | 4.9 | 5.3 | 4.9 |
| 40-60 | 5.8 | 5.0 | 4.8 |
| 50-70 | 7.3 | 5.3 | 5.5 |
| 60-80 | 10.2 | 6.6 | 7.3 |
| 70-90 | — | 9.3 | — |

**FUEL CONSUMPTION**

| | mpg | mpg | mpg |
|---|---|---|---|
| Overall | 22.2 | 21.7 | 18.7 |
| Touring | — | 28.2 | — |

# GIANT TEST

## Triumph Stag    Scimitar GTE    Datsun 260Z

SHOPPING FOR A SPORTING CAR COSTING between £3000 and £3500? Individuality ranks high among your priorities, and so does strong, torquey performance. You want a fifth gear for effortless and economical long-range cruising, and a measure of practicality as well—enough to allow a couple of kids or the occasional in-law to ride in the back. You have a good choice: there are three popular cars that fulfil all these needs, scoring especially highly on individuality. They have so much individuality, in fact, that each has its own special something to stand apart from its opponents. For instance, the recently-arrived Datsun 260Z 2+2 is a low and smooth 2+2 with a convenient hatchback, the Reliant Scimitar GTE is a sort of sporting estate car (*very* sporting), and the Triumph Stag is a 2+2 with a detach-able roof. These special features—each a rarity—give these cars unique appeal. There is just enough overlap in their purpose and the way they carry it out to make them true competitors.

After the latest price rises, the Reliant Scimitar GTE Overdrive is the cheapest of the trio at £3337. The Stag (with hardtop) is just a few pounds more at £3342, while the 260Z 2+2 is up another £100 at £3435. When the 260Z was released about six months ago, it was very much more expensive than the other two. Now the dreaded inflation has levelled things up.

## STYLING, ENGINEERING

The Datsun is a development of the original 240Z which Nissan introduced five years ago,

the 2+2 configuration coming at the same time as the first major mechanical change: upgrading of the in-line, six-cylinder engine from 2.4 to 2.6litres. Styling was altered greatly for the 2+2—rather, a fistful of inches were slotted into the wheelbase and the roof-line was altered to suit. The result is a longer car that is visually better balanced than the 260Z two-seater, although the ruggedly handsome look is still there.

As before, the tailgate lifts up to reveal a large luggage platform behind the seats: the rear ones fold down to give even more carrying capacity. Although the Datsun looks a big car, it is, in fact, fractionally shorter than the Stag, the largest of our group.

Stag styling is completely different from the Datsun: because it has a removable roof, it is built on more conventional lines

# GIANT TEST

## Triumph Stag

*Despite* Motocar's *insistence that it's a latterday Lotus Elan, the Stag is in fact not a very sporting vehicle. It behaves tolerably well in corners but its limits are low. If they are transgressed, surprises follow—fast*

## Scimitar GTE

*In these photographs the GTE is going 10mph faster than the Stag (above), yet it is rolling less and still has a bit in hand. Its handling is very progressive, and for such a commodious vehicle it is surprisingly 'chuckable' when driven hard*

## Datsun 260Z

*The beefy Datsun reminds us of certain British sports cars of the '50s and early '60s. It feels strong, even crude, but in fact behaves remarkably well and, thanks to radically improved Japanese-built tyres, now offers outstanding adhesion wet or dry*

with a normal boot following on after the cockpit—in brief, a Triumph 2000 coupe with a lift-off roof. An unusual feature is the integral roll-over bar.

The Scimitar is different again. Thoroughly unique in the motoring world now that Volvo's imitator 1800ES has died a welcome death, it is essentially a two-door estate car that has been neatly styled to make it both very appealing to the eye and highly practical. It is, no doubt, the car that inspired the increased practicality of the Capri II. A further point where the Scimitar differs from the Datsun and Stag is in body construction: while they have unitary chassis/body construction, it uses a steel chassis with separate fibreglass body. That the GTE's styling is effective becomes apparent when you stand it alongside the Stag: although introduced two years later, the Stag looks a good deal older with its ornate front and rear treatment and coke bottle flanks.

There are as many variations in the mechanics of the three cars as there are in their styling. The Stag is powered by a V8, the GTE by a V6 and the 260Z by an in-line six. The Stag and GTE both have four-speed gearboxes with electric overdrive; the Datsun has a five-speed gearbox. The GTE has a live axle; the Stag and Datsun have independent rear ends. The Datsun isn't a highly-advanced car mechanically, but it does have enough interesting features to rate it as above average. The in-line six has an overhead cam and twin carburettors, the gearbox has five speeds and the rear suspension is independent. All three have disc brakes at the front with drums at the rear, and they all have servo assistance. The Datsun alone has a rear pressure-limiting valve.

The Stag's 2997cc V8 is the largest engine —by a mere 3cc from the GTE's Ford V6. More than 400cc behind comes the Datsun's 2565cc in-line six which, like the other two, has oversquare cylinder dimensions. While the GTE V6 has a super-staid cast-iron block and cylinder heads, the Stag and Datsun engines have alloy heads on iron blocks. Their valves are operated by overhead cams as well, compared with stodgy pushrods in the V6 (which uses a solitary twin choke carburettor against the twin setups of the other two). Power outputs? The Datsun engine might be the smallest, but it gives the most power: 151DINbhp at 6000rpm, with 147lb/ft at 4400rpm. While the Stag V8 doesn't have quite as much power (146DIN at 5700rpm) it has, not surprisingly, a good deal more torque—167lb/ft at 3500rpm. The GTE's V6 has an even greater discrepancy between bhp and torque, giving 135DINbhp at a tame 5000rpm and 172lb/ft at 3000rpm. Final product of the gearing is a very high 25.9mph/1000rpm for the Reliant, 24.1 for the Triumph and 22.6 for the 260Z.

## PERFORMANCE

If you're expecting the mythical might of the V8 to let the Stag kick up its heels at the other two cars, you will be disappointed. It's the slowest of the three by quite some margin. Not only is it the slowest against the watch, but it feels that way too, since the burbly V8 spreads its torque right over its rev range and makes the acceleration process seem unfussed. Mated to such high final gearing, it is an engine well-suited to effortless cruising at ton-up speeds—a point the Triumph engineers no doubt favoured over outright acceleration, although by most standards the Stag is, of course, a quick car.

But if the Stag power plant is thoroughly unfussed at speeds above 100mph, then it is matched rev-for-rev by the V6 in the GTE. On the one hand, the overdrive gearing in the Reliant lets it cruise with even fewer revs for a given road speed; on the other, the intermediate gearing and the hefty torque mean that the GTE can blow the pants off the Stag while it's getting there. It manages just to shade the Datsun too, although the difference between their times is hardly worth worrying about. Years of refinement have turned the Ford V6 into a smooth, pleasant engine with excellent flexibility, although it still isn't really happy to be cranked right out: it gets noisy and a little harsh beyond 5000rpm. You drive the GTE on torque and the wide range of gearing available, and never need to worry about lack of revs.

The Datsun's straight six is a lusty old thing that can never be accused of having too little torque for the job in hand. Datsun even want us to think the engine is capable of as many as 7000rpm too, but the redline on the tacho dial is in reality only for show; although the engine will, in fact, pull that far, it becomes far too harsh and noisy to tempt you more than once or twice. Power drops off sharply after 6000rpm anyway, so there's just no point in trying for the extra 1000rpm. For fast touring, the 260Z engine has to turn over quicker than the V8 and V6; up to 4000rpm (90mph in fifth) it is fine, but beyond that it begins to rumble and intrude.

However, the six is almost as flexible as the other engines and has a notable advantage in fuel consumption to back up its solid acceleration. Driven as hard as it will possibly go in all conditions, we could not get its fuel consumption to drop below 23mpg—an impressive figure. Cruising at 85 to 90mph for hours on end it returns a steady 27.9mpg. The Reliant is only slightly less impressive, dropping to 21.6mpg when thrashed and giving around 28mpg during fast cruising in overdrive top. The Stag logs 21mpg as its worst, but is very hard pushed to give better than 25mpg at cruising speeds—a pity, since the Stag V8 is easily the most pleasant of the three engines.

## RUNNING COSTS

Tot up the total fuel consumption for 12,000 miles running and the difference between the three cars is not great. Just on 23mpg is the figure most Stag owners will return most often, so that means a year's fuel bill will be around £283. GTE buyers should average 24mpg, which means an annual cost of around £270. At an average 25mpg, the Datsun's yearly tally will be £259 (all three run on four-star fuel, although you can get away with 3-star in the 260Z).

Maintenance is a straight-forward matter for each of the test trio, with the Triumph having the edge because of a 6000-mile interval between servicing compared with 3000 miles for the others.

## HANDLING & ROADHOLDING

For some time we have been scratching our heads in amazement over a *Motocar* statement that the Stag has a handling/ride compromise almost as good as that of the Lotus Elan (which is extinct, anyway). Granted, it rides quite nicely; but if you're thinking of the Stag as a really sporting car with deft handling, you're way off beam. The power steering is still far too light and has practically no feel; corners taken at any real pace leave the driver with a lot of uncertainty as to what is going to happen. In fact, because the roadholding is so modest the Stag begins to understeer strongly. With enough speed up, it is then prone to total contradiction, with a lightning switch to oversteer of often-alarming proportions. Once again, the non-communicative steering is of no help in dealing with this, and the driver feels remarkably insecure and lacking in control over the car. We must point out, should you think that we are talking of wildly excessive speeds, that in the corners we were using for our action photography, the Stag was at its limit at just on 60mph whereas the Datsun and GTE were going through at 70mph with perfect manners. So the Stag is *not* a driver's car; it needs to be taken into bends relatively slowly then guided quietly through. It is far more a boulevard vehicle than a sports car.

For a car with a live rear axle, the Reliant Scimitar has outstanding road-holding and handling. The only fault that can be found in its behaviour is some bumpsteer and the back will also joggle a little off line over bumps in the middle of bends taken near the limit. Apart from that, the car is totally trustworthy, signalling its intentions clearly and having such finely balanced handling that at the very limit it switches from the merest understeer to the mildest oversteer. The sorting could hardly be better.

The Datsun displays the same sort of handling excellence, with a similar (if not greater) level of roadholding. Because the steering is heavy and reaction-prone, the car feels as if it is a strong understeerer. It isn't, and it puts up very fast times through bends. However, it too is prone to bumpsteer and the wheel kicks strongly in your hands. One of the things that most impressed us about the Datsun was its tenacious grip on wet roads. Even using all the power in first and second, we rarely managed to break the tail away. A combination of the efficiency of the rear end and the new compounds used for the Japanese Bridgestone radials is the reason.

## DRIVER APPEAL

The Stag's V8 engine is easily its most outstanding and appealing component. The ride is good, but apart from that and the fact that the car can be a hardtop, a soft-top or roofless, it isn't rewarding or especially enjoyable to drive. You soon become disenchanted with the pleasures of the engine when the thing won't go round corners with any sense of security. Around town, the Stag is quite happy and pleasant because the steering is so effortless, the vision is so good and flexibility of the V8 is made to order for slow-moving traffic.

At first the Datsun seems a crude and harsh vehicle. Care must be taken with the clutch or else the drive train will answer with clunks and jerks. The engine, as it revs, feels more robust than anything else, and the steering is hefty. Yet, when you've lived with the Z for a while you begin to like it because of its rather crude, hefty character. It feels like a modern-day adaptation of the old Austin-Healey 3000. There is also the fact that it really does handle well, and that the roadholding can be trusted implicitly. The five-speed gearbox has well-chosen ratios and the change is good. When really flying in a series of bends, the balance becomes apparent, and the car can then be driven enjoyably on the throttle in a way that the Stag can't hope to match.

The Reliant offers similar enjoyment during spirited driving: perhaps even more so because it does not require such a firm hand on the wheel. At low speeds, the steering is fairly heavy, but it lightens quickly with speed and has excellent feel. Despite the bumpsteer on poor surfaces, and the occasional tail hop, it feels a more refined car than the Datsun. Of course, the four-speed gearbox, with its two overdrive ratios (third and top), means that there is a gear for every possible situation. Or you can simply play about with them all if you're feeling so inclined.

All three cars have good instrument and control layout. The Stag is perhaps the least impressive, since its steering wheel (which does adjust, alone in this trio) obscures the instruments in certain positions. The instrument glasses also reflect. Both the Datsun and GTE have excellent instrument displays, although they are laid out in different styles. Very American, the Datsun's dash has deep nacelles which house the instruments. Even if it looks gimmicky, it does work well and the dials are clear. Minor instruments include water thermometer, oil pressure and temperature (in one nacelle), ammeter and fuel (combined in another) and a clock. In the Stag, the minor instruments include a battery condition gauge, but surprisingly no oil pressure dial. Contrasting with the Datsun, the Stag's instruments are set into a wooden panel, and the GTE's fascia looks almost as traditionally British, even if it doesn't have wood facing. Thoughtfully, the GTE instruments are laid out from left to right in order

## Triumph Stag

*Single-ohc-per-bank V8 is a tightish fit but delivers ample torque as smoothly as one might wish. Cockpit is in keeping, not well integrated but almost luxuriously equipped, and with an adjustable steering wheel. Padded bench in back is for occasional use*

## Scimitar GTE

*Accessibility suffers because the Scimitar's Ford V6 engine is squeezed right back under the bulkhead in the interests of good weight distribution. Interior is a masterpiece of uncluttered planning, flexibly arranged to take two, three or four adults plus their baggage*

## Datsun 260Z

*Plenty of space around the 260Z's simple, inline engine for routine access. Interior is uncompromisingly sporting, yet really does have space behind the front seats for two small adults or large children. Luggage room is quite generous, too*

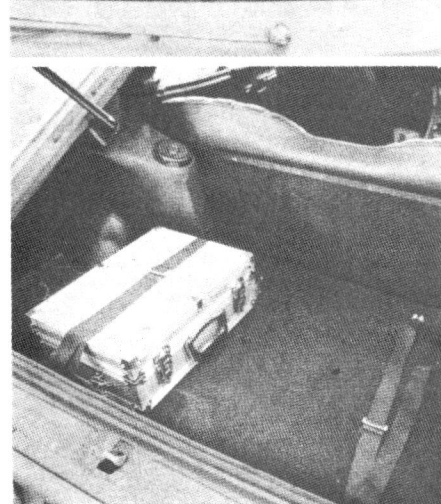

*Triumph Stag (top) has the smallest boot, but is by no means inadequate and has the advantage of being covered and lockable so that valuables are out of sight. Scimitar (centre) features a novel rear seat which folds half at a time. Datsun rear seat folds more conventionally*

of importance, and they're all easy to read. Pedals are fine (as in the Stag and Datsun, there is a footrest), the gearlever is easy to reach and works easily and the overdrive switch juts out from the dash so that the right index finger can flick it without the hand having to move from the wheel. The overdrive comes in and out very quickly and is unusually smooth. Minor switchgear is well located on the edge of the dash, to the left of the wheel, but because there are wipers front and rear, washers and lights, it can be confusing. The only thing that really does detract from the GTE driver's pleasure is the nasty steering wheel: it flexes under pressure in the middle of bends and looks cheap.

Pedal layout is equally good in the Datsun and Stag, and so are the gearlever locations. Both have twin stalks for the minor controls, but the Datsun's are more pleasant to use because they work more smoothly.

## COMFORT

The Stag wins hands-down for ride, especially around town. Its suspension is soft and soaks up bumps with a relaxed, easy style that it retains on motorways and most backroads as well. The seats have excellent squabs but the cushions are fractionally too flat. The driving position itself is good, thanks only to the adjustable wheel. But no matter how comfortable the Stag may be as a long-range motorway car, it is let down by frightful wind noise. The rush around the window frames starts to drown the radio at 45mph, and from then on just gets steadily worse until at 80mph you give up and turn the radio off. At higher speeds, conversation is difficult and at 100mph becomes tiring. So this point, together with the tricky handling and inadequate roadholding, means the Stag is more suited to boulevard purposes than anything else. Although it is the largest of the three cars, the Stag uses its space poorly by comparison with the GTE and 260Z, although two, even three kids can squat on the padded back shelf. Equipment and finish is good (overdrive and power steering are standard) and the car has an air of some luxury. The integrated roll-over bar offers permanent protection as well as providing vital body stiffness. The car comes with either soft-top or hardtop or both if you pay extra money. While the tops both fit well and keep out water and draughts with equal efficiency, the soft-top is hard to erect and the hardtop needs two men to lift it on and off. A good point of the ragtop's function is that it folds down into a well behind the cockpit and is concealed by a metal cover. Boot space is poor and getting the spare wheel out from under the boot floor is difficult.

The Datsun 260Z's interior is among the best-planned of any car. As a 2+2 it is a resounding success, with the available space being utilised so well that two small adults (say two women of around 5ft 3in) can travel there for hundreds of miles in some comfort. In that respect alone, the Datsun is rare

among its 2+2 contemporaries. But one of the sacrifices made to achieve this carrying capacity is the thinning of the front seats so that their squabs are hard and lack the necessary lumbar support for long trips. Although it has an independent rear end, ride isn't one of the Datsun's finer points. There seems to be enough spring travel, but the damping is inadequate and at low speeds the ride is harsh and juddery over bumps. Going quickly on open roads, the ride becomes quite comfortable. In all, the Datsun shows enough crudeness in its ride to point to it being thoroughly underdeveloped: the Japanese just don't yet understand about a good ride.

The Stag—and the GTE as well—are well-equipped, but they can't compete with the Datsun. Bringing the others up to an equivalent level means that the price moves in the Datsun's favour. The car is full of impressive detail points, too. Points like a ventilation system that directs cold air onto the face while sending hot air to the feet; trouble light on an extension lead mounted under the bonnet; straps to hold luggage in place, rubber under the fuel filler cap so the paint won't be damaged and so on. With the rear seat folded, the luggage area in the 260Z is vast, although there is nowhere to hide valuables. Under-bonnet accessibility is excellent, with all the ancilliaries laid out for easy reach—things are a lot more cluttered under the Stag bonnet, and with the V6 so far back in the Reliant's bay it can be difficult to work on during tune-ups.

The Scimitar is another firm rider at low speeds, covering uneven surfaces with a deal of knobbiness but not thumping so loudly as the Datsun when it strikes hefty bumps. With speed it, too, becomes thoroughly comfortable although the limits of the rather basic suspension are shown up by bumps in fast corners. Seats are comfortable, but they need higher backrests. As in the Datsun, foot space in the front is formidable and all the creature comforts are present. Space in the rear is better than in the Datsun and adults can travel there in comfort so long as their knees are tucked either side of the front squabs. Of course, the Scimitar's rear seats can be folded flat, one seat at a time, to provide the optimum seating/load space combinations. With both seats down, there is enough platform space to cart as much as in a small conventional estate car. Like the Datsun, the rear window lifts up on self-supporting struts and incorporates electric demisting, as with the Stag. Alone, the GTE has a rear-window washer and wiper—a necessary touch. Reliant's approach to the interior appointments has been so thorough that you now get a fold-out cover that clips into place to hide luggage.

## SAFETY

Roadholding and handling of the GTE, and especially the Datsun, are strong safety factors unto themselves, whereas the Stag

# FOR

## Triumph Stag

*Triumph Stag has clear instructions for raising the roof, big red lights warn when doors are open at night and the steering wheel is adjustable*

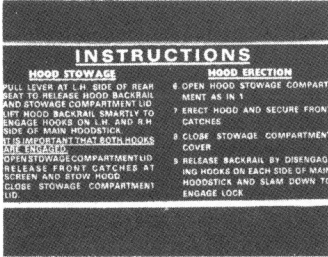

**INSTRUCTIONS**

**HOOD STOWAGE** | **HOOD ERECTION**

## Scimitar GTE

*Scimitar GTE gets a fold-out cover to hide valuable luggage, switching in overdrive is fingertip work and a powerful wash-wipe system keeps the rear window clean*

## Datsun 260Z

*Datsun 260Z lists among its excellent detailing an under-bonnet trouble light, a fuel cap that can't be lost or damage the paintwork and straps that hold luggage in place*

# AGAINST

## Triumph Stag

*Triumph Stag's spare wheel is hard to extract, checking battery levels is difficult and the steering wheel rim obstructs the view of the instruments*

## Scimitar GTE

*Scimitar GTE has four halogen headlights but on our car, at least, poor adjustment made them ineffective; wheel is cheap-looking and bends, fuel filler doesn't lock*

## Datsun 260Z

*Datsun 260Z's luggage must be disturbed to change a wheel, the tinny hubcaps carve up fingers trying to remove them and the seats are too thin for day-long comfort*

131

# GIANT TEST

can't be called a truly stable car if driven briskly. It just doesn't have enough road-holding and handling to come up to the standards of the others. But that hefty rollbar is a first-class safety feature and so are the strong headlamps and good wipers. Vision in all directions is excellent and the seatbelts are well-mounted and comfortable. Heating is good, but ventilation on really hot days isn't potent enough (presumably, you have to take the roof off). A dual-line braking system is standard.

Dual-circuit brakes come in the Datsun too, and so does that pressure-limiting device. But the 260Z's brakes can be made to fade all too easily in high speed stops, so one good point is balanced out by a bad point. Seat-belt mountings are not perfect in the Datsun and some people find the belts uncomfortable. Headrests are standard though, and there is good padding on the steering wheel boss. A laminated windscreen is standard.

The Scimitar has no sharp cowls inside the cabin to cause injury, and has good seat-belts. Its windscreen is laminated and it too has four halogen headlights (although the Datsun only has two, they are big ones). Glass is Sundym all round, and a full rollbar (concealed under the trim) is part of the basic construction.

## CONCLUSIONS

After a great deal of soul-searching, we've decided that we would settle for the Scimitar. It is an extremely versatile car with a great deal of appeal stemming from its originality. The only real faults are occasional bumpsteer and its slightly harsh ride at town speeds. You can easily live with that in the knowledge that the car behaves so nicely at high speed and is so pleasant to drive quickly.

On the other hand, we could be as happy with the 260Z—a car with superior roadholding overall (and a good deal of potential for developing the suspension) and faultless handling. The Datsun does indeed have a very robust, strong sort of character that can be off-putting at first. But you learn to like it the more you drive the car: it has the same sort of hairy-chested, vintage appeal of the old Austin Healey sixes. Suspension noise, engine harshness above 4500rpm and seats that aren't quite comfortable enough are the only significant faults, although the steering is really too heavy for women to manage around town. The amazing equipment level is a strong selling point.

A poor third, the Stag is outclassed by the GTE and 260Z for open road performance. It does have a fine engine, a good transmission, comfort, svelte looks and a lot of dash when the roof is removed. These are points that will make it a very good car indeed for the Highgate ladies to use on shopping trips, or for your commuting office manager to impress his friends. Meanwhile, it's left to Scimitar and 260Z to look after the real drivers. ●

### Scimitar GTE
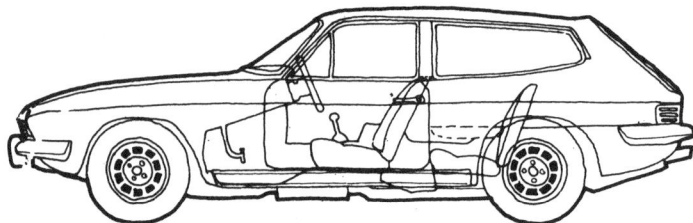

### Datsun 260Z
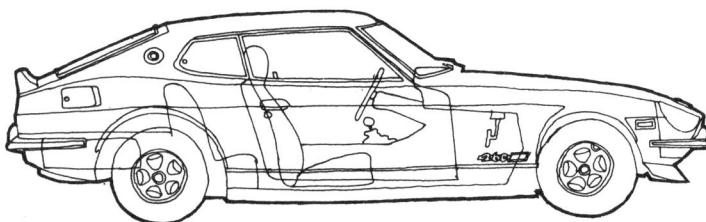

### Triumph Stag
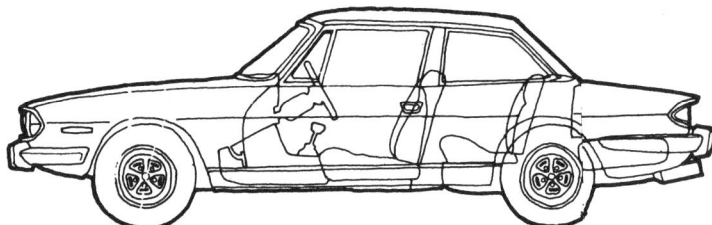

SPEEDS IN GEARS

|  | FIRST | SECOND | THIRD | FOURTH | FIFTH |
|---|---|---|---|---|---|
| Scimitar | 0-40 | 0-65 | 15-90 | 20-106 | 25-122 |
| Datsun | 0-47 | 0-72 | 18-104 | 21-117 | 27-125 |
| Stag | 0-42 | 0-60 | 10-91 | 15-118 | 20-120 |

PERFORMANCE

|  | 0-30 | 0-40 | 0-50 | 0-60 | 0-70 | 0-80 | 0-90 |
|---|---|---|---|---|---|---|---|
| Scimitar | 3.1 | 4.5 | 6.6 | 8.8 | 11.7 | 15.6 | 20.2 |
| Datsun | 3.1 | 4.6 | 6.7 | 8.9 | 12.2 | 15.3 | 19.8 |
| Stag | 3.5 | 5.4 | 7.5 | 9.8 | 13.2 | 17.1 | 22.3 |

FUEL CONSUMPTION

Scimitar 22-28mpg, Datsun 23-28mpg, Stag 21-25mpg

# A Pair of British V8 Cars

QUITE by chance I was able to try two very varied British sporting cars, both powered by V8 engines, emanating from different sources. With over 80,000 miles on the E-type Jaguar's speedometer I decided it might be a good thing if the factory had a look round it and gave it "a decoke and valve grind" so to speak, before setting off on another six months of continental touring. While this was being done Andrew Whyte, Jaguar's enthusiastic Publicity Manager, kindly loaned me a car from the British Leyland pool, which turned out to be a Triumph Stag, with hard-top. In the course of using this car for general work I found myself due to head towards Malvern, in Worcestershire, to attend the Annual Dinner and prize-giving of the Morgan Sports Car Club. Where possible I like to "be dressed for the occasion" so I got in touch with Peter Morgan, the President of the Club and the Managing Director of the Morgan Motor Company, and suggested it might create a favourable impression if I arrived in a Morgan Plus Eight, which he thought was a splendid idea. As I had never driven the latest type of Morgan sports car I arranged to borrow one for the day, to find out about Morgan motoring, and finish up at the Abbey Hotel in Malvern for the evening function. The result was that I was able to sample two British sporting cars with V8 engines over the same terrain.

Both of these cars are of orthodox front-engined layout, both using V8 engines, but there the similarity ends. The Stag is built by Triumph on a production line and extends little towards the title of sports car, while the Morgans are hand-built one at a time, and are pure sport from front to rear. The Stag is powered by Triumph's 3-litre V8 engine derived from the basic design of the 4-cylinder engine built for Saab, and the car is complete in every way, from heated rear window to electric side windows, all mod-cons. in the way of warning lights, illuminated instructions on what to do and what not to do, an overdrive behind its four-speed gearbox, four seats and a well-equipped luxurious finish. Consequently there is quite a bit of weight for the 3-litre V8 to propel in Stag form, whereas the Morgan is the complete opposite, having few mod-cons., little in the way of creature comforts, apart from perspex side screens, and a button-on hood, with two seats and a small luggage space. As it is powered by the Rover 3½-litre V8 engine, its performance goes without saying. As it charges up steep hills in third gear, accelerating all the while, you are conscious that there is not much weight for the 3½-litre alloy engine to propel, and when you stand on the brakes, with discs on the front and drums on the rear, it stops very abruptly and once again you are conscious of there not being too much weight to stop.

On paper these two cars would appear to have a lot in common, but in practice they are as chalk to cheese, or as the VSCC once described two sporting cars of the 1920s, plum cake or chocolate eclair, both very pleasing to the palate, depending on what mood you happen to be in. The controls of the Triumph Stag are a nightmare to anyone like myself who is not ergonomically adjusted and has only recently discovered what the publicity boys mean by ergonomics, so that I continually had the wipers going when I meant to turn right, and squirted the washers when I meant to blow the horn; this was due to the two stalks protruding from the steering column, one to the left to operate the wipers and squirts and one to the right to operate the indicators, headlamp flasher, headlamp dip and horn. The four-speed gearbox is simple enough and the overdrive switch is on the top of the lever, and high-speed cruising in overdrive fourth gear is very restful, but invariably when I snicked into third gear on reflexes I had forgotten about the overdrive switch and overdrive third gear was not what I wanted for the conditions prevailing. A good four-speed or an even better five-speed gearbox is my real answer; I have never liked overdrives, viewing them as an admission of failure in the basic design of the car. However, driven in a relaxed and gentle manner the overdrive did help enormously on fuel consumption, but with all the Stag's weight and the not very beefy engine, performance tails away alarmingly in overdrive fourth gear. On the Morgan Plus Eight there are no sops to ergonomics; what there is, is functional, and there isn't much of it. The Rover four-speed gearbox is just that, with reasonable ratios, but nothing to rave about, nor is the actual gearchange, but the "beef" of the Rover engine means that you can do most things in third and fourth, the engine whizzing round to 5,000 r.p.m. very happily, at which speed things happen quite quickly. Whereas the Triumph V8 does not feel particularly pleasant at over 5,000 r.p.m., the Rover needs a constant eye on the tachometer it is so rev.-happy and it would be hard to imagine two V8 engines with such dissimilar characteristics. The Triumph engine emits a woolly sort of exhaust note, never being very convincing about being on eight cylinders, while the Rover gives off a very deliberate V8 boom, encouraged by Morgan's sporty exhaust system.

The suspension on the Stag is a strange mixture, for over undulating bumpy roads it is superb, while on billiard table surfaces it does not seem to be able to make up its mind what it ought to be doing. The independent suspension to all the four wheels is soft, with long travel, and on the sort of surface that would make a good special-stage in a rally, it is truly outstanding, with 50-60 m.p.h. and more being most impressive, but that same speed on a good main road is not outstanding, the car seeming to want to get into a wobbling motion, and it gets no better, or no worse, at speeds well over 100 m.p.h. The steering is power-assisted and suffers from imparting no "feed-back" whatsoever, so that you do not really know what the front wheels are doing until they have done it. Under normal touring conditions this is no hardship, but it becomes tiresome in a side wind at speed for you are always making corrections too late, and it may be this feature which causes the suspension to impart the wobbling motion described. On poor surfaces the suspension rides the bumps so well, with the wheels following the undulations, that the steering feels all right, and equally it feels fine when pounded over special-stage surfaces, but for circuit-type surfaces and cornering it is unpleasant. In complete contrast the Morgan Plus Eight is pure vintage, with its hard independent front suspension by vertical sliders and coil springs and its rigid rear axle on semi-elliptic springs, but on normal roads it is remarkably predictable, and the steering, while being heavy, tells you all you want to know about the behaviour of the front wheels. That the Morgan has the ability to bound over bad surfaces is clearly demonstrated by the number of Morgan owners who run their cars in the Land's End and Edinburgh trials. Leaping and bouncing over the rough stuff without falling apart is one thing, but travelling swiftly over bad surfaces without alarm and despondency is another thing altogether.

While the Stag is a nice, easy and restful car to drive, that covers the ground pretty quickly without any fuss or strain, it does not provide any particular type of FUN. The Morgan, on the other hand, has got to be FUN from the word go and its performance covers up its shortcomings. It comes provided with a tonneau cover, and when this is buttoned up and the doors are shut you suddenly realise there is no way of getting in, for there are no external door handles and the tonneau fittings are concealed by the closed doors. Enquiring of a group of fanatical Morgan owners on the best way of getting into the car, I was told to undo the central zip fastener, climb over the spare wheel and clamber into the car from the centre! You have got to have a sense of humour to be a Morgan owner. While the Stag wafts silently along in a smooth and elegant manner, unflurried and undramatically, encouraging the owner to listen to the radio and observe the speed limits, the Morgan brings out all the sport in the owner, encouraging him to charge from corner to corner in a pretty unruly manner, powering round the swerves for the sheer joy of motoring.

The Morgan must be the last stronghold of what motoring used to be all about, while the Stag is what the Department of the Environment would like us all to accept as a Sports Car, or Executive's Car of Slightly Sporting Demeanour, but nothing unruly or out of step with modern-day thinking. Fortunately it takes all sorts to make a world, so they need all sorts of motor cars, and in the Stag and the Plus Eight we have an interesting pair, both powered by interesting V8 engines. If you are carefree and able to enjoy yourself without inhibitions then the Plus Eight is for you; if you are responsible, with responsibilities and all that they involve, and may only be permitted to give the appearance of enjoying yourself, then your car is the Stag. I have never managed to get my priorities fully sorted out, so I enjoyed both of them, but for opposing reasons.—D.S.J.

*An Executive's Car of Slightly Sporting Demeanour.*

# STAG-SWEET

# AND FLEET

STYLISHLY individual and sophisticated in personality the Triumph Stag is a true gentleman's sporting carriage. It's also well-mannered, comfortable, thoroughly British, and we loved it.

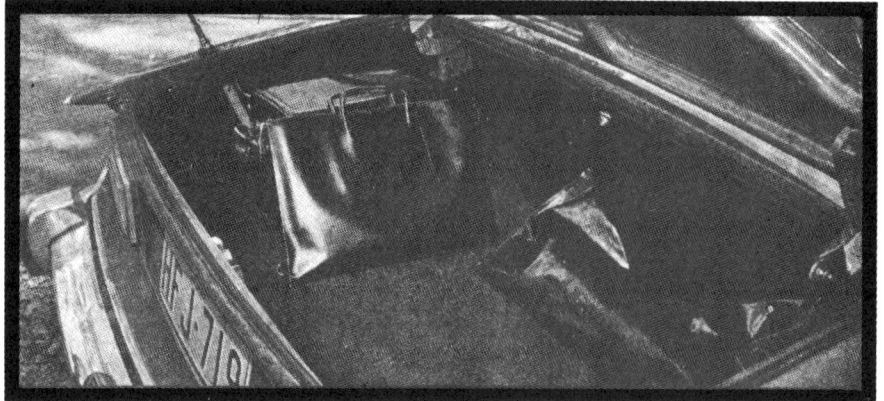

RIGHT: It's not what you'd call a large boot, but for two people touring there is ample accommodation behind the front seats to supplement the boot. Note carpeted finish and large toolkit.
BELOW: Handling is stable and predictable, but the power steering has a slightly dead feel at speeds over 110km/h. The optional mags and wide track give Stag a wide stance on the road.

◄ Although Stag comes complete with a lift-off hardtop we conducted most of this test with the top removed. As an open car the Stag was delightful, with noticeable lack of buffeting and wind roar. The car's forte is highway cruising and at constant touring speeds Stag can return up to 25mpg.

TRIUMPH'S Stag is still a relative mystery to many Australians — the car was in dramatically short supply at its introduction in July 1970. There were hardly enough for Brits with the ready cash, let alone the Aussie export market.

With its current $10,000-plus pricetag here, it's a safe bet Stags will retain their mystique, and exclusivity. It has been in relatively popular demand, but now with a new thrust by Leyland Australia, aimed at the top end of the luxury market, this car and a number of other luxury Leyland cars will be seen more often on Australian roads. The Stag *is a luxury* car — in the truest sense of the word — it costs a lot of money, it's basically only a two-plus-two and consequently it's a luxury to be able to own one. It's not quite as self-indulgent as some touring coupes, there's ample accommodation for four over short distances, but it is really an excellent piece of luxury *personal* transport.

The styling caused mixed reactions during our test period. Some said it was dated, some said it was confused and awkward and many drooled over is Michelotti-inspired lines.

The big styling points of the car are its lift-off hardtop and the massive integrated roll-bar. Everything else is basically the same as the current Triumph sedans. The hardware is identical, as is the mechanical layout, the only real difference being the V8 engine.

The 2997cc V8 powerplant was built exclusively for the Stag. Basically it is two Dolomite four cylinder engines arranged at 90degrees. The output (109kW) is adequate to push the car along at respectable touring speeds, with plenty in reserve. The test car ran down the standing 400m under 18seconds and that's quite impressive.

In Drive the excellent Borg-Warner 35 auto transmission changes up at 62km/h and 105km/h and this was quite acceptable. The ratios might not look very sporting at first glance, but in practice the gearbox ratios are well-matched to the V8's stepless torque curve. There is nothing we would like changed in this regard. Although sporting buffs would probably prefer the four-speed manual

# STAG

gearbox, we enjoyed using the BW box in typically sporty conditions.

When British Leyland introduced the Stag they endowed it with a "British enthusiast drives across Europe" image and heavily promoted the car's ability to cruise swiftly and easily along continental autobahns and mountain passes. Indeed the Stag handles such conditions with ease and comfort and for that express purpose it is a delight.

The handling is precise and predictable, although the power steering can give a slightly 'dead' feel under certain circumstances. It is nowhere near as good as the system which BL use on the Jaguar XJ sedans. We doubt whether the Stag really needs power steering, but the unit fitted at present could do with some serious attention if Triumph claim to offer pin-sharp response.

The all-independent suspension front McPherson struts/rear trailing arms is supple and positive. The ride was comfortable and yet the driver has ample control when he calls on the car to perform quickly over twisty territory. For a nose-heavy car the Stag is quite surprising in its reaction. Most times it understeers, but tends to become neutral quite quickly. It will oversteer in certain circumstances, but it's not in the least vicious. It is a feature of the Stag's rear suspension design that the tail should eventually break away with ultimate cornering pressures, but a gentle squeeze on the throttle makes the tail settle down nicely and allows the driver to power smoothly out of the corner.

The Stag is reasonably stable, but can be affected by heavy side gusts. Wet weather driving requires a little more concentration and we warn that there shouldn't be too much uncontrolled squeezing of the throttle in greasy conditions. In the wet the tail simply breaks away and the Stag is quite a handful when you lose it.

At first the ride felt a little too firm, but combined with the excellent front seats we decided after a short while that the Stag is a great long distance touring car in the typical British manner.

The brakes on the test car were a little disappointing, but overseas reports tend to confirm that braking is not the Stag's forte. The brakes work well enough, but we encountered lock-up on a number of occasions during test, and the deceleration rates were what we'd expect, but nothing to write home about. Pedal pressures were moderate, but the brake efficiency 'went-off' rather quickly and the brakes got very 'wiffy' shortly after our five maximum stops from 100km/h.

We believe the test car would have performed better with a harder pad material, but say again that the brakes performed alright, we just felt that a car with this performance potential should have a more effective set of stoppers. The handbrake was just so-so.

The driving position is very good, for short or tall drivers. The seats offer a wide range of adjustment and are very comfortable. The relationship of seats to wheel, dash and windshield is excellent. The steering column is adjustable for rake and the leather-bound wheel was pleasant to use. The column controls are normal Triumph hardware and convenient in operation. The remainder of the dashboard controls were quite

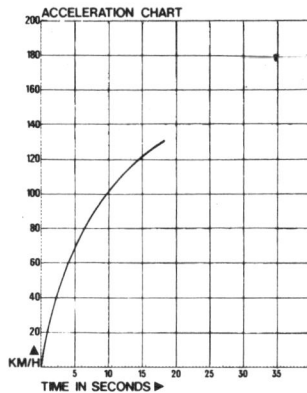

LEFT: Stag's V8 is a totally unique engine — it is used only in the Triumph sporty car. The 2997cc engine puts out a healthy 109kW, but returns a welcome 25mpg.
ABOVE: Optional on the test car, but standard on the newer models, the racy five-spoke mags give the car a stable footing for sporty driving.

## ROAD TEST DATA & SPECIFICATIONS

Manufacturer: .. British Leyland Motor Corporation, Coventry, UK
Make/Model: ............................... Triumph Stag
Body type: ............................ 2-door convertible
Test car supplied by: ....................... BLMC, Coventry

### ENGINE
Location: ............................................ Front
Cylinders: .................... Eight, 90degree vee-formation
Bore & Stroke: ...................... 86mm x 64.5mm
Capacity: ........................................ 2997cc
Compression: ................................... 9.25 to 1
Aspiration: ............. Twin Stromberg 175CDS carburettors
Fuel pump: ..................................... Electric SU
Fuel recommended: ............ Maximum octane (90 plus)
Valve gear: ......................................... SOHC
Maximum power: ................... 108.8Kw @ 5700rpm
Maximum torque: ................ 226.4Nm @ 3500rpm

### TRANSMISSION
Type/locations: ..... Borg-Warner 35 automatic transmission, T-bar shift
Driving wheels: .................................... Rear

| Gear | Direct Ratio |
|---|---|
| 1st | 2.390 |
| 2nd | 1.450 |
| 3rd | 1.000 |
| Final drive | 3.7 to 1 |

### SUSPENSION
Front suspension: .... Independent by McPherson struts with anti-roll bar
Rear suspension: ........ Independent by Semi-trailing arms, coil springs
Shock absorbers: ............................... Telescopic
Wheels: ........................... 5.5J x 14 4 Stud
Tyres: ................................... 185 x 14

### STEERING
Type: ............. Rack and pinion, power assisted, adjustable
Turns lock to lock: .................................... 2.8
Wheel diameter: ................................. 36.2cm
Turning circle: ...................................... 10.4m

### BRAKES
Front: ..................... Disc. Diameter: 26.8cm
Rear: ..................... Drum. Diameter: 22.9cm
Servo assistance: ............................... Standard

### DIMENSIONS AND WEIGHT
Wheelbase: ..................................... 254.0cm
Overall length: ................................ 442.0cm
        width: .................................. 161.2cm
        height: ................................ 125.8cm
Track, front: ................................. 133.7cm
        rear: .................................. 134.6cm
Ground clearance: ............................ 10.2cm
Kerb weight: ................................. 1265kg

well-placed and the electric window switches very handy — they're located at the trailing edge of the centre console, adjacent to the handbrake.

The transmission shifter is also stock Triumph and features a stepped gate, it works okay, but the quadrant lighting is poor at night.

The ventilation is exceptionally good (especially with the roof off) and we were surprised that a European design could be so adequate for Australian conditions. The dashboard centre and side vents give a good flow of air and direction control is simple. The heater controls are well-placed and easy to operate. The quarter windows in the doors are operated by a knurled rotary knob and supplement the fresh air supply.

Also impressive is the quietness of the Stag. The motor is smooth and mechanically quiet, and wind noise is apparent, but not offensive. There was some flutter around the upper A-pillar, but a lot of that had more to do with carefully re-fitting the hardtop in its proper position.

The hardtop removal and replacement is definitely a two-person operation, but the actual fitting system is quite simple, once you get used to carefully lining-up the hardtop locking slots and getting everything square before dropping it into place. We can't in all honesty say it was draught and water-free. The test car's hardtop leaked and allowed wind to whistle around the occupant's ears, but we are assured this was a particular fault with the test car and has now been rectified and should not be seen as a chronic complaint.

The soft-top hides under a well-designed panel which runs around the rear of the passenger compartment. The auxiliary soft-top is a great idea and we were very impressed with the quality of the top and the simplicity of raising and lowering it.

Back seat comfort (actually comfort isn't the right word) is marginal, and it's not a long distance pew by any regards. However it has pleasant seat cushions and backrest and for short distances (and short people) it is entirely adequate.

It's our opinion the Stag really is a *grand tourer* — it's auto travel in the grand, luxurious manner. The Stag is stylish and comfortable and completes its duties without fuss and bother.

It is a practical car in many respects. The rubber-lined recess on the top of the dashboard is great for cigarette packets, a ball-point pen, roadmap or the like. The glovebox is large and well-placed and the doors feature handy (and capacious) map pockets. The controls are very handy and *driving* the car is not a difficult task at all.

The unique integrated rollbar is finished in zip-on vinyl and of course became part of the design due to US crash regulations. It serves another purpose, increasing the rigidity of the body structure — always a problem with open cars. By using this concept BL engineers were able to cut down weight appreciably, the Stag weighs a neat 1265kg, giving a good power to weight ratio of 11.6kg/kW.

Despite its high pricetag we believe the Stag has a definite future in Australia. Its suspension seems to cope pretty well with the various road surfaces and it's reasonably economical. It's comfortable and offers the luxury car buyer everything he would want and because of its styling it has a unique appeal.

ABOVE LEFT: The pedals are a good size and well-located. The rubber pads are big enough for the driver not to make a mistake when he wants to stop in a hurry. CENTRE: Access to the rear seat is good, but it's definitely a short-distance affair. RIGHT: General finish is a very high standard and the car is very comfortable over most surfaces. As a high-speed touring machine it is smooth and swift.

## CAPACITIES AND EQUIPMENT

Fuel tank: . . . . . . . . . . . . . . . . . . . . . . . . . . . . . . . . . . . . . . . .64litres
Cooling system: . . . . . . . . . . . . . . . . . . . . . . . . . . . . . . . . .10.4litres
Engine sump: . . . . . . . . . . . . . . . . . . . . . . . . . . . . . . . . . . 4.5litres
Battery: . . . . . . . . . . . . . . . . . . . . . . . . . . . . . . . . . .12V 56AH
Alternator: . . . . . . . . . . . . . . . . . . . . . . . . . . . . . . . . .Lucas 45A

# PERFORMANCE

**FUEL CONSUMPTION**

|  | Litres/100km | (MPG) |
|---|---|---|
| Average for test: | 14.1 | (20.2) |
| Best recorded: | 13.9 | (20.7) |
| Possible best (under optimum conditions): | 11.1 | (25) |

**ACCELERATION**

|  | Drive | Held |
|---|---|---|
| 0-40km/h | 2.7 | 2.7 |
| 0-60km/h | 4.5 | 4.5 |
| 0-80km/h | 7.0 | 6.9s |
| 0-100km/h | 10.4 | 10.6s |
| 0-110km/h | 12.1 | 11.9s |
| 0-120km/h | 14.0 | 13.7s |
| 0-130km/h | 18.7 | 18.6s |

**OVERTAKING TIMES (holding gears)**

| km/h | 2nd | Drive |
|---|---|---|
| 40-70 | 3.9 | 3.8 |
| 50-80 | 2.9 | 2.9 |

|  |  |  |
|---|---|---|
| 60-100 | 5.8 | 5.8 |
| 80-100 | 3.7 | 3.7 |
| 100-130 |  | 7.9 |

**STANDING 400M**

Average: . . . . . . . . . . . . . . . . . . . . . . . . . . . . . . . . . . . . . . . . .17.8
Best Run: . . . . . . . . . . . . . . . . . . . . . . . . . . . . . . . . . . . . . . . .17.6

**SPEEDS IN GEARS**

|  | Drive | Held | rpm |
|---|---|---|---|
| 1st | 62 | 110 | 6000 |
| 2nd | 105 | 150 | 6000 |
| 3rd |  | 188 | 5600 |

**BRAKING**

Three maximum stops from 50km/h:

| Stop | G-force | Pedal pressure (kg) | Distance (M) |
|---|---|---|---|
| 1 | .84 | 20 | 11.7 |
| 2 | .77 | 38 | 12.8 |
| 3 | .94 | 20 | 10.4 |

Five maximum stops from 100 km/h:

| Stop | G-force | Pedal pressure (kg) | Distance (M) |
|---|---|---|---|
| 1 | .80 | 20 | 49 |
| 2 | .75 | 15 | 52 |
| 3 | .89 | 15 | 44 |
| 4 | .87 | 18 | 45 |
| 5 | .96 | 20 | 41 |

**SPEEDO CORRECTIONS**

| Indicated: | 60 | 80 | 100 | 110 |
|---|---|---|---|---|
| Actual: | 62 | 83 | 105 | 116 |

# STAG

Perhaps it could be fitted with 'air' as standard equipment, but Leyland Australia do not consider the factory job capable of offsetting Aussie temperatures to their complete satisfaction, so they bring in Stags without air conditioning and recommend certain local companies to fit an integrated system for between $600 and $700. That's pretty good, considering you get a local unit which can cope with the climate, and, it's fully-integrated.

There are no dramatic changes to the standard equipment levels on Stag, except that the next shipment of cars will feature the nifty mag wheels as standard equipment rather than as an option.

Leyland Australia report that between 450 and 500 Stags have been delivered to Australian owners over the past five years and expect to sell more than 400 in the next year — so that's a good indication of the increasing popularity of this fine English touring car.

You'll gather we liked it, and we did.

The Stag does everything it's supposed to and is a smart, sophisticated machine — we just hope Leyland Australia is keeping a close eye on demand, because we think it will skyrocket as this relatively mysterious stranger makes more new friends down under. ⓗ

---

## COST SCHEDULE

**Make/Model:** . Triumph Stag Automatic

**Pricing** (basic): . . . . . . . . . . . . $10,210
(as tested): . . . . . . . . . $10,414

**Options** (prices):
Alloy wheels: . . . . . . . . . . . . . . . .$185
Stereo cassette/radio: . . . . . . . . .$199

**Registration:** . . . . . . . . . . . . . $153.15

**Insurance Category:** . . . . . . . . . . . . . . 4
Rates quoted below are for drivers over 25 with 60 percent no-claim bonus and where the car is under hire-purchase. This is the minimum premium level — decreasing rates of experience and lower age groups may have varying excesses and possible premium loadings.
Non-tariff: . . . . . . . . . . . . . . . . . .$341
Tariff: . . . . . . . . . . . . . . . . . . . . .$377
N.R.M.A.: . . . . . . . . . . $410 (Group 8)

**Warranty:** . . . . 12 months or 20,000 km

**Service:**
Initial service is free. This covers the first 1500 km and includes inspection and adjustments. Oil used is charged for.

**Other Services:**
Lubrication and maintenance services every 5000 km. Parts and labour is charged.

**Spare Parts:**
(recommended cost breakdown)
Disc Pads (set of four): . . . . . . . . $27.84
Muffler (front): . . . . . . . . . . . . $32.00
Muffler (rear): . . . . . . . . . . . . . .$41.95
Windscreen: . . . $37.20, $119 laminated
Shock Absorbers: Front: . .$46.20 (strut)
Rear: . . . . . . . $23.50
Headlamp Assembly: . . . . . . . . . .$46.00
Taillamp Assembly: . . . . . . . . . .$17.00
Bumpers: Front: . . . . . . . . . . . .$162.00
Rear: . . . . . . . . . . . . .$70.00
Front Guard: . . . . . . . . . . . . . .$104.00

# 10,000 MILES ON TRIUMPH STAG

by C. John French

This originally started off as a Running Report requested by the Editor (whom God preserve) when I had completed 7000 miles in the Mark 2 Stag. However, with the privilege he shares with the opposite sex of changing his mind at short notice, it appeared that 7000 miles was too far for a Running Report (and rightly so for a gentleman advancing in years), so I threw away the deathless prose I had already submitted and started again, when the car had completed 10,000 miles which apparently makes this a long-term test.

Before proceeding farther, you should know that this report is contributed by the Publishing Director of this highly reliable journal, and not by one of those blasé sophisticated young men who are usually responsible for such affairs. This means, of course, that the views expressed are those of an individual whose journalistic talents are generally confined to judging others, and who is at last free to show his knowledge and skill as a tester and journalist.

The reason for selecting the Stag as my personal car can be traced to the Road Test in *Motor* in September 1973. I borrowed the test car one evening and with the hood down, motored home to delectable sunny Surrey, and for the first time in many, many years, asked my ever-loving wife if she would care for a blow in the car after dinner. Not being a violent man, she took my meaning immediately and we actually went driving for pleasure in the fresh air in what had almost ceased to exist—a Grand Touring car. Regular readers may remember that the French Blue road test car had Borg-Warner automatic transmission and it was this version, with the addition of a hard top, that Henlys delivered in January 1974. In its hard top role, the car looks very distinguished indeed to my eyes.

The car, driven exclusively by me, has now done just over 10,000 miles, and has proved a source of considerable pleasure from the moment it arrived. It was also regarded with considerable interest, not to say envy, by the rest of the staff and the great British motoring public. Only recently did a gentleman (obviously) say there was only one thing wrong with it; that it was in my and not his possession.

I remember now that the Editor did ask me for some facts and I have kept a meticulous record of the car's behaviour.

For the first six months of its life I left the hard top on and used the car mainly for commuting 34 miles a day. During this period the only faults were a failed bulb on the automatic console and the flasher unit, both replaced at the first free service. Since then the car has given yeoman service.

The automatic transmission and the power steering take most of the fatigue out of commuting in London, and the acceleration on kick-down has I think surprised various other enterprising motorists, left astern at very short notice. The change up in automatic drive with your foot hard down on the accelerator takes place at 40 mph from first to second, and at 70 mph into top; such full throttle changes are marked by a distinct jerk through. Driven normally, the changes are almost imperceptible but in 30 mph zones one has to lift off gently to engage top, which the box is reluctant to do under 40 mph.

During the petrol shortage I was forced on to a 3-star (91 octane) petrol but I could not detect pinking nor any other trouble. I stayed on 3-star in the interests of economy for some time but for high-speed touring (100 mph or more) abroad, there was always the fear of detonation which you couldn't hear, so I now use 4-star again.

Going to Le Mans through France I did notice that the temperature after the first hour was moving nearer to the red danger area, but as I had been holding 6000 rpm on 3-star petrol for some time, perhaps that's not so surprising.

The oil consumption has been negligible—in 10,000 miles, apart from the two service oil changes, it has consumed only 2½ pints. The petrol consumption remains fairly constant between 17-19 mpg when used only in London, but improves to 22-25 mpg on other longer journeys. The average mpg for 10,000 miles is 19.2 which, considering the high town mileage, seems to me very reasonable and

is almost the same as the overall figure of our original road test car. Some of these figures were obtained while the hard top was in position and although I thought they would improve with it off, this has not proved the case—probably the extra drag offsets the lighter weight.

This leads me into a very interesting (and in retrospect amusing) saga—the removal of the hard top and the erection of the hood. Taking the hard top off requires two people but is not too difficult. However, the emergence of the hood for the first time is an altogether different matter.

Carefully following the instructions in the handbook and with the aid of two strong men, we were able to erect it in the manner prescribed—ie engage the front catch pins and after releasing the spring-loaded catches at each side of the frame, then snap (splendid word) the rear plunger into the body catch. Being determined to do things alone, I then lowered the hood again, and with the aforementioned help, not only raised it again but lowered it too. The evening being sunny, I motored home open-aired.

Now, at this time I was living alone (my wife having departed on one of her numerous holidays) and having read the nasty weather forecast for the following day, I decided that the hood must be

| | Long term test Stag | Road test Stag |
|---|---|---|
| **MAXIMUM SPEED** | | |
| Lap | 115.2 | 112.4 |
| Best quarter mile | 120.0 | 116.8 |
| **ACCELERATION** | | |
| 0-30 | 4.8 | 3.6 |
| 0-40 | 6.8 | 5.2 |
| 0-50 | 9.8 | 7.5 |
| 0-60 | 12.2 | 9.9 |
| 0-70 | 15.3 | 12.8 |
| 0-80 | 19.5 | 17.0 |
| 0-90 | 26.4 | 22.4 |
| 0-100 | 34.7 | 30.2 |
| Standing quarter | 18.7 | 17.6 |
| Standing km | 33.2 | 32.2 |
| **ACCELERATION IN KICK-DOWN** | | |
| 30-50 | 5.0 | 3.9 |
| 40-60 | 5.4 | 4.7 |
| 50-70 | 5.5 | 5.3 |
| 60-80 | 7.3 | 7.1 |
| 70-90 | 11.1 | 9.6 |
| 80-100 | 15.2 | 13.2 |
| **FUEL CONSUMPTION** | | |
| Overall | | 18.9 |
| Touring | 22.1 | 21.7 |

## Fuel consumption

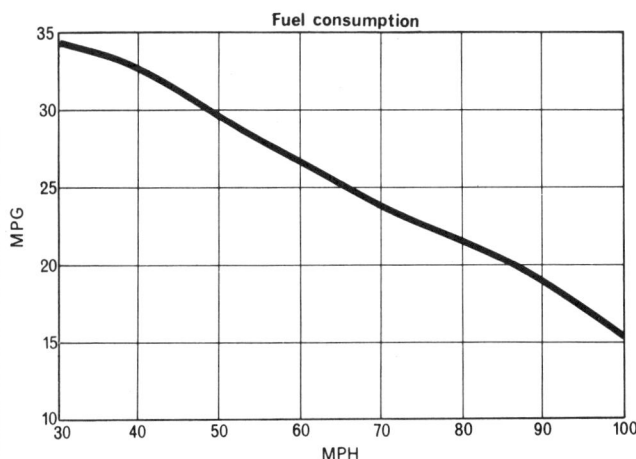

## Tyre wear

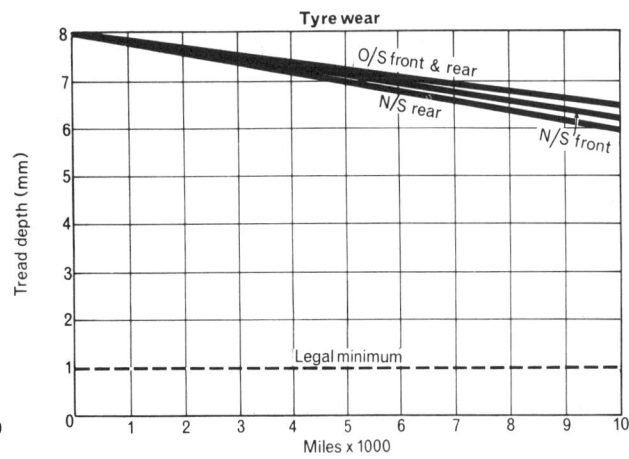

Fuel consumption graph shows the benefits of obeying the 70 mph limit. Tyre wear graph reveals remarkably low tyre wear for a car of this type

The Stag's "very distinguished" profile with hard top in place. It requires two to remove the top, a conversion which had no noticeable effect on the fuel consumption

erected – by myself. I keep reminding myself to tell the manufacturers that unless you are a professional strongman, it is unlikely that the hood can be erected alone except by engaging the *rear* catch first and *then* dealing with the front catches where one can exert some leverage.

At one period whilst trying to follow their instructions, I was standing on a kitchen stool with sweat pouring off me and an evil look on my face, stamping with my foot, endeavouring to engage the rear catch but completely without success. Having gone through their drill four times, I then tried the method I have outlined, that is to engage the rear catch *first* and the front one *last*. The whole operation was then completed in two minutes. The neighbours continue, as before, to regard me strangely as they pass, and mutter "something to do with publishing, I believe".

It is a very efficient hood and there have been no noticeable draughts or leaks in heavy rain. On my particular car, I did for a time experience minor difficulty in engaging one of the tonneau catches which house the hood, but Henlys have now adjusted the cable and it all works very well. In fact the hood is so easily raised and lowered when you know how, that at Le Mans this year, in lovely weather, we lowered the hood to travel from the hotel to the track and then popped it back on in a trice. There is a picture of the car outside the Hotel des Hunaudieres on the Mulsanne Straight which shows how pleasant an open car can be on such occasions (and in fact makes me feel restless to get back).

The pleasure of open-air motoring has another benefit not immediately obvious: you are not tempted to drive very fast because the exhilaration is quite enough at 70 mph. At speeds above this, the buffeting around the back of the neck becomes uncomfortable.

The suspension is firm but not harsh, and improves with a full load. When pushed into a corner, using the power to help the car round, an unwanted kick-down change can cause oversteer, but one

soon allows for this and there always seems to be an ample reserve of adhesion on the Michelin XAS radials, when entering one of those corners which appears as though it will never end. Nevertheless, the rear-end does rock when going really fast through winding main-road turns.

The power-assisted brakes, discs at the front and self-adjusting drums at the back, have performed very well. I wore out a set of front pads in the first 3500 miles but a harder set of replacement Ferodos are still wearing well.

One word of warning. I thought I'd read the handbook thoroughly but overlooked the manufacturer's instruction not to raise or lower the electrically operated windows without the engine running. Consequently I left myself with a flat battery at the Lygon Arms, Broadway, and I would like to thank the garage there who started me with jump leads and would not accept a fee.

When the Stag was delivered I had fitted a Lucas audio LS120 push-button radio, described in full in *Motor*, June 19, 1974; as it has proved very satisfactory, I agree

with the assessment of it in that issue.

In a long motoring life one has many experiences. I remember in my very first car, a pre-war Standard 10, managing a complete spin on ice, finishing up still proceeding in the right direction. In an Austin utility during the war I had a slight collision with an excessively large steam locomotive during the blackout on the docks at Marseilles. I have been rolled over by an over-enthusiastic French chauffeur in a Delage sedanca-de-ville (on the only hairpin on the pass from Marseille to Toulon with a ditch instead of a precipice). Inevitably, I suppose I was likely to have a fire sometime as well, and it happened at 9,300 miles in the Stag – a burning moment to remind you how important it is to have a fire extinguisher.

The fire was caused by the positive lead from the battery chafing on the pipe feeding the power steering pump. On a run back from Brighton, I noticed that the engine appeared to be running hot so I opened the bonnet lid and had quite a nasty turn – about 4-5 inches of the positive lead was white hot with a small flame from the smouldering insulation. Amazingly

a small hole had also been rubbed in the oil pipe and with the admission of the air by raising the lid, I suddenly had flames a foot high.

Fortunately I had in the boot an absolutely splendid device called the Tyrenfire extinguisher marketed by East London Hydraulics Ltd., Tyre Division, 46 Spurstowe Road, Hackney, London E8. This is a $CO_2$ extinguisher so easy to operate that I read the instructions on the top as I ran from the boot back to the engine. By a simple turn of the nozzle top, the fire was immediately extinguished; turning the nozzle in the opposite direction stopped the flow of $CO_2$, leaving an ample charge still available. Fortunately, the incidental damage was virtually nil. This extinguisher can be easily re-charged and can also be used for inflating flat tyres and is a very worthwhile investment at £7 to £9 according to size.

I notice that on the latest Stag there is a clip on the positive lead to prevent any movement; in their repair to the electrics of my car, Henlys taped the lead up to save me from burning before my time.

With 4-star petrol at its present 54p per gallon, it would cost £274.11 for 10,000 miles for the fuel, to use the Stag as I have done, mainly in urban areas, probably 25 per cent cheaper in the country.

Summing up, I'd like to dwell on the effect the Stag has had on me, which amuses my friends – I have grown my hair longer, wear more extraordinary ties and racy hats and generally behave as though I have shed 20 years, which can only mean that I am enjoying the car immensely. One does not look out of place in the commuter stream and can out-dash the coxiest of county show gentry when desired – in all, a very pleasing dual purpose Grand Touring car. It is also very pleasant to motor to the office on fine mornings with the hood down – it gives one the impression of proceeding on a day of fun instead of the usual nose-to-the-grindstone stuff.

If permitted, I will give you more information of the Stag's progress at a later mileage.

Unique as a Grand Tourer? The Stag in picturesque surroundings with its top in place. Left: the car's dual personality symbolised by its driver's selection of headgear. Below right: tasting the open air delights on the way to Le Mans

**Wind in the hair for $2500 a seat.**

# TRIUMPH STAG
## —fresh air for four

Triumph's classy Stag drop-top is not an all-out performance car, nor is it really a soft and silent interstate cruiser. But PETER ROBINSON found it a pleasant car, just the same.

TRIUMPH'S Stag holds a unique place in the motoring world. There is nothing like it in America, Europe or Britain and even the Japanese, who pride themselves on finding and filling gaps in the market, don't have an open touring car that attempts to carry four adults.

The Stag is not a sports car — it never really pretended to be one — but it makes a fine boulevarde cruiser that's easy to drive, pleasant to be in and capable of winning smiles from long-legged blondes who previously ignored you. And although the specification is mildly exciting and the equipment level is high, the Stag has one major fault that prevents it being a genuine Grand Touring car in which an inter-capital city dash becomes merely a moment behind the wheel, undertaken for its own sake.

That fault — excessive wind noise — makes us wonder about the writer of the Stag's brochure which proclaims positively about the "loveliness of the long distance tourer" and "built to take the continent in its stride". The rest of the Stag is up to these romantic illusions. But poor sealing of the superbly finished but extremely heavy hardtop allows not only the wind to roar but rain water into the cabin. And

*Right:*
*Stag corners well enough to suit its performance potential, but at the limit the independent rear end makes the handling rather twitchy. Similarity to Triumph sedans is obvious in front-end styling.*

*Left:*
*Stag roll bar is not detachable — accounts for better rigidity of this car than other drop-top Triumphs WHEELS has driven. Rear seat room is adequate for small people.*

# TRIUMPH STAG

the soft-top, again cleverly designed and neatly built, buffets and also generates plenty of wind noise.

So for us it was hardtop off and soft-top down for the duration and the Stag was at its excellent best in this form, but then we didn't drive it to Melbourne . . .

Triumph developed the Stag almost as an afterthought to the 2000/2.5 sedan range. Michelotti styled the car as a one-off for a motor show, but before the car reached the show Triumph's top executives saw the car and rushed it off to Coventry in Britain. At that time it was still powered by the 2-litre six but the V8 engine was already on the stocks — simply by grafting two Dolomite fours together — and it was decided to use this to give the Stag the performance it deserved.

Although the styling might suggest otherwise the roll bar was designed as an integral part of the car from the start. The bar was actually going to be bolted into position so people could take it off if they wanted to. Then the Triumph engineers discovered that the body need addition torsional rigidity and another bar from the windscreen back was added and it was in that form it reached production. The roll bar has two functions, it protects the occupants of the car in the case of a roll over and also reduces scuttle shake — something that plagues the old TR6.

It's five years since the Stag was first released in Britain and four years since I drove the first example to come to Australia. Since then, of course, the Stag has found a comfortable niche in the Australian market and sells so well there is an eight week waiting list. Since Leyland Australia took over distribution of the car from Australian Motor Industries and dropped its local manufacturing program, it has been taking a greater interest in its fully imported models, hence this road test.

The latest models are more refined and clearly show that the Stag has been the subject of much detail modification since it was released. There has been a reduction in assistance for the power steering, the steering wheel is smaller in diameter, the brake system is dual circuit, there's an intermittent wiper and a number of other minor changes, including a very slight increase in the power output as a result of rationalising the combustion chambers and pistons with the four cylinder Dolomite range.

*Right:*
*Stag engine is understressed and smooth three-litre V8 with single overhead cam on each bank. It is virtually two Triumph Dolomite Sprint engines grafted together.*

*Centre:*
*Cockpit is pleasant and quite well finished. Instruments are comprehensive and easy to read and column-mounted controls work well.*

*Far right:*
*Boot is uncluttered and nicely trimmed, but rather small and shallow for four people's luggage. Spare under boot floor.*

Briefly the Stag is a fairly long (4420 mm — 14 ft 5.75 in.) but relatively narrow (1612 mm — 5 ft 3.5 in.) two plus two convertible, powered by a mildly developed 3-litre V8 engine. It's a heavy car at 1275 kg (2807 lb) and although the standard car comes with a four-speed gearbox and overdrive most are sold with a three-speed automatic which is more in character with the general softness of the car.

The small V8 — which has yet to be used in any other Leyland product — produces only 109 kW (146 bhp) at 5700 rpm and so the performance, while effortless, is well below the standard expected from a car with sporting pretensions and a V8. Certainly the manual version would be quicker but the automatic put down standing 400 m (quarter mile if you didn't know) in a best of 17.9 seconds and averaged 18.1 seconds.

The engine is redlined at an optimistic 6500 rpm but had trouble reaching 6200 in low and 6000 in intermediate, in fact our figures show there is no advantage in running the engine out to more than 5500. Leaving the transmission in drive range the Stag produced standing 400 m times equal to those achieved by running it right out. In drive it changed up at 4000 rpm from low to intermediate and 4250 rpm from intermediate to drive.

Triumph uses the Borg-Warner type 35 transmission in the Stag but on the test car it didn't perform as well as we expected. Under full throttle acceleration it is smooth, but on light accelerator applications it clunks into higher gears and is anything but smooth on the way down again. This jerky action spoils the Stag unless the driver concentrates on the throttle openings to reduce this tendency to a miniumum.

The transmission is controlled by a short T-bar with a dog-leg change pattern which makes it difficult to know which gear the car is in. However, it is virtually impossible to select neutral inadvertently and the selector can be pulled back into intermediate very quickly. The kickdown is responsive at middle range speeds but is reluctant to

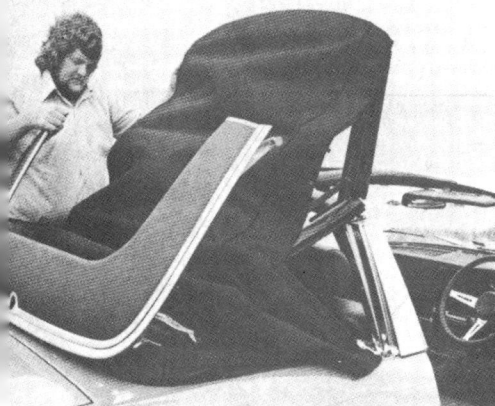

# TRIUMPH STAG

pick up low at more than a crawl.

Compared with the previous Stag we took for a short impressions run four years ago, the biggest single improvements are in the power steering and the smaller steering wheel, which have improved the feel and responsiveness of the steering. It is only when driving very quickly that the steering becomes too light and loses the ability to convey to the driver just what is happening to the front wheels. In most driving situations, the powersteering is exactly the way it should be, it just takes some time to get used to its quirks.

The Stag sits on fat 185 HR 14 Michelin XAS radial — or Avons of the same size — and this together with its independent rear suspension should give it outstanding roadholding and handling. Up to a point it does, but as greater confirmation that it is not a pure sports car, the handling and roadholding becomes a little nervous when you really start to drive it hard. Understeer is the basic steering characteristic and it can reach quite high proportions although the limits of adhesion are high.

There is a surprising lack of throttle steer control which detracts from the overall driver appeal of the car, at least in spirited motoring. Over bumpy surfaces the Stag wiggles around at the back end and develops a strange lurch when driven quickly through a series of S-bends that suggests the old Triumph 2000 problem of spline lock-up is still present.

If the roadholding has its limitations the ride is excellent because there is little road noise and pitch is well controlled. Although there is some body roll, it never affects the handling. The Stag has good brakes except in adverse conditions when the pedal gets spongy and needs to be pumped to get back to full efficiency.

The Stag really shines in its finish

*Soft-top is a masterpiece of convenience engineering. To lower it you simply (A) unclip it at the rear and fold forward to the roll bar, (B) raise hood storage bin cover behind rear seats, (C) concertina the hood into storage compartment and (D), drop the cover and the job's done. All in 30 seconds.*

and equipment level. The cockpit is typically Triumph with two large instruments in front of the driver for the tachometer and speedo and two smaller dials on either side for the minor instruments. Excellent steering column controls — they are also used on the Marina — look after lights, wipers and horn. The other controls are on the console.

Heating and ventilation is complete but the small circular vents at the extremes of the dashboard lack flow through when the hardtop is in place. Of course, when the roof is off/down there is no real need for additional ventilation the open cockpit provides plenty of fresh air without creating too much turbulence. Travel in the Stag with the hardtop removed — a difficult two-man job because it is so heavy — is extremely pleasant and with the side windows raised there is little wind roar.

Triumph has ensured a good driving position by making the steering wheel adjustable for both rake and height and there is a handle at the front of the seat cushion to alter the tilt of the seat. We found its range limited and our drivers still complained that the seat cushion was too flat.

The small rear seat is fine for children and can be used by adults for short journeys, but head and leg room is restricted. The boot too is rather small, although it is well finished in pile carpet and sensibly shaped.

The lavish equipment includes electric windows and the hardtop is standard in Australia, along with power steering. There's also a clock and map reading light, and the hardtop even has a heated rear window.

The Stag's role in life is simple. It's for middle aged men searching for their lost sports car youth; it's for women who love the wind in their hair but don't want it ruffled; it's for nights at the Opera House and general slipping between fashionable shops in fashionable suburbs.

It's not a European style GT car with taut handling and roadholding, it's not for the ultimate performance buyer or somebody looking for a modern open two-seater but its faults are minor compared with what the Stag achieves. *

# TRIUMPH STAG

## SPECIFICATIONS

**MAKE** . . . . . . . . . . . . . . . . . . . . . . . . . . . . TRIUMPH
**MODEL** . . . . . . . . . . . . . . . . . . . . . . . . . . . . . . Stag
**BODY TYPE** . . . . . . . . . . . . . . . . . . . . . . .Convertible
**PRICE, As tested** . . . . . . . . . . . . . . . . . . . . . . $10,043
**OPTIONS FITTED** . . . . . . . .Hardtop, stereo-cassette player
**ENGINE:**
Cylinders . . . . . . . . . . . . . . . . . . . . . . . . . . . . . . Eight
Valves . . . . . . . . . . . . . . . . . . . . . . . . . . . . . . Overhead
Carburettor . . . . . . . . . . . . . . . . . . Two Stromberg CDS
Fuel pump . . . . . . . . . . . . . . . . . . . . . . . . . . . Electrical
Bore x stroke . . . . . . . . .86 x 64.5 mm (3.385 x 2.539 in.)
Capacity . . . . . . . . . . . . . . . .2.997 litres (182.9 cu in.)
Power, at 5700 rpm . . . . . . . . . . . . . .109 kW (146 bhp)
Torque, at 3500 rpm . . . . . . . . . . . .226 N-m (167 lb/ft)
**TRANSMISSION:**
Type . . . . . . . . . . . . . . . . . . . Three speed fully automatic
Gear lever location . . . . . . . . . . . . . . . . . . .Central console
**RATIOS:**

| | Gearbox | Overall | km/h per 1000 rpm | mph per 1000 rpm |
|---|---|---|---|---|
| First | 2.39:1 | 8.84:1 | 14.2 | 8.4 |
| Second | 1.45:1 | 5.37:1 | 22.0 | 13.7 |
| Third | 1.00:1 | 3.70:1 | 32.2 | 20.0 |
| Final Drive | 3.7:1 | | | |

**CHASSIS AND RUNNING GEAR:**
Construction . . . . . . . . . . . . . . . . . . . . . . . . . . . Unitary
Suspension, front . . . . . . . . . Independent, MacPherson struts,
coil springs, anti-roll bar
Suspension, rear . . . . . Independent, semi-trailing arms, coil springs
Dampers . . . . . . . . . . . . . . . . . . . . . . . . . . . Telescopic
Steering type . . . . . . . . . . . . . .Power assisted rack and pinion

Turns lock to lock . . . . . . . . . . . . . . . . . . . . . . . . . .3.0
Turning circle . . . . . . . . . . . . . . . . . 10.2 m (33.6 ft)
Steering wheel diam . . . . . . . . . . . . . 368 mm (14.5 in.)
Brakes, type . . . . . . . . . . Disc/drum, power assisted
**DIMENSIONS:**
Wheelbase . . . . . . . . . . . . . . . . . 2540 mm (100 in.)
Track, front . . . . . . . . . . . . . . . . 1337 mm (52.6 in.)
Track, rear . . . . . . . . . . . . . . . . . 1346 mm (53.0 in.)
Length . . . . . . . . . . .4420 mm (14 ft 5.75 in.)
Width . . . . . . . . . . . . . . . . 1612 mm (5 ft 3.5 in.)
Height . . . . . . . . . . . . . . . . 1258 mm (4 ft 1.5 in.)
Fuel tank capacity . . . . . . . . .58 litres (12.75 gallons)
Kerb mass (weight) . . . . . . . . . . 1275 kg (2807 lb)
**TYRES:**
Size . . . . . . . . . . . . . . . . . . . . . 185 HR14
Make fitted . . . . . . . . . . . . . . . . . Michelin XAS
**GROUND CLEARANCE:** . . . . . . . . . . . .102 mm (4.0 in.)

## PERFORMANCE

**TEST CONDITIONS:**
Weather . . . . . . . . . . . . . . . . . . . . . . . . . .Fine, cool
Surface . . . . . . . . . . . . . . . . . . Castlereagh Dragstrip
Load . . . . . . . . . . . . . . . . . . . . . . . . .Two people
Fuel . . . . . . . . . . . . . . . . . . . . . . . . . . . Premium
**SPEEDOMETER ERROR:**

| Indicated km/h | 50 | 70 | 90 | 110 | 130 | 150 |
|---|---|---|---|---|---|---|
| Actual km/h | 52 | 71 | 90 | 110 | 128 | 148 |

**FUEL CONSUMPTION ON TEST: Distance and Conditions.**
Check one . . . . . 7.9 km/l over 235 km (20.3 mpg over 146 miles)
Check two . . . . .6.7 km/l over 262 km (18.9 mpg over 163 miles)
**MAXIMUM SPEEDS:**
Fastest run . . . . . . . . . . . . . . . . . . . . 186 km/h (115 mph)
Average all runs . . . . . . . . . . . . . . . . . 184 km/h (114 mph)
**IN GEARS:**

| | Drive | | Held (See text) | |
|---|---|---|---|---|
| First . . | 47 km/h (35 mph) | 4000 rpm | 88 km/h (55 mph) | 6200 rpm |
| Second | 94 km/h (60 mph) | 4250 rpm | 132 km/h (82 mph) | 6000 rpm |
| Third | 184 km/h (114 mph) | 5700 rpm | | |

**ACCELERATION: Through the gears:**
0-50 km/h . . . . . . . . . . . . . . . . . . . . . . . . . . . . . .4.6 sec
0-70 km/h . . . . . . . . . . . . . . . . . . . . . . . . . . . . . .6.7 sec
0-90 km/h . . . . . . . . . . . . . . . . . . . . . . . . . . . .10.0 sec
0-110 km/h . . . . . . . . . . . . . . . . . . . . . . . . . . .13.7 sec
0-130 km/h . . . . . . . . . . . . . . . . . . . . . . . . . . .19.5 sec
0-150 km/h . . . . . . . . . . . . . . . . . . . . . . . . . . .27.4 sec
**KICKDOWN: In the gears:**
30-60 km/h . . . . . . . . . . . . . . . . . . . . . . . . . . . .2.8 sec
40-70 km/h . . . . . . . . . . . . . . . . . . . . . . . . . . . .3.2 sec
50-80 km/h . . . . . . . . . . . . . . . . . . . . . . . . . . . .3.7 sec
60-90 km/h . . . . . . . . . . . . . . . . . . . . . . . . . . . .4.4 sec
70-100 km/h . . . . . . . . . . . . . . . . . . . . . . . . . . .4.6 sec
80-110 km/h . . . . . . . . . . . . . . . . . . . . . . . . . . .5.7 sec
90-120 km/h . . . . . . . . . . . . . . . . . . . . . . . . . . .6.2 sec
100-130 km/h . . . . . . . . . . . . . . . . . . . . . . . . . .7.8 sec
110-140 km/h . . . . . . . . . . . . . . . . . . . . . . . . . .8.6 sec
120-150 km/h . . . . . . . . . . . . . . . . . . . . . . . . .10.8 sec
**STANDING START – 0-400 m (0-¼ mile)**
Fastest run . . . . . . . . . . . . . . . . . . . . . . . . . . .17.9 sec
Average all runs . . . . . . . . . . . . . . . . . . . . . . . . .18.1 sec

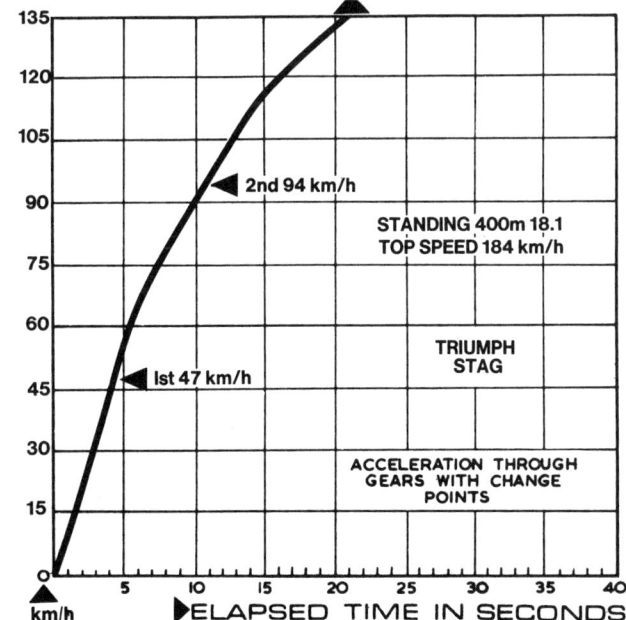

2nd 94 km/h

STANDING 400m 18.1
TOP SPEED 184 km/h

TRIUMPH
STAG

1st 47 km/h

ACCELERATION THROUGH
GEARS WITH CHANGE
POINTS

km/h ►ELAPSED TIME IN SECONDS

# 20 000 MILES ON TRIUMPH STAG

**" Motor's " Publishing Director, John French, continues the tale of his widely travelled two plus two**

It seems a far cry from *Motor* w/e September 21, 1974 when I reported on the progress at 10,000 miles of my Triumph Stag on long-term test. For a start 4-star petrol costs were then calculated at 54p per gallon, now the price is 73p.

The second 10,000 miles confirms my original opinion of the car as one of considerable attraction, particularly in appearance, and with a distinct character and temperament. On the opposite page you will find a "Service Diary" item relating to a bearing failure which the engine sustained shortly after the 10,000 mile report. Apart from this incident which, at the time, caused me some considerable alarums and excursions as I was en route from the Parish Show to Spain for a holiday, the car had performed admirably.

At the time of the 10,000 mile report I said that the Borg-Warner automatic box was "lazy," ie that it showed a reluctance to change up or down gently. As the car was still under warranty a gentle carp at Borg-Warner via Triumph has caused them to bend their skills to this problem and the gearbox now provides a smooth upwards change under both steady and fierce acceleration.

An annoying clonk now occurs only when proceeding in Drive and the box in intermediate gear, though when lifting off, the box will change into top at much too low a speed, with a jerk. This occurs only in town driving and usually where there is a short distance between corners. With a little foresight you can engage "hold" in intermediate until the second corner is rounded.

The Michelin XAS radials are unchanged and look good for at least another 8,000-10,000 miles (30,000 miles—not bad for such a car). The power-assisted brakes continue to serve excellently, though I am on my third set of disc pads. Regular readers will remember that the original ones were too soft and had to be replaced at 3000 miles; the harder second set were replaced at just over 18,000 miles. This means that the second set lasted for 15,000 miles and the linings for the rear drum brakes remain unchanged but will probably need replacing at the next service, coming up within the next month.

On my return from France in October I replaced the hood with the hardtop for winter usage without difficulty; again it has proved draught-free and rainproof. Having reached the allegedly good weather period in the United Kingdom only recently (on May 1) I once again removed the hardtop to its place in the garage and the hood has emerged from the tonneau, still in first class condition. It is put up and down without trouble to suit the vagaries of our changeable climate and creates remarkably little extra noise. In the last report I commented on the buffeting which occurred with the hood down at high speeds but since then I've found this is reduced if the side windows are lowered.

Apart from the normal service changes, oil consumption is negligible. The engine and ride are commendably smooth and the lines of the body, in my view, have not dated in the five years since the car's introduction. It still elicits admiration—the attendant at the local supermarket car park says he sells it regularly at least four times a week. The average mpg including a large percentage of driving in town, with very few long journeys to assist, has averaged 19.1 mpg but on a recent run to Thruxton and back it averaged 22.5 mpg.

The front adjustable seats have provided comfort for all shapes and sizes of passengers but other than for children—or two small adults—the rear seats are uncomfortable. As I commented in the previous report, a load in the back does improve the ride though. No major replacements

# SERVICE DIARY

## STAG-GER

The two pictures (below) show how a combination of minor faults can have dramatic effects. The car was a Stag (see story opposite) which had always shown a slight increase in temperature at sustained high cruising speeds. While still under guarantee it received dealer attention to cure a noisy timing chain, but on re-assembly it was alleged subsequently to have been incorrectly re-timed—but only slightly.

During routine servicing a new oil filter of proprietary manufacture was fitted, recommended for use in the Stag (many others had been fitted in previous Stags without causing any trouble) although BLMC pointed out later

**Damage to a replacement oil filter on the staff Stag led to failure of two big ends . . . .**

**. . . which in turn caused two pistons to hit the valves, as the indents testify**

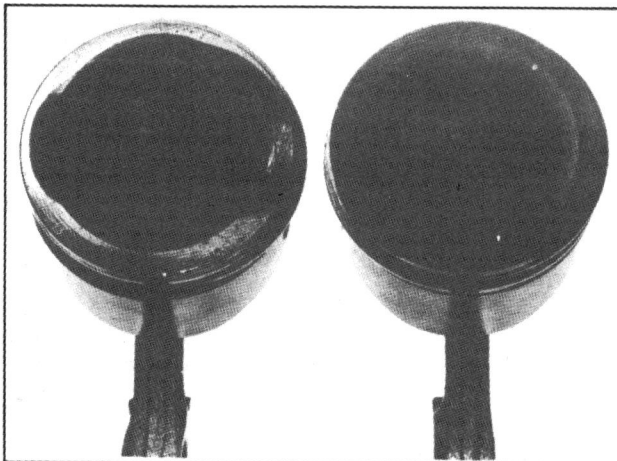

have been required—only the occasional bulb for stop lights but nothing more, and there is no sign of deterioration through rust or corrosion on the chromium plating.

I have rectified one minor irritation. The plug leads from the cylinder bank on the opposite side to the distributor pass over the rocker box and are meant to be held clear by a plastic gate which is not tall enough. The leads, therefore, were being gently cooked as the rocker box warmed up, giving off a hot plastic odour, now cured by lifting the leads higher and taping them to the bulkhead.

The performance, alas, can no longer be used legally in the higher speed range but the acceleration is still lively and at the present legal maximum the engine is so smooth and understressed that it feels it could go on forever.

The Stag still gives me considerable pleasure not only to drive but to look at, and it will be a pity if in BLMC's rationalisation programme this body shape should disappear.

Summing up, despite the service errors which caused the engine failure in France, I took the Stag to Le Mans and went on from there to Spain. As I intend to continue with it as my personal transport, and if the Editor so desires, I may be able to give you a further year's history at this time in 1976.

that these particular filters were not on *their* recommended list.

The sequence of events leading up to the failure of two big ends causing the pistons to hit the valves appears to have been as follows. A minor rise in temperature, no doubt aided by the re-timing error, was accentuated when the car had been cruised for approximately 100 miles on the Autoroute du Sud at between 90-100 miles per hour. The pressure of the overheated oil damaged the top of the filter element and the sealing gasket, as the picture shows. A series of foreign bodies arriving at the bearings caused the trouble, but fortunately the rise in temperature was spotted in time before even more expensive damage was done.

Fram, who manufactured the filter, had had no similar experience in four years of manufacture. But at the time of the correspondence (last February), stated that even one failure such as this was unacceptable to them. Pending a re-design they withdrew this filter from their catalogue for Stag application—a commendable act on their part.

Finally, Triumph traced the minor overheating to a slight casting obstruction in a waterflow channel, the removal of which has effected a cure.

## FLAT

Four minutes flat (if you'll pardon the pun) to fix a puncture—and the cost? 88p, which I think is expensive.

It was all done properly, too. When the source of the puncture was discovered the tyre was removed from the wheel, a clearance hole made, a dab of adhesive stuck on the inside, and a mushroom-headed plug pulled through. The tyre was then replaced on the rim, inflated, the protruding end of the plug was cut off (as shown) and the wheel was placed in a tray of water to check for leaks.

Of course, you could say that you pay for such service, but against that the repairers (a tyre factor, not a garage) had the wheel for 48 hours. I reckon these people must make quite a

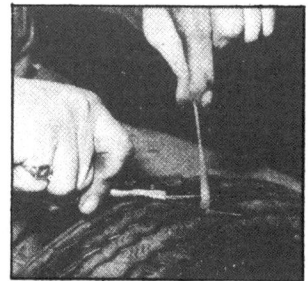

**Are the tyre people charging too much for fitting and repairs?**

good living: the firm in question employs two fitters, both of them fully occupied: at 10 repairs to the hour it means £9 per hour per man, or £72 a day.

On one occasion I saw a Triumph drive in: it was jacked up, all the wheels were fitted with new tyres, the wheels replaced and the driver was away 35 minutes later. Now, with new tyres at (say) £11 each, or £55 the five, and remembering that the factor works at about 33⅓ per cent discount, that gives him just over £18 for about half an hour's work. And alignment and balancing are extra. . . .

## BANG

With so much tight street parking these days it is quite common to see a car reversing into a spot, and then see the car behind give a shudder—occasionally accompanied by the tinkle of glass. Yet for some very strange reason the drivers seldom look fore or aft to see if they have done any damage to their own or the other car. They must know they have hit, yet they walk away quite nonchalantly.

I for one now know where so many of these little dings and dents on wings and bumpers come from: recently, after one of these performances (you can't call them accidents) I checked the two cars. I was amused to find that the offending driver had come off considerably worse, for what she hadn't realised was that the 3.4 Jaguar she had reversed into was fitted with a towing bracket!

As a matter of interest I measured the height of various bumpers from the ground to the top of the bumper bar. The figures were:

| | |
|---|---|
| Austin 1800 | 16½in |
| Ford Corsair | 19in |
| Ford Cortina | 23in |
| Hillman Hunter | 21in |
| Austin Maxi | 18in |
| Singer Gazelle | 16in |
| Austin Mini | 16in |
| Jaguar 3.4 | 16in |
| Triumph Herald | 19½in |

All measurements are with the vehicle unladen.

Four of the cars had towing brackets standing 4-4½in proud of the bumper, three of them had reversing lamps broken, again because they protruded beyond the bumper, and none had overriders.

149

# Multi-point Stag
By Maurice A. Smith

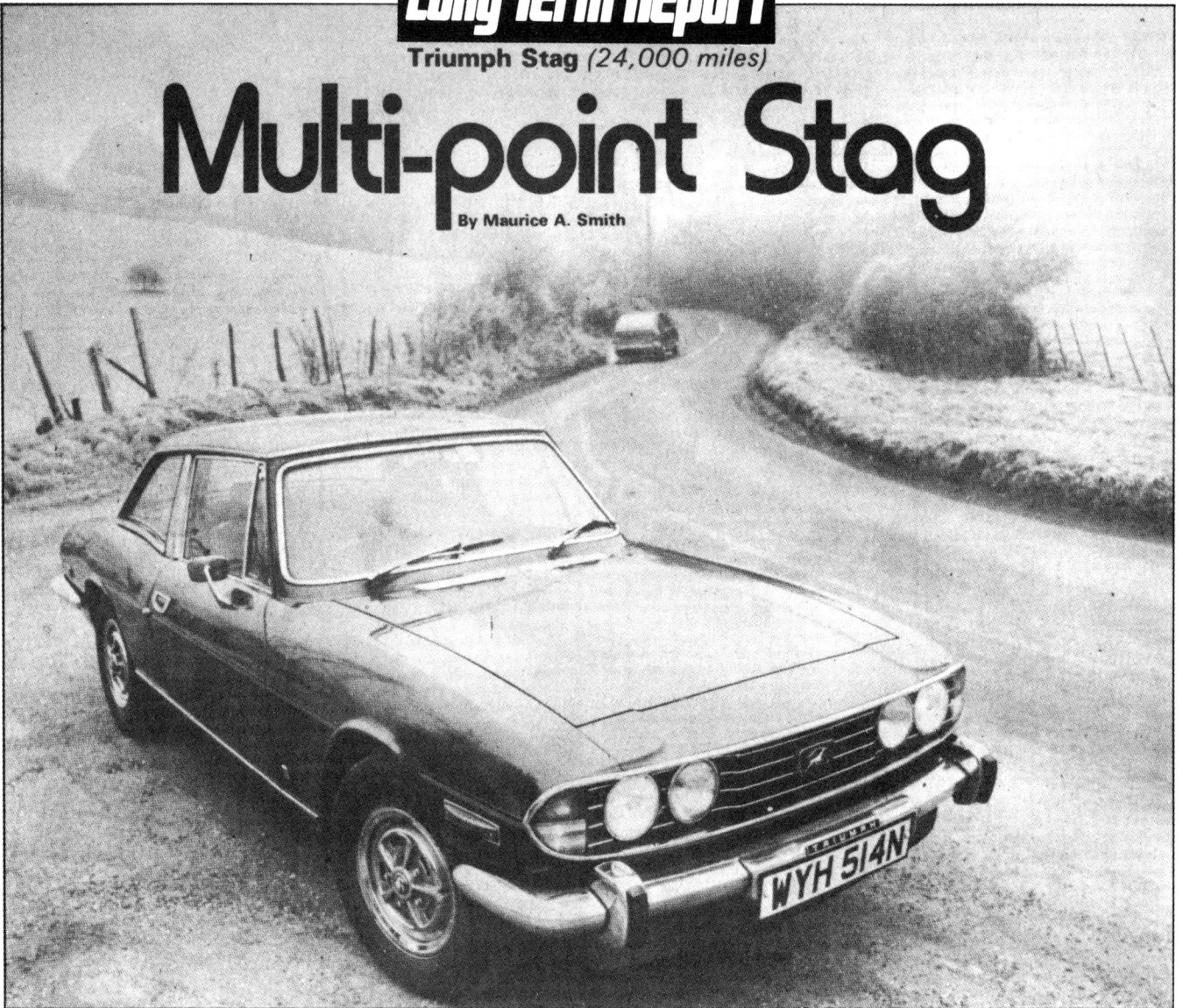

STAGS HAVE CARVED a definite niche for themsleves in the diminishing list of out-of-the-ordinary British models. Whether it is the niche originally foreseen for them is open to question. The fortunes of the model have swung around as a result mainly of external influences by no means peculiar to Stags. They came on to the market late in June 1970 at what now sounds a very modest price of between £2,000 and £2,500 according to equipment and extras. Now you have to pay nearly £6,000 for the same thing which seems expensive.

On seeing their first Stag some people said "ah, a big TR" which Stags are not, and they went on to complain about handling and performance shortcomings by 3-litre sports car standards. The jumble of speed limits, high petrol costs, road congestion and anti-motoring propaganda in the intervening 6½ years have taken some of the steam out of performance appeal and the majority of those who operate Stags found, and still find, the acceleration, speed and handling very adequate. Whether Triumph knew they were introducing a car which would also have real woman appeal is another question to be considered. Certainly it is a very popular "wife's car" for those who can afford it.

What a pity that British Leyland did their too familiar trick of introducing an attractive new model and then failing to produce them. Very many sales in Europe and the UK were lost through failure to deliver or even to quote a firm delivery date, and now it is too late since the production days of Stags must be numbered.

Turning from the general to the particular, I have heard of a few Stags that have "blown up". Cylinder head gaskets, bearings, pistons, even a broken rod have been mentioned as causes. Looking closely at these, it seems more than coincidental that a few of the owners have been the sort of "out-to-prove-something" drivers who might be expected to blow up cars and put rods through crankcase sides. Let's say it is just good luck that our two Stags have stayed together, the second one now on 26,000 miles, remaining sweet and responsive and always starting at the second short churn, whether sun-baked or ice-covered.

Oil consumption is negligible but the oil looks dirty by the time a change is due. The pressure water cooling system needs topping up most months and one wonders where the water goes. Like Jaguars and sometimes other large cars such as Rolls-Royces and Mercedes, the Stag has a steamy exhaust most of the time, said by some to be owed to condensation in long and elaborate exhaust systems. I wonder whether the suspect Stag cylinder head gaskets are allowing slight seepage of water into the cylinders. Would this occur more when the engine is hot or when cold? Such small quantities, if they exist, do not show up in the oil or on the plugs.

It seems likely that failure to check water level or notice symptoms of shortage may have led to the engine overheating reported by a few owners. If water is short the heater matrix goes cold before the temperature gauge gives any indication of abnormal temperature. Overheating is first indicated when idling or crawling in thick traffic and there is a return towards normal as soon as the engine and water pump are revved up. This may not be noticed by a driver concentrating on the traffic and obscuring the gauge with his left hand on the steering wheel. It is cured when the expansion bottle is replenished.

Talking about climatic extremes, the Stag's heater will really cook you if you want it to and can include plenty of face level cold air while your legs are being toasted. Full air conditioning was an option on early models but was not well regarded. Last summer in Britain was a real Stag summer and the soft top was folded away in its neat flush fitting box for months on end. The soft top has a zip round its rear window allowing it to fold open and flat. The rear quarters are blind and special care is needed when moving into and across traffic lanes.

The detachable hard top comes

as near to being a coupé conversion for the winter as you can hope for and includes a heated rear window which plugs itself in when the hard top is attached. Extractor rear quarter lights are fitted.

Like Harold Wilson I have said it before and I will say it again; the worst sin on the part of one's car is to pack up suddenly and leave you stranded. There are degrees of sin depending on where, when and how it does it. Our Stag sinned a little bit. Petrol starvation leading to momentary instances of cutting out gave a hint of trouble to come. I chose not to take the hint at once and did not pull into the next Leyland garage. As a result while I was driving into London to get service on the following day, the electric petrol pump in the boot finally died and left me stranded on the Albert Embankment. I had to be towed in and the replacement unit cost £16.50. Otherwise we have enjoyed over 24,000 miles in 2½ years without any trouble of note. The car has had regular services more or less at the specified times, there have been a few minor adjustments and bulb replacements.

The most expensive trouble to date was with the alternator which wore down its brushes last August at 22,580 miles, causing minor internal damage by the time the red no-charge warning light came on. Strictly speaking this might have been anticipated and a check made at the 18,000 miles service. In addition a repair rather than a replacement alternator should have been possible except that there is a built-in regulator which suffered and there was the time out of use to be considered if a repair were to be made. The cost of new parts, plus labour charges was a reasonable £22.54 from Autocar Electrical, who are good friends but in no way connected with *Autocar* magazine.

At the very beginning of August

(21,036 miles) a minor but illegal fault ocurred — you have guessed it, electrics again. The indicator flasher unit became sluggish and sometimes gave up all together. A replacement for this small sealed unit cost £5.40 trade price. Perhaps they are gold plated. The battery is still good and very seldom needs topping up.

Checking back on the maintence records I see a note of eight new sparking plugs at 16,098 miles. This could have waited until the 18,000 mile service but the earlier time was convenient for sparing the car. Otherwise the entries are routine.

There are lots of good points about this Stag, some already mentioned in two earlier stories but worth summarising in what might be a final report on this particular car (although we don't plan to sell it yet). First there is the high degree of comfort in the broad sense, owed to good quality, properly shaped seats with three modes of adjustment and in addition to the two-way adjustable steering column. The very effective heating and fresh air supply have already been mentioned. Forward view is above average and aided in rain by the two speed and intermittent wipers which sweep full arcs, the

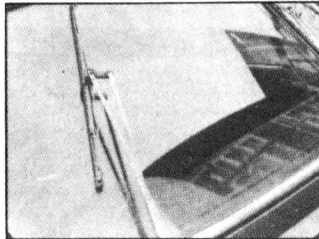

blade on the driver's side being articulated to move parallel to the screen pillar and wipe the last sector. Head room is adequate for people up to about 5ft 10 in but limited for those over 6ft. It always amuses me to watch an "anti" chap

of 6ft. proving his point that he cannot sit in a Stag at all and then next day another "pro" chap of similar size proving that he can tuck in perfectly comfortably with a bit of careful seat adjustment. As my aunty used to say "there is none so blind as those who do not want to see".

Touches of convenience and class are provided by screw-opening front quarter-lights (or ashtrays as some might say) and electrically operated windows — which reminds me that the mechanism on the driver's side is getting a bit lumpy in operation, which may be ominous.

The original Michelin tyres are still fitted having been rotated corner to corner when scuffing was becoming obvious on the outside

treads of the front wheels. Stags have a good lock and light, power-assisted steering so drivers tended to turn abruptly when manoeuvring — as tyre squeal on dry tarmac or ruts in gravel drives indicate. Long tyre life must be owed in part to the big wheels which of course carry more rubber than small ones as well as having a larger footprint on the road.

If I can afford eight (or more) cylinders I shall continue to do so. The smoothness, the idling, the starting, even the exhaust note are, or should be, that bit better than with a four or six. Thinking of exhausts, here is another good feature of the later Stags: To date there have been no leaks or rattles and no section has needed replacement in 2½ years.

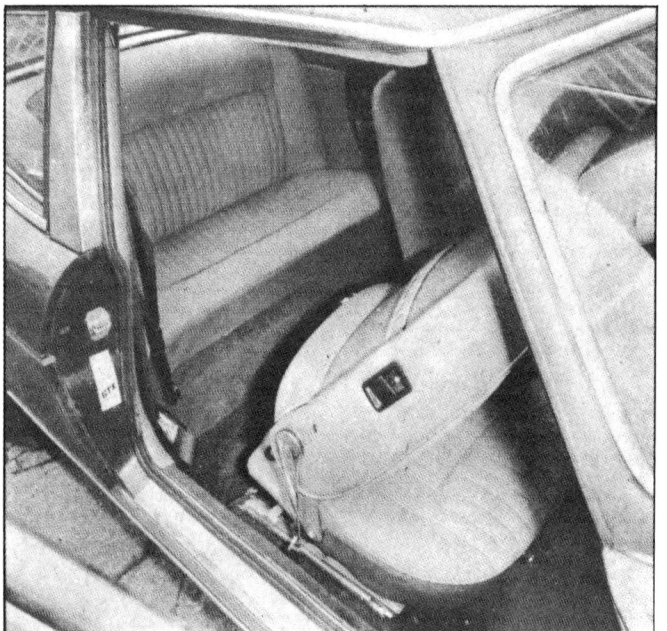

*Above. The roll-over bars do not detract from the open car feeling, and front seat passengers are well protected from wind buffeting. Above right. Comfortable, three-way adjustable seats with good shoulder support and compact head restraint. Just enough room at the back (right) to call it an occasional four seater, and reasonably easy to climb in*

Pictures may indicate that this long-term car has non-standard brushed nylon covered seats "borrowed" from the Triumph saloon line and not usually fitted to open models. I have preferred sitting on this material having experienced the alternative vinyl on my earlier Stag, and it has stood up well. It is more prone to picking up dirt and fluff but has not been difficult to brush clean. Another non-standard feature on this Stag is the black air dam under the nose. It was fitted ex-TR6 for four reasons; to improve appearance (the Stag nose appearing too high); to improve top speeds by reducing drag; to reduce front-end lift and improve directional stability at high speeds (Stags get lighter in front at around 100 mph or over and are then more susceptible to gusts, having little keel surface); to try to increase air flow through the radiator and engine compartment (Stags have been known to overheat in extreme conditions, particularly if towing a trailer). The first point is a matter of opinion, the other three seem to have been modestly effective. Some 3-5 mph higher than average top speed was recorded when the car was fully run in at about 12,000 miles. The 123 mph best figures indicated should be about 116 mph true. This compares with the *Autotest* Stag automatic maximum of 112 mph.

Luxury fittings obviously cost money and when included as standard items cumulatively account in part not only for the attraction but also the high price of the Stag today. Steering column adjustment, opening quarter lights, electric windows, three-speed wipers with articulated blade — these have already been mentioned. Some other details are the indicator repeaters on the sides of the front wings; the powerful twin halogen headlamps; viscous fan coupling; lockable flush petrol filler cap; red door-open warning

## Maximum speeds

| Gear | mph | | kph | | rpm | |
|------|-----|-----|-----|-----|-----|-----|
| | LT | RT | LT | RT | LT | RT |
| Top (mean) | 115 | 112 | 185 | 180 | 5,750 | 5,660 |
| (best) | 117 | 113 | 188 | 182 | 5,850 | 5,710 |
| Inter | 89 | 89 | 143 | 143 | 6,500 | 6,500 |
| Low | 54 | 54 | 87 | 87 | 6,500 | 6,500 |

## Acceleration

| True mph | Time (sec) | | Speedo mph | |
|----------|-----|-----|-----|-----|
| | LT | RT | LT | RT |
| 30 | 4.2 | 4.1 | 31 | 29 |
| 40 | 6.0 | 5.8 | 41 | 39 |
| 50 | 8.1 | 7.9 | 52 | 50 |
| 60 | 10.7 | 10.4 | 62 | 61 |
| 70 | 14.7 | 14.2 | 73 | 72 |
| 80 | 19.2 | 18.6 | 84 | 84 |
| 90 | 25.7 | 24.9 | 94 | 95 |
| 100 | 35.7 | 34.5 | 105 | 107 |

| | | | |
|---|---|---|---|
| Standing ¼-mile | LT | 18.3 sec | 77 mph |
| | RT | 17.9 sec | 78 mph |
| kilometre | LT | 33.7 sec | 97 mph |
| | RT | 32.6 sec | 98 mph |

| mph | Top | | Inter | | Low | |
|-----|-----|-----|-----|-----|-----|-----|
| | LT | RT | RT | LT | LT | RT |
| 10-30 | — | — | — | — | 3.1 | 3.0 |
| 20-40 | — | — | — | — | 3.2 | 3.0 |
| 30-50 | — | — | — | — | 3.8 | 3.7 |
| 40-60 | — | — | 5.5 | 5.3 | — | — |
| 50-70 | — | — | 6.1 | 5.9 | — | — |
| 60-80 | 8.9 | 8.6 | 7.6 | 7.4 | — | — |
| 70-90 | 10.2 | 9.9 | — | — | — | — |
| 80-100 | 15.3 | 14.8 | — | — | — | — |

## Consumption

**Overall mpg:** LT   20.8 (13.6 litres/100km)
RT   17.2 (16.4 litres/100km)

*Note: "RT" denotes performance figures for Triumph Stag Automatic tested in* Autocar *of 10 June 1971*

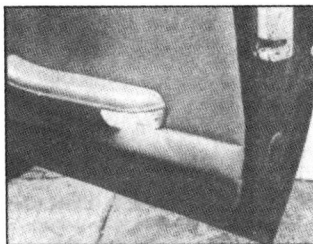

lamps in the arm rests; dipping day/night interior mirror; thick rubber inlays in the bumper over-riders and framed front number plate; and independent boot lamp and an electric clock which continues to keep good time.

What would I now want included if there were to be a Mk 3 Stag, assuming that only details could be changed at this late stage in its life? In no special order, I would change

the design and position of the segmented warning lamps dial which I find ugly and usually obscured by the driver's left hand. The upper seat belt attachments (or guides) are dreadful. They cause the straps to rub the seat back; they tangle and jam and seldom take up tension.

The rear head room in the hard top is too limited and if even another inch could be found it would be valuable. The soft top could do with transparent quarter panels for better rear vision. The styling strips along the side of the Mk 2 car should be replaced by proper rubbing strips with rubber inlays. A second filler cap on the port side would often be convenient. The plastic clips intended to hold down the floor panel of the boot are miserable fragile little objects which should be improved.

I would expect that with some better dampers the initial harshness, often felt when the car is rolling over coarse road surfaces, could now be improved. This is not to be regarded as a serious shortcoming and the ride in general is very comfortable and quiet.

There are no complaints about the way paintwork, trim and carpets are standing up, and so far there are no rust pocks like the two or three which came through the paintwork on our first Stag. In fact, after a clean and polish the car can still look practically new.

Good, low mileage secondhand Stags can now be had for between £3,000-£4,000. This may sound a lot but compared with other cars of a broadly similar type — 3000 Capri, Scimitar, Datsun 260Z — and including lesser models bought new in this price range, they are well worth considering. In my experience Stags will provide very pleasing and versatile driving with, of course, the additional rare pleasure for those that want it, of open-car motoring. □

*Left. Engine compartment keeps pretty clean and the equipment is accessible. Above, a battery mounting in the nose is inaccessible but cool. It seldom needs topping up. Power steering pump may need to move before battery can be removed, but luckily this has not yet been necessary*

*Left. The hard top gives closed car comfort but there is some extra wind noise from the screen top joint. Below. Four-way adjustment of the steering column. in-out and up-down*

*Above. Specially fitted TR6 air dam, which apparently brings aerodynamic advantages*

*Below. The official Witter tow attachment is convenient but prominent. Not the solid bumper section*

## Cost of ownership

| Running costs | Life in Miles | Cost per 10,000 miles £ |
|---|---|---|
| One gallon of 4-star, average cost today 80p | 19.2 | 416.66 |
| One pint of top-up oil, average cost today 38p. | 4,000 | 0.95 |
| Front disc brake pads (set of 4) | 35,000 | 2.54 |
| Rear brake linings (set of 4) | 16,780 | 5.88 |
| tyres (front pair) | 30,000 | 9.56 |
| tyres (rear pair) | 30,000 | 9.56 |
| Service (main interval and actual cost incurred) | 6,000 | 22.99 |
| †Repairs | | 47.54 |
| **Total** | | 515.68 |
| **Running cost per mile:** | **5.1p** | |
| **Approx. standing charges per year** | | |
| Insurance | | 89.50 |
| Tax | | 40.00 |
| **Depreciation** | | |
| Price when new (less soft top and radio) | | £3,158.00 |
| Trade in cash value (approx) | | £2,800.00 |
| Typical advertised price (current) | | £3,250.00 |
| **Total cost per mile (based on cash value)** | **10.0p** | |

'Insurance cost is for 60 year old driver, living in Surrey

†Repair costs include alternator, fuel pump, indicator flasher unit, bulbs and one wiper blade.

## Specification

**Engine:** 9 deg V8, 86×64.5mm (3.39×2.54in.), 2,997 c.c. (182.9 cu.in.); CR 9.3 to .1; ohc; 2×Stromberg carbs, 146 bhp (DIN) at 3.500 rpm; max torque 167 lb.ft. (23.1 mkg) at 3.500 rpm.

**Transmission:** Front engine, rear drive. Automatic, overall ratios 3.70-8.51, 5.37-12.36, 8.84-20.40 rev 7.73-17.80. Top gear mph/1,000 rpm 19.8.

**Suspension:** ifs, MacPherson struts, lower links, telescopic dampers, anti-roll bar. Rear, independent, semi-trailing arms, coil springs, telescopic dampers. Steering, rack and pinion (power assisted).

**Brakes:** hydraulic dual circuit (servo), 10.6in. front discs, 9.0×2¾in. rear drums.

**Dimensions:** Wheelbase, 8ft 4in. (254 cm); front track 4ft 4¼in. (133 cm), rear track 4ft 5in. (135 cm). Overall length, 14ft 5¾in. (441 cm), width 5ft 3½in. (161 cm), height 4ft 1½in. (126 cm). Turning circle 34ft (10.4 m). Unladen weight 2,720 lb (1,235 kg). Max payload 1,000 lb (454 kg).

**Others:** Tyres 185HR-14 in.; 5½in. rims; Fuel 12¾ gal (58 litres); service interval 6,000 miles. Warranty period 12 months/unlimited mileage.

# Triumph Stag

THE SHAPE of the Triumph Stag is essentially the work of the Italian stylist Michelotti, who was of course responsible for the looks of several other Triumphs before. It was conceived in 1966 (before the birth of British Leyland) as a sports-coupé using the Triumph 2000 floor-pan and drive. Several notable changes occurred whilst the idea was still on paper, including a shortening of the wheelbase, a new nose and tail — which Triumph liked enough to incorporate on the 2000 (and 2.5PI) to make the longer Mk.2 versions (before the Stag was announced) — and the substitution of a 90-degree vee-8 version of the single ohc engine Triumph supplied originally to Saab for the 99. Originally the swept volume of the new engine was to have been 2½ litres, but this was increased to 3.0 with a slightly bigger bore but much shorter

stroke than the Saab unit, giving very over-square dimensions — 86x64.5mm, 2,997 c.c.

The Stag is of course a luxury sports-car, with all of the comforts of a good 2 + 2 saloon, provided that you don't ride in the back, for any great distance at any rate. It bears some resemblances to the same sort of cars produced by Mercedes-Benz at the time, including a hefty steel detachable hard top (you need at least a strong friend to help you remove it) or convertible soft top. Since Triumph were original-enough thinkers to fit an elaborate forward-braced roll-protection bar (principally to meet expected American safety requirements), it is a little puzzling that they didn't evolve the steel folding roof that we still need as the answer to the wind-noise and thief-proofing weaknesses of the fabric soft top. The inside is handsomely equipped,

with Triumph's idea of a wooden facia, very convenient fingertip controls for major accessories, electric lifts for the windows, and manual or Borg-Warner type 35 three-speed automatic transmission. Overdrive was an option at first. Clutch and brakes were bigger than on the 2.5PI, whilst the suspension — MacPherson strut front, semi-trailing arm independent at the back — was similar to the Triumph saloons, though with different rates for the coil springs. The 2.5PI gearbox had a higher first gear, but the final drive ratio (3.7) was set midway between the 2000 and 2.5PI ones (4.1 and 3.45 respectively), which, going by our Road Test of 30 July 1970, gave the car a slightly low-geared top and a reasonably long overdrive — though judging by the fact that maximum speed was achieved in overdrive (115 mean, 117 best

against direct top's 113 mph) it could usefully be longer still.

The engine looked a quite advanced design, with a cast iron block and light alloy heads, each with single overhead camshafts driven by chains rather than exposed belts, a two-plane crankshaft, a jack-shaft and skew-gear-driven water pump in mid-block, and wedge-shaped combustion chambers fed by two SU-principle Stromberg carburettors inside the vee. Triumph took some trouble to make the unit easy to work on; you can for example remove the camshafts without disturbing the sprockets.

It developed a claimed 145 bhp at 5,500 rpm (the 2.5PI six produced 142 bhp in TR6 guise) and maximum torque, appreciably more than the 2.5PI, was 170 lb. ft. 3,500 rpm. Overhead camshafts, especially at and before the Stag's 1970

*Stag could be had manual or automatic, with hard or soft top. The soft top version had a rigid roll-over bar for safety (above). V8 engine fitted easily under bonnet (below) and facia was well equipped and of high quality (bottom) with a full range of instruments and face-level ventilation.*

## Performance data

| Road Tested in Autocar of | Stag manual (O/d) 10 June 1971 | Stag auto 10 June 1971 |
|---|---|---|
| **Mean max speed (mph)** | 116 | 112 |
| **Acceleration (sec)** | | |
| 0-30 | 3.5 | 4.1 |
| 0-40 | 5.1 | 5.8 |
| 0-50 | 7.1 | 7.9 |
| 0-60 | 9.3 | 10.4 |
| 0-70 | 12.7 | 14.2 |
| 0-80 | 16.5 | 18.6 |
| 0-90 | 21.8 | 24.9 |
| 0-100 | 29.2 | 34.5 |
| | | |
| **Standing ¼-mile (sec** | 18.2 | 17.9 |
| **Top gear** | | |
| 10-30 | — | — |
| 20-40 | 10.8 | — |
| 30-50 | 11.1 | — |
| 40-60 | 11.3 | — |
| 50-70 | 11.6 | — |
| 60-80 | 13.2 | 8.6 |
| 70-90 | 16.1 | 9.9 |
| 80-100 | 21.9 | 14.8 |
| **Overall mpg** | 20.6 | 17.2 |
| **Typical mpg** | 25 | 20 |
| **Dimensions** | | |
| Length | 14ft 5¾in | |
| Width | 5ft 3½in | |
| height | 4ft 3½in | |
| kerb weight (cwt) | 25.1 (25.3 auto) | |

launch, were usually expected to bring, amongst other things, a better top end to an engine's performance, but as with the Dolomite engine from which the Stag one is derived, this was not so on this car. Smoothness and a pleasant if not at all exceptional spread of power were its best characteristics; there was none of the quite exciting urge high up of the 2.5PI. Certainly a little of that impression is due to the extra weight of the Stag (which is 2½cwt heavier than the saloon), but the rest is the fault of the engine which never felt as eager as its size and specification declared it ought to be.

Triumph's not very subtle power-assisted steering was standard. It was (and remains) entirely satisfactory for most users, being light and quite "quick" in a car which doesn't roll too much ordinarily and which has reasonably good roadholding. Its failings when compared with other systems were a lack of any feel and a curious and highly distinctive "notch" effect at the straight-ahead position. Turning the rather large steering wheel (adjustable though in both tilt and telescope senses) either side was accompanied by a noticeable — if small — increase in effort needed, so that at first one tended to steer a little jerkily. Ultimate roadholding is limited by the inevitable camber-change effects caused by fact that semi-trailing arms are not pure trailing arms.

## What to look for

Having been around for only a relatively short time, and also since Triumph seem to have learnt more than some other makers about avoiding it, rust and body rot do not seem to have been noted so far. Details like fittings may have deteriorated superficially, but we have not heard of any serious cases of decay on the Stag to date. That doesn't mean of course that such troubles may start to appear at any time, so a proper inspection both on top and underneath for rust is still very worthwhile; the best protected of cars will deteriorate fast if it has suffered crash damage which hasn't been repaired properly. The same applies to stone damage at both the front and behind wheels; sills can suffer in the hands of the show-off and the clumsy.

The suspension has a good reputation amongst dealers too. The usual wear of pivots occurs, but not to any unusual extent; test for this in the usual way, by shaking each wheel sideways whilst a helper keeps the brakes on — if there is play, a bush may need replacing; if there is play with the brakes off, then suspect a wheel bearing.

A peculiarity common to any of the Triumphs using semi-trailing arm rear suspension is a curious lurch sideways from the back when you go from drive to neutral or over-run — during a gearchange for instance. It doesn't always happen, but when it does it is caused by the splines of the telescopic drive shafts sticking and then freeing, thereby locking, then releasing the rear suspension. Ball splines avoid this problem, but they are much more expensive than the Stag's plain splines. Triumph have always told us that this is a lubricating problem that should not arise, though they admit that it sometimes does. It isn't a thing to worry about however.

The gearbox has a reputation for synchromesh that wears out, and a strong weakness for gearlever buzzes after a lot of use. Cases are reported of a metallic ping from the propellor shaft rear companion. flange where it couples to the quill shaft of the differential, due to looseness. But no one we spoke to mentioned any true unreliability of the transmission. Final drives have a good reputation.

The power steering can leak, but to no greater extent than on any other power-steered car. Inspect the gaiters on the rack for signs of oil leaks. Oddly, several dealers told us that repairing power steering (with new seals and so on) has always proved rather unsatisfactory, though they find it hard to explain why.

Brakes troubles are rare. The hold of the convertible can leak and if it has not been used for a long time can prove very awkward indeed to erect. Once one has managed to lift the weighty hard

*Boot was large, with a wide opening (above); folding the aft section of the soft top was somewhat complicated (right) but back seat was quite roomy despite the width swallowed up by seat well, which meant no proper elbow rests (bottom). Thick roll-over bar supported belts and housed interior lights.*

top (which comes with a heated rear glass) into place, the clips securing it (very like the similarly excellent ones on the Mercedes 280 SL) should work well.

The engine of the Stag has, at any rate in older cars, proved to be the great weakness, to such an extent that there is a tendency to christen the car the "Triumph Snag." Aluminium alloy heads are ideal things in theory, but if they are not designed or made correctly, they can give a disproportionate amount of trouble. The Stag heads have a bad reputation for warping in service, often at only around 25,000 miles, thereby blowing gaskets which if they are not dealt with quickly can all too easily lead to damaged head faces needing skim-

ming if they are to be rescued. The trouble is that there is very little meat to be skimmed; up to 0.010in. is officially allowed for removal (one dealer told us he had got away with 0.020in. but did so unhappily), but this isn't always enough of course. Corrosion is another problem if whoever maintains the car has not used a corrosion-inhibiting coolant as he should have done. Head bolts then seize as they have done on many aluminium engines in the past. New heads each cost well over £200.

On earlier models there was a history of oil pressure bothers leading to crank journal wear, but

this has reportedly been cured. Triumph themselves say that the head troubles are a thing of the past too, but any prospective buyer must feel some disquiet knowing that there is at least one London-based firm who, we are told, will fit you a Rover vee-8 in place of the Stag unit. And he may also be aware that all the indications are that Leyland rationalisation will almost certainly mean the dropping of (at any rate) the engine in favour of the Rover engine. A pity, because if properly developed both the engine and less so the car could be made into something much more positively successful. □

*Left: Hard top could be detached but was massive affair needing two men to lift off or replace; the process took some time.*

## Approximate selling prices

| Price range | Stag (manual) | Stag (auto) |
|---|---|---|
| £1,300-£1,400 | 1970 | 1970 |
| £1,400-£1,500 | 1971 | — |
| £1,500-£1,600 | — | 1971 |
| £1,600-£2,000 | 1972 | — |
| £2,000-£2,200 | — | 1972 |
| £2,200-£2,550 | 1973 | — |
| £2,550-£2,700 | — | 1973 |
| £3,000-£3,200 | 1974 | — |
| £3.200-£3,500 | — | 1974 |
| £3,500-£4,000 | 1975 | — |
| £4,000-£4,500 | — | 1975 |
| £4,500-£4,750 | 1976 | — |
| £4,750-£5,000 | — | 1976 |

## Spares prices

| | |
|---|---|
| Short engine (exchange) | £335.88 |
| Cylinder head (one, new) | £407.16 |
| Gearbox assembly (exchange with overdrive) | £140.40 |
| Clutch pressure plate (exchange) | £15.66 |
| Clutch driven plate (exchange) | £10.21 |
| Automatic gearbox assy, with convertor (exchange) | £123.12 |
| Propellor shaft universal joint repair kit (each) | £5.77 |
| Final-drive assembly (new) | £130.68 |
| Brake pads — front (set, new) | £9.94 |
| Brake shoes — rear (set, exchange) | £9.72 |
| Suspension dampers— front (pair) | £49.68 |
| Suspension dampers — rear (pair) | £19.44 |
| Radiator assembly (exchange) | £43.50 |
| Alternator (exchange) | £28.08 |
| Starter Motor (new) | £76.10 |
| Front wing panel | £41.85 |
| Bumper, front (new) | £56.16 |
| Bumper, rear (new) | £75.28 |
| Windscreen toughened | £22.68 |
| Windscreen, laminated | £26.19 |
| Exhaust system complete | £67.07 |

All the above prices include VAT at 8 per cent.

## Milestones and chassis identification

| | Chassis No. |
|---|---|
| **June 1970:** 2 + 2 convertible / hard top based on Triumph 2000 / 2.5PI introduced, with new vee-engine based on four-cylinder design first seen sold to Saab (later in Triumph's Dolomite and derivatives). Distinctive anti-crash bar provided. Power steering, electric windows standard. Overdrive for manual box and Borg-Warner automatic transmission and air conditioning optional. Detachable hard top version, with soft top as extra. | 2 BW |
| **October 1972:** Overdrive on manual version standard | — |
| **February 1973:** Matt black sills and tail panel, re-styled instruments, courtesy lamp in roll bar | 20001 |
| **January 1974:** Hazard warning and seat belt warning lamps fitted | 31153 |
| **March 1975:** Air conditioning option dropped | — |
| **January 1976:** Alloy wheels, tinted glass and laminated screen now standard | 40793 |

# To say the Stag

... came about by accident is a gross over-simplification, but if it hadn't been for the friendship between two men it might have been a totally different animal. Mike McCarthy talked to Harry Webster, formerly Chief Engineer at Triumph

HARRY WEBSTER first met Giovanni Michelotti back in 1957, when the Italian stylist was working for the coachbuilder Vignale: he was responsible for the facelift of the Standard Vanguard, a model which in fact became known as the Vignale Vanguard.

"I used to go and see 'Micho' every other weekend or so for, oh, five or six years or more ... My missus used to play hell with me ...! I used to set sail (after the races) from Coventry to London, fly to Paris, then change and fly on to Turin — it was the only way I could get there on a Saturday. I used to work with him, often on Saturday night and certainly on Sunday, then set off home on Monday morning — there's a direct flight back: it was very convenient ...

"'Micho's' first major job was the Herald, and he went on to design most of what was to become the Triumph range — Spitfires, Vitesses, GT6s, and so on ...'"

Having completed the Triumph 2000 and 1300, Michelotti asked Harry if he could have a 2000 chassis for a one-off show special. "These styling boys, they're proud fellows, they always want to show you what they can do, show off a bit ... Anyway, he asked me for a 2000 chassis so that he could put a one-off body on it and put it in the Turin show that year(1966 — MM).

"I agreed to let him have the chassis, but on one condition — that if we liked it we would pay him for it and take it back to Coventry ...'"

It was on one of his trips to Italy that Harry saw what Michelotti had done. He had created a smooth, sleek convertible featuring a wide, horizontal grille behind which hid the headlights (fronted by electrically-operated flaps), wire wheels and a sharply cut off tail again with wide horizontal tail-lights, which echoed the shape of the grille particularly at the edges where the side lamps came to a point. Interestingly enough one feature that was to make the production car so distinctive didn't appear on the prototype

— the T-bar over the passenger compartment.

"Well, when I went to see it, I liked it, took it back, and set the wheels in motion for production ...'"

The other major difference between the prototype and the production car was the engine. As first envisaged, the Stag was to be equipped with the 2.5 litre fuel-injected straight six engine from the TR and 2.5PI. However, in 1963 a clever Triumph engineer, Lewis Dawtrey, had drawn up a report, proposing Triumph's future engine plans. It boiled down to basically two: a slant-four, originally of 1500 cc but capable of enlargement to 2 litres, and by "ordering in" the other half of the four cylinders, a V-8. When Spen King, now Deputy Chairman, BL Technology, took over as Chief Engineer at Triumph after Harry's appointment to the Austin-Morris division following the Leyland/BMC merger, he, too, liked the car but felt it could take a lot more power: he therefore delayed the introduction of the car until the V8 was ready. This put the weight up slightly, and, though the six (since the car was basically a 2000) slotted straight in, the engine bay had to be widened to take the V8. This led to a lot more modifications until eventually there were very few (if any) common body parts between the Stag and the big Triumph saloons. And it also meant that the car didn't see the light of day until 1970.

The T-bar was another story again. "After 'Micho' finished the prototype we made some hand-built prototypes ourselves — and suffered from the most enormous scuttle-shake! Oh boy, it was horrid! You almost had to try and catch the steering wheel, if you know what I mean! The torsional stiffness of the body had gone to hell, of course, and the only way to get it back was to join up the A and B posts (*the structural members fore and aft of the door — MM*) with a good torsional box across the top, and that's exactly what the T-bar does''. So it wasn't just brought in to help with roll-over

regulations or for styling? "Oh no! Of course it helped with the roll-over conditions, but it's very much part of the structure, I can assure you!''.

Were there many other problems? "Not while I was in charge, remembering it was to have the six cylinder engine — after all most of the suspension, engine, transmission and so on had been sorted in the saloon. One of the interesting problems was the convertible top: it had to disappear completely into a bin behind the seats with a lid on so that we could fit the hard top. The way the hood folded gave us some headaches ...''

So the Stag came about because Harry liked what Michelotti had done

— but what was the rationale behind the car? "Triumph realised back then that we couldn't compete with the big boys — we had to find a niche or market gap to fill, and that was where the Stag came in. It was aimed at the "young executive", someone who'd gone through the motor bike/sports car/family and family saloon, and wanted something different, something sporty but with creature comforts. Hell, I wanted one myself at the time!''

Then the surprise of the day: "I'm getting one myself, actually. My wife wants me to have one when I retire — something to play about with ...''

How's that for confidence in your own product?

Above: Harry Webster (on the left) with his successor at Triumph, Spen King. They were responsible for the birth of the Stag.

Left: the Stag as envisaged by Michelotti, *sans* T-bar and side ornamentation but with wire wheels and hidden headlights

## Little deer ...
■ Proud owners of Triumph Stags may like to know that a model of their car, to 1/43rd scale. is available.

# Past prejudice

Triumph's Stag could have been a contender against Mercedes' mighty SL in the seventies, but reliability was not on its side. After two days' driving, Martin Buckley says the gap is closing as Leyland's Cinderella matures with age

The Triumph Stag looked set for greatness in 1971. It was a great idea: what else could you buy with open four-seater practicality, hard top coupé capability and the wuffly sophistication of a home-grown V8 engine at such a bargain price? Leyland's answer to the SL, they called it, and, but for a terrible reliability record, that wouldn't sound laughable today. It was a fine car, the Stag, but badly executed.

It took more than a decade for the big Triumph to regain its credibility. With engine problems licked – a Stag shouldn't overheat or run its bearings these days – its many good qualities as a GT car can shine through.

To find out just *how* good a Stag can be we thought a run down to the west country would be a good idea – in convoy with a Mercedes Benz 350SL. As far as we know, the pair have never been extensively compared in print, because their prices were so far apart when new. Could *you* have justified paying twice as much for a Mercedes 350SL?

On Sunday afternoon I collect the Mercedes from The Autodrome's King's Cross showroom. It's a 1973 350SL, an automatic like

**Our car waits for the off at Surrey Stags, above, and below, on Lyme Regis quay**

almost all these open Mercs, with straight bodywork but lots of miles under its belt. It's up for just under ten grand. The doors still thunk shut like a vault and the engine starts on the first twist of the key; but for a few rattles and some smoke it goes well. Driving through town it's familiar: those broad-shouldered looks don't rate a second glance, but then the shape has been around so long and so many were sold that it isn't a surprise. They only stopped building cars that looked just like this a couple of years ago. Myself, I like neither the stockbroker belt/bank robber's wife connotations nor the high waisted, stubby proportions.

Without an owner's handbook it takes me 20 minutes to figure out the release for the flush-fitting steel tonneau cover and put the hood up, but it locks into place tightly on the screen rail with the help of a tyre-lever type implement in the glovebox.

I'm up at 7am and by 7.30 the SL is heading out along the A3 towards Haslemere in Surrey. That's where Graham Munt's Surrey Stags is based and, by the time I'm pointing the SL's nose on to his forecourt at 8.30, his team – Stuart and Bob – are already hard at work. Graham got into Stags 11 years ago after buying a cheap one as a young man.

"That car cleaned me out, but I've been working on them ever since." Graham reckons there is nothing too difficult about the Stag, but the engine doesn't suffer fools gladly. "The oil and cooling systems are marginal, any problems will soon show up. Regular oil and water changes are the most important thing." He also points out that it won't take kindly to constant 100mph motorway cruising like the Mercedes either. Graham deals in and works on the cars at the top end of the Stag market – £10,000-plus – and sells most to professional people in the 35-55 age bracket who want a sporting open car with four seats. At that point one of Graham's satisfied local customers burbles down the high street in a white Stag, waving as he goes.

Graham points me in the direction of a superb MkII Stag in the corner, the white car I'll be driving over the next couple of days. It turns out not to be one of Graham's Stags but one belonging to a friend, Roger Morrish. It's an original, low-mileage machine for sale at £16,000. These MkII Stags have accent stripes on their flanks, a matt black tail panel and sills

but keep the steel wheels. Like 70 per cent of Stags, this has the Borg Warner self-shifter. The Stag is still a common sight in Britain as most of the production stayed here. It's a prettier car than the SL, I think, especially with the too-bulky hardtop removed. The proportions are right, the coke bottle 'hips' above the rear wheels adding visual character, although the fat screen pillars, derived from the 2000 saloon, jar.

By the time I've cast an eye over the car, photographer Simon Childs has arrived. We sling his bags into the SL's deep, narrow boot, say our goodbyes and head out along the A3, stopping after 10 minutes at a Little Chef to take some breakfast on board.

Already the Stag is making a good impression. Even with the heavily three-quarter blind-spotted hood erect it's much less gloomy than the Mercedes inside. The scuttle is low, the slender, leather-bound helm steeply angled (although it can be adjusted to rest on your thighs if need be), the driving position in the PVC-trimmed seats easy to tailor. Triumph did a good line in dashboard in the seventies and the Stag had the best of them, with a fine spread of circular instruments set back in dull-finish timber, plus an all-systems-go, multi-warning light dial that BMW claimed as its own later. The powerful blower clears the screen quickly, and the power windows slide up and down quietly.

We're off again after 30 minutes, out into a gloomy morning that threatens rain, heading along the A3, Stag following SL through Grayshot, Liphook and other twee southern villages until we hit Petersfield and pick up the A272 to Winchester. The rain's coming down now and we're both taking it easy, keeping the speed down to 60.

Out on the M27 the 200bhp SL has the legs on the 145bhp Stag – but only just. Both are happy at 85 but the Triumph is far from quiet. Short gearing for the three-speed auto means the crisp V8 has to work hard in top, the 19.8mph-per-1000rpm final drive corresponding to a fussy 4000rpm at 80mph. The rag-top buffets and hisses too, making the radio near useless.

The A31 is busy and, with no overtaking opportunity, I lose the Merc for a while so

First stop – breakfast at Little Chef, A3

Crab-catching still a living, Lulworth Cove

Simon has to wait for me near a roundabout. The weather is still miserable and neither car is in its element with the hood up.

We're heading down the A352 now, through Wareham on the way to the first photo-stop at Lulworth Cove. On the B-road to Lulworth Simon guns the Merc for the first time, sending a smoke signal of blue fumes out of the exhaust, and I follow his lead. Road tests criticised the Stag's servo steering but I find the rack and pinion effectively loaded and pleasantly accurate, although it does lighten rather than get heavier the faster you

Merc familiar, owing to long production

go, leading to over-compensation as the understeer turns into pleasant neutrality. It keels over some through tighter turns, but not enough to put me off, and the semi-trailing arms have a surprising amount of grip in hand. But there is a rubbery, disjointed feeling to the Stag's chassis I can't put my finger on, and an odd wiggle from the back-end when I come off or on the power. That's a common Stag ailment, it turns out, due to the grease in the driveshaft splines drying up.

At Lulworth we drive the cars down a single track road to the Cove, where there's a cafe and a launching ramp into the blue-green water. On the beach a fisherman is busying himself with a net and, on the quayside, a basket of still-live crabs are stirring: they'll soon be in a pot of Shipham's paste, I muse.

The cafe is full of kids on a school trip, who take little notice of Simon taking strange pictures of me drinking coffee and eating ice cream. I find beauty spots like this depressing and not a little sad on such a rotten day. I'd like to return in better weather.

Off on the A35 to Lyme Regis now, and we're surprised to be going through mist as we drive down the steep incline into the town. These days it's best known as the place where *The French Lieutenant's Woman* was filmed (the writer John Fowles lives here) but it's always been popular as a place '...for sea air and bathing' as the dated little guide book puts it, a change from the spa towns of Bath or Cheltenham. Jane Austen wrote, and set, *Persuasion* here in 1804.

It's made up mostly of Georgian buildings and narrow streets but its most famous feature is the Cobb, a stone harbour wall going back to at least the 1300s that hugs the

SL interior neat; black plastic oppressive

Buckley dips oil at fuel stop in Dartmouth

Classic dash in Stag; roll bar aids rigidity

Dipstick a stretch in Stag; V8 now reliable

boats in the harbour like a protective, parental arm. It's easily big enough to drive along too, so we take the cars to the end and set up a shot outside some fishing sheds. The rain still pours, but nobody bothers us.

There are more bleak, rainy miles ahead on the road to Exeter and I think how much nicer the Stag would be with the manual overdrive box to take advantage of the 3-litre V8's torque. It's a lovely engine though, this V8. Smooth, flexible, free-spinning and powerful enough to give this hefty car eager if not startling performance. At this moment it's hard to believe it ever gave problems.

On the outskirts of Exeter I spot an intriguing El Camino style Triumph 2000 pick-up outside what looks like an old gate-house. The owner's garden is filled with crumbling 2000s, a car the Stag shares its Michelotti styling touches with.

Heading down the A379 we hit a village called Starcross, home of Brunel's atmospheric Railway Museum. It's housed in a grade one listed building and I'm keen to go in and find out more. Sadly it's closed and, after reading the blurb on the board outside, I'm none the wiser.

The day is closing in now and we want to get to Dartmouth before nightfall. The weather is clearing up as we drive through Torbay. With its bingo halls and a flash-looking pier it seems to be the west country equivalent of Blackpool. The last time I was here was 1970, as a toddler.

We hit Dartmouth at 6pm and head for the Royal Castle hotel, an old coaching inn that Simon has stayed in before. It started life as two houses built by rich merchants in the 1630s: they joined forces in 1782 to become the Castle Inn. It became the Royal Castle Hotel when the Prince of Wales began to frequent it in the late 19th century, and since those days all kinds of celebrities have stayed here: Cary Grant, Gregory Peck, Diana Dors and Faye Dunaway among them. More

Castle Hotel, Dartmouth, 'famous since Drake first sailed', now yours for £30 a night

Brunel's Atmospheric Railway; closed

Buckley studies the next day's route

161

Stag a prettier car than the SL, although, with 145bhp to the Merc's 200, it's not as fast

recently it was used as part of the set of an Agatha Christie thriller. I'm sufficiently star-struck by all this to deem the £30-a-night bed and breakfast within budget, so we check in and get our heads down for an hour.

We have a drink in the bar at 8 and eat at 8.30. I have turbans of salmon, which is good but barely fills a hole. Simon admits to not being a fan of the Mercedes simply because it isn't his type of car, but then he drives a Renault GTA everyday and before that a Caterham Super Sprint 1700.

The quickest way across the bay at Dart-mouth is by the Kingswear ferry so we decide to head off on the foot passenger ferry to investigate the possibilities of a picture on board while the cars are being taken across. "No problem" says the cheery man in the ferry office (who has done this with Simon before). We pay our bill at the Royal Castle, load our bags and lower the hoods, a simple operation on both cars once you know the ropes, although the two-stage movement of the Stag's canvas takes a little longer. I notice the Triumph's beefy roll-bar for the first time – which helps make it feel so stiff for an open car – and the generous-by-GT-standards rear seats. This is a good package.

Driving through the narrow streets to the landing point I get the full force of the Stag's bellowing caricature of a bent eight exhaust note. The ferry consists of a recently built 'float', capable of taking six cars, lashed with ropes to a little tug, which dates from the thirties. There's been a ferry service operating continuously here since the 1860s. Now, there's a crossing every 20 minutes or so.

After a petrol and oil stop we're leaving Dartmouth by mid-morning and driving along the coast road. There's a blue sky for the first time, so the hoods stay down. We're on the 'English Riviera' now – and this is perfect Stag country.

After a stop for more shots we swap cars. The SL driver's seat has sagged so I feel almost as though I have to peer through the spokes of the big wheel. It's a depressingly bleak land-scape inside the SL, with little to relieve the black moulded plastic, and I miss the Stag's electric windows and friendly ambience.

There's greater initial urgency to the 350's acceleration, to the accompaniment of a tappetty growl. The three-speed auto has a more gentle change than the thumpy Borg Warner slush-box in the Stag but, oddly, the Merc's soggy-feeling four discs can't match the disc/drum retardation of the Triumph.

Perhaps the worst omission in the SL is that, unlike the Triumph, it doesn't have any rear seat: it was twice the price of the Stag, but half as practical.

We haven't gone much further up the road when we come to Torcross and one of the most fascinating sights of the trip – a floating tank. It was a British adaptation of the American Sherman called the DD (Duplex Drive) which used an inflatable canvas screen to allow the tank to float ashore, driven by a pair of propellers that folded up on dry land where the tank could then operate normally, driven by its tracks. This one was sunk during practice operations in the war and stayed there until 1984 when it was rescued by local man Kenneth Small, after negotiations with the US Army. Apparently he had it running

SL and Stag share a ride across Dartmouth bay on the Kingswear ferry; up to six cars can be pulled across at a time by the ancient tug

162

together with no problem; so well that I put the pistons in as well that day.'

Because it was winter, Roger brought the heads and valves into the house and stood them in front of the gas fire so they got to an even temperature. 'I had the Workmate in the lounge and shimmed the heads up sitting in front of the fire and the television,' he recalls. Lesley had no objection to this. 'At least it meant I saw him occasionally,' she says philosophically.

Roger sent the Borg-Warner three-speed automatic gearbox away to Liveridge Autos to be reconditioned, and by the end of January the running gear was ready to go back in, complete with an electronic ignition pack and a new five-core radiator. Before installation he

resprayed the engine compartment and, with some help from neighbours, the car was back together the following month.

Roger and Lesley never really consider

> **'Roger brought the heads inside and stood them in front of the fire'**

their Stag to have had an 'off the road' restoration, as all work carried out on the car has been done over successive winters.

Every summer the Triumph has been back on the road, covering many miles attending shows at Stratford, Trentham Gardens and Woodvale.

Therefore, it was November of 1991 before they decided to turn their attention to the car's bodywork, where quite a surprise awaited them. 'When putting the tonneau cover on I'd noticed some pop rivets and showed them to someone who said they shouldn't be there. You couldn't see them at all on the outside,' Roger explains.

When he began taking the paint off the rear wings, however, he discovered why. A wheelarch repair section had been riveted in place on top of the old arch and the join hidden with filler. In places this was

someone had pop-rivetted a repair panel over rusty metal. Roger had to learn how to MIG weld, and fast!

166

---

shortly afterwards and drove it to its present site, where it serves overlooking the sea, as a D-Day war memorial.

Bigbury-on-Sea is the next stop, but on the way we stumble across a convincing-looking vintage sports car on a filling station fore-court in Aveton Gifford. It has very Aston Martin Ulster proportions but turns out to be Fergus, built up the road in Churchstow and based on a Morris Marina of all things. Tope Body Repairs is next door to the filling station and inside Charlie Tope and Dave Stirling are restoring a Stag and a TR4, apparently to a good standard. In the corner there's a sound 2 + 2 E-type, fresh in from the States.

The run to Bigbury gives me a chance to open up the Merc for the first time. It feels a more substantial, grown-up car than the Stag, heavy-footed and clumsy but vicelessly safe. It squats heavily coming out of tight bends but the roll is well checked, the assisted steering aiming the car accurately. Surprisingly, the ride is rather lumpy and the shell has definite scuttle shake, something the Stag hardly suffered from at all. It's a competent but charmless car to drive quickly.

Burgh Island is what we have come to Bigbury-on-Sea for, a small piece of land surrounded by sea when the tide is in, with a 1929 art deco hotel built on it. We consider driving the cars across the beach to the hotel but think better of it. Good job too, because the bloke behind the check-in desk tells us that the beach belongs to the Duke of Cornwall and you need permission from the Duchy to drive on it (a giant sea tractor takes guests across when the tide is in). We decide to have a look around this amazing building.

It was designed by Matthew Dawson for the industralist Archibald Nettlefold in 1929 and quickly became a haunt of the racy set of the thirties: Agatha Christie, Edward, Prince of Wales and Wallis Simpson, Mountbatten and Noel Coward all stayed here. The interior is the most impressive part, with its magnificent art deco ceilings and glasswork. The Ganges room is the actual captain's cabin of the Ganges, the last sailing vessel to be the Flagship of the British navy, built in 1821. The cabin was bought by Nettlefold when the ship was broken up in 1930.

Burgh Island fell into disrepair in the postwar years and was bought by Beatrice and Tony Porter in 1985. They have brought it back to near-original order with authentic art deco furniture in every room: the originals were burnt in a massive beach bonfire a few years ago. At £73 a night it's expensive, but I make a mental note to come back someday.

We take one last shot of the cars with Burgh Island in the distance and start off home, taking a less convoluted route through Glastonbury, and eventually ending up on the M3. By 8pm I am handing the keys of the Stag back to Graham, well pleased, even surprised, with how the once unloved car has put up over the last 48 hours. It still can't match the Mercedes in much of what it does these days – the intrinsic quality isn't there – but, with reliability on its side, it comes much closer than you might at first imagine. I liked the Triumph Stag but found myself unmoved by the Mercedes SL.

Thanks to The Autodrome (tel: 071-833 2076) for the loan of the SL, and Surrey Stags (tel: 0428 658427) for the Triumph.

Merc's bulk shows against lithe Stag as Buckley examines 'swimming Sherman'

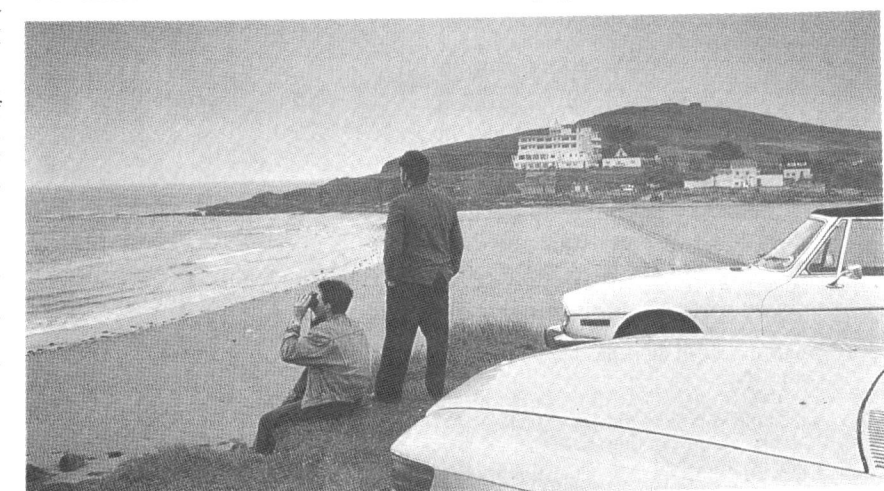

Buckley ponders the winner on ferry

Dartmouth passenger ferry, left, and roofs up against drizzle as Buckley writes

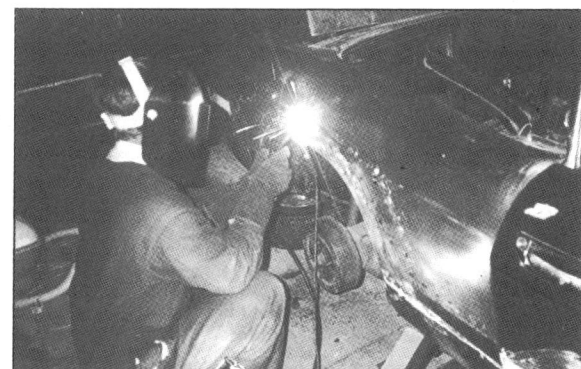

Last stop is Burgh Island, with art-deco hotel; you drive across the beach to reach it

163

# Stag in stages

Drive through summer, restore all winter is Roger and Lesley Phillips' secret to happy classic car ownership. Zoë Harrison relates how they rebuilt their Triumph Stag one piece at a time

Photos by **Zoë Harrison**

Rebuilt engine and gearbox are slotted back in with the help of a good friend!

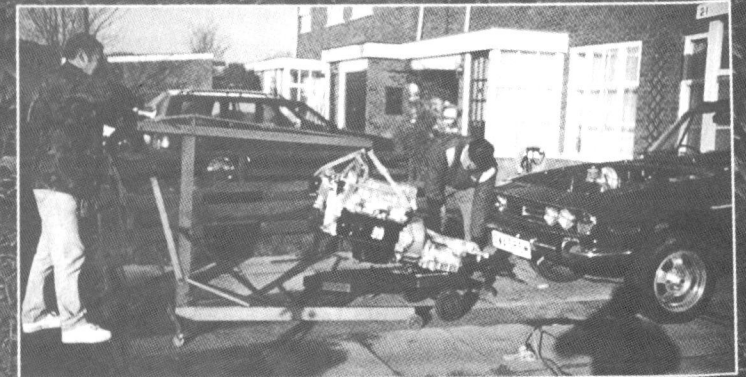

It can't be denied that whenever anyone carries out a survey of popular classics, the Triumph Stag will always feature highly. They're a versatile car, roomy, well-equipped and benefiting from a 146bhp, 3.0-litre V8.

On the other side of the coin, that alloy V8 engine does have a reputation for overheating, and people tend to question their reliability. Roger and Lesley Phillips are keen to dispel the myths surrounding their immaculate 1973 Stag. In fact, the biggest problem they have is keeping below the 5000 miles a year agreed with their insurance company.

The Phillips bought their Stag in August 1988. Roger had not particularly been after a Triumph and it was Lesley who put him on to the marque. 'We were just looking for an open-top sports car and I'd been thinking of a Westfield Seven, but once I'd seen a Stag, I knew that's what I wanted.'

Lesley had her own very practical reasons for suggesting the car – the fact the driver's seat adjusts vertically as well as horizontally, and the steering column position can be altered as well. 'I'm fairly short in the leg and there are a lot of cars I can't actually drive,' she admits. 'I can either reach the pedals, or see over the top of the dashboard, but not both. I don't have that problem with the Stag. And we like V8s, which reduces the choice quite a bit.'

The general appearance of the Triumph really appealed to them, as well as its practicality. 'There are not many sports cars you can properly fit four people in,' Roger adds. 'It's a bit of a squash, but at least you can get adults in.' His in-laws were taken for a spin recently, to prove this point. 'We only did about 30 miles, but they are both over 80...'

The Stag was bought as a running restoration project with 96,000 miles on the clock. As far as they're aware no major engine work had been carried out. There were some immediate jobs that needed

**Stag has never been off the road for more than a few months at a time since the Phillips bought it in 1988**

**Below, dashboard was re-veneered by Roger Phillips using a secondhand item picked up for £15**

**Below right, engine was meticulously rebuilt over one winter and has proved totally reliable ever since**

doing to keep the Triumph on the road. 'In the first week it had new timing chains, rear shock absorbers and the rear wheel cylinders and brake shoes were changed,' Roger recalls. Some welding was also required on the rear subframe mounts.

That was that until October 1990, by which time the Stag then had another 10,000 miles under its wheels. Roger decided the moment to tackle his first V8 engine rebuild had arrived. 'Everyone had frightened me with stories about not being able to get the cylinder heads off,' he recalls. So, before he took the engine out, he had a go at taking all the studs out, to prepare himself for the job in hand. 'I didn't know whether to rebuild this one or buy a reconditioned engine.'

To his surprise, it only took 1½ hours one Sunday morning to remove all the

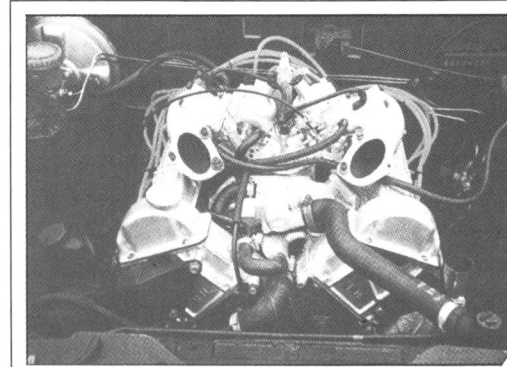

in one hand a[...]
other,' Lesley [...]

The mach[...]
the 2997c[...]
Chesman [...]
Road in [...]
block [...]
heads [...]
The [...]
30t[...]
c[...]

studs. 'I couldn't believe it, they came out so easily,' he adds. A friend at the garage where Roger handles the accounts help[...] take the engine and gearbox out of the [...] but he has done much of the other w[...] himself at home. 'With a workshop manua[...]

Right, working conditions inside the Phillips' single garage were definitely on the cramped side!

Below, extensive body work was needed around the rear arches once the paint had been stripped

up to a quarter of an inch thick, stretching along the rear wings and onto the door. Roger admits that whoever did the filling was a master, but it would have taken less effort just to do the job properly in the first place. 'They'd rubbed it down to the shape of the car just right,' Lesley adds. 'No-one noticed it was wider than it should have been.'

As soon as Roger drilled the rivets out, the new arch fell off. It turned out the original problem was a relatively small amount of rot around that area, so he used part of the replacement panel to repair the old. 'I'd never done any welding before, but my neighbour had a MIG and I asked him if he'd do it for me.' The answer Roger got was not quite what he expected. The neighbour lent him the gear, showed him how to get started, and left him to it. 'I ended up lying underneath the car putting the flames out as the underseal caught fire,' Lesley recalls. Ah, happy days...

Small patches were necessary in the footwells and boot floor, but nothing major. Roger stripped the whole car down to bare metal, mainly using a hot air gun and a chisel, until it was ready to be taken to the paint shop, Stratford Car Bodies on Western Road. 'That's the only time it's been on a trailer since we bought it,' he declares. Although the Stag's original colour was blue, it was red when the Phillips bought it, and Pimento red it has stayed. Two-pack paint was used for durability.

Various bits of brightwork were refurbished while the Triumph was off the road, including the bumpers, rear lights, petrol cap and various engine parts. These were done by A G Warren on Banbury Street in Birmingham.

By April 1992 the Triumph was back on the road again. This time for a run to Euro Disney as well as attending Benson and Hedges events around the country.

The braking system has been completely rebuilt during a winter break. Roger fitted some second-hand calipers after one seized, plus new discs, seals, shoes, pipes and

Proper Stag alloys have replaced the Wolfrace wheels seen far left, opposite

**Roger and Lesley Phillips find their Stag comfortable, fast and a true four-seater**

hoses. All the work he did himself was carried out in their small one-car garage at home, which is barely wide enough to get the doors open. 'We had to jack it up and push it right over until it was just about touching one wall to work on it,' he remembers.

He has made some interesting minor modifications as he's gone along, including putting in an extra bonnet release cable. 'If it breaks, you've no way of getting in there except by punching a hole in the inner wing,' he explains. Likewise, he's modified the way the rear seat fastens in case the tonneau release cable snaps.

Roger's latest project has been to make a new veneered dashboard, which he did using a secondhand dashboard picked up for £15. He took over the dining room table for about six weeks while he stuck veneer to the dash, rubbed it down and varnished it. He ended up using 1200 grade wet and dry paper and thinning the last coat of varnish right down to avoid brush marks. 'People see it now and ask where I bought it,' he adds.

Over the last couple of years the Phillips' Stag has successfully completed journeys all round this country, and numerous trips to Europe as well. In April this year they did a quick 1600-mile hop round France,

Belgium, Holland and Germany, all without a hitch.

They do much of their travelling abroad with Classic Car Tours, based in Alfred Square, Deal, in Kent. This is a specialist firm, catering for individuals or clubs, which organises runs or tailor-made holidays for classic car owners to see Europe. On some runs a 'modern' back-up vehicle is even provided.

Not that the Phillips' car has ever needed

rescuing. The only problem encountered so far was when Roger replaced the radiator cap. The old one was weaker than he thought and the new cap built up too much pressure in the cooling system, blowing out the water pump seals.

Roger puts the reliability exhibited by the car so far partly down to meticulous reassembly work on the engine. He changes the oil religiously every 2000 miles and flushes the cooling system out, using a good quality glycol-based anti-freeze to help prevent corrosion within the engine.

Roger and Lesley certainly have no qualms about using the Stag as a proper Grand Tourer, for which it is very well suited. As they put it, 'It has power steering and electric windows, the seat adjusts so much that if you can't find a comfortable driving position in this then you have very little hope of finding it in anything, and it can carry four adults. The V8 motor has bags of torque and will happily sit at 30mph or cruise at 80mph on the Autobahn. What more can you ask for?' ∎

*Thanks to Roger and Janet Hammersly at Sambourne Hall Guest House, Sambourne, near Alcester, for the photo location.*

**Stag was trailered to the paint-shop – the only time since the restoration began**

# SEASIDER

On the only part of the English east coast that faces west, we discovered a Stag in a colour which almost matches the broad sweeps of sand that make up the local shoreline. Story by Tony Beadle, photography by Mike Key

**Above top:** Coastal track provided an ideal setting for convertible Stag - even in the winter!

**Left:** Black soft tops were standard on '76 models, this one is due for renewal fairly soon but is still effective in keeping the weather out when necessary

**Above:** Stylised grille badge

The North Norfolk village of Heacham sits alongside The Wash, just a couple of miles below the bustling holiday resort of Hunstanton and about six miles away from the royal residence at Sandringham. Close to the shore, behind the tall hump of the sea defence dunes, sprawl hundreds of mobile homes and caravans like a giant flock of seagulls at rest. Inland, despite the construction of many modern bungalows, the traditional centre of the village remains virtually unchanged, with its narrow streets, tiny terraced cottages and old fashioned shops. Further to the east, alongside the main road, is the world famous Caley Mill lavender farm, probably Heacham's most notable landmark and a popular tourist attraction.

In a quiet part of the village, hidden well away from the holiday traffic congestion, can be found the home of

# SEASIDER

Triumph enthusiasts Les and Sharon May and their three children. Although ignored by the thousands of holiday-makers who visit the area every summer, the May residence would seem, even from a casual glance, to be an obvious point of interest to any Triumph lover, if only because of the 2.5 P.I. estate (which Les calls his 'one day' car) and Sharon's Mk 1 2000 saloon parked in the front garden.

But it is neither of these two vehicles that is the focus of our attention (special though they both may be in their own way) but the 1976 Topaz Yellow Stag that Les has driven and

> "...before that it had been in the same hands since it was first sold by Hall's of Finchley."

restored over the past three years. This is not the first Stag that the 40-year old heating engineer has owned, but his previous experience proved less than happy, "I had another Stag for about two years before this one," said Les. "It was in primer when I bought it, and I didn't realise just how bad it was until the rust started to come through. When I worked out how much money it was going to cost to put right I got rid of it and bought a Vauxhall Senator!"

However, the desire for a Stag wouldn't go away, and in March 1992 he started at the top of a list of six cars to look at. "This was the first one I saw, in Thamesmead, east of London and

that was it," Les recalled. Although MLN 565P came with very little history, he was able to establish that he is, in fact, only the fourth owner of the car from new. The previous two custodians had each kept the car for a couple of years, and before that it had been in the same hands since it was first sold by Hall's of Finchley. Built in January 1976, the Stag had been looked after quite well, and was still powered by the original V8 engine.

After enjoying driving the Stag around that first summer, inevitably the dreaded metal decay started to come through in all the usual places - around the rear wheel arches and along the seam between rear quarters and sills in particular - and so Les decided on a bare metal respray. Removing the existing paint revealed the full extent of the rust, plus the fact that at some time the Stag had been involved in a prang resulting in damage to the front nearside corner. Doing 99% of the work himself, Les was fortunate in being able to have the use of a space in the corner of the workshop of Tommy's Body Repair of Heacham and he wanted to thank Adrian Hayward for this.

New outer wings, sills, rear wheel arch repair panels and outriggers are only part of the list of components that were accumulated for the rebuild and even then, some small patches had to be made up to fix the floor. The doors however, only had a tiny bit of rust on the bottom which was sandblasted out. Although to anyone else the fin-

**Above top left:** Refurbished interior looks better than new. Black with wooden dash and trim is a classic combination, deep pile carpet was first used in '76 models as part of Leyland standardisation process
**Above left:** Front wings and sills were replaced and all rust cut out prior to bare metal respray in matching colour to the original factory Topaz Yellow
**Far right:** Engine compartment before and after restoration
**Right:** The first coat of primer is sprayed on

# SEASIDER

ished job looks first rate, Les himself is disappointed with the rear arches. "One side had already been filled and I didn't have another car to refer to," he explained. "And to me they look a bit squarish - it's a real bug bear of mine." Once the bodywork renovation was complete it was sprayed with PPG two-pack to match the original factory Topaz Yellow colour.

But the Stag refurbishment went more than skin deep, and as well as having the radiator re-cored, a new power steering rack (from the Stag Owners Club spares department) and new bushes for the suspension were installed. The engine didn't have to come out, but Les fitted a new timing chain set and replaced the oil pump as well as having the carburettors over-hauled while he was at it. In all, the parts came from a variety of sources: Rimmer Bros of Lincoln supplied most of the panels; S.N.G. Barratt of Wolverhampton, James Paddock of Chester and the now defunct Northants Stag

*"...like any such project, just keeping it in pristine condition is an ongoing task."*

Centre provided most of the rest, together with the aforementioned SOC Spares. Chromework was replated by S.B. Products in Peterborough, and the aluminium sill covers were re-anodised by Electro Plating Services of Glasgow because Les was working up there on a contract at the time.

There are 68,000 believed genuine miles registered on the clock and Les May reckons to add a thousand or so more each year getting to as many shows as he can. So, what does it perform like now after all his work? Unfortunately, the rapidly fading light of a dull, grey winter day meant my experience of this Stag was limited to some all too brief excursions along narrow country lanes as we searched for a suitable photo location. Even so, with the top down and the heater on to combat the cold, it all proved utterly enjoyable - the glorious V8 exhaust note reverberated off the stone buildings as the Borg Warner automatic shifted cleanly through the gears and we took the sharpest of corners in our stride. It was an intoxicating feeling, and one can easily see how the media enthusiasm for the Stag was generated when it was announced.

Defined as a high-performance luxury grand tourer and christened 'The Car of The Seventies' by some writers the Stag was compared to prestige continental makes with the promise of Ferrari or Porsche performance at a cheap price. Intended for an October 1969 Motor Show introduction, the Stag was eventually launched on 9th June 1970, but the delay didn't affect the optimism expressed about the car's future. The journalists loved it, and predictions of up to 14,000 units a year being produced were freely bandied about. The *Autocar* concluded their Stag appraisal in July 1970 with the words 'We liked it so much in fact that we shall be adding one to our long term test fleet as soon as we can get delivery.' Symptomatic of the problems that were to beset the Stag it was June of the following year before the *Autocar* was able to report the first of their experiences with a long term test vehicle.

Three years and 27,000 miles later, the *Autocar* staff were still sufficiently enamoured of the Stag to talk about buying an identical replacement - but availability remained a problem, with up to a six month long waiting list quoted. This situation was blamed on a series of production problems, but with only 5,504 cars delivered from the factory in '73 many potential customers gave up the wait (especially in overseas markets) and went elsewhere.

By the time Les May's Stag came off the assembly line in 1976, the annual figure had dropped to 3,110 units and year-to-year changes were minimal: deep-pile carpet for the interior; five spoke alloy wheels, tinted glass and stainless sill covers on the outside. And at the end of the road, in June 1977, total Stag production for the eight years of its life was just under 26,000 - a small fraction of the numbers anticipated at the beginning of the decade.

As far as Les is concerned, it's a long way from the end of the road for MLN 565P - there's plenty of work still to be done, a new convertible hood for one, plus some detailing on the removable hardtop - but like any such project, just keeping it in pristine condition is an ongoing task. Recognition of his efforts came when he won 'Best Home Restoration' at the Sandringham Classic Car Show but for Les the enjoyment comes from preserving a car that in his words, "Is regarded as something of an underdog in classic car circles, and always get a lot of stick from the 'Quentin Willson' brigade!" That sort of criticism is best ignored, for there's no doubt that despite the production difficulties that crippled the car's chance of the sales success it deserved, the Stag was conceived and built in the true Triumph tradition of quality and performance.

## FACTFILE

### 1976 TRIUMPH STAG
2+2 Grand Tourer convertible

**Engine:**
SOHC V8, cast iron block, aluminium cylinder heads
*Capacity:* 2997cc (182.9 cu.in.)
*Bore and stroke:* 86 x 64.5mm (3.39 x 2.54 in.)
*Compression ratio:* 9.25:1
*Maximum power:* 146bhp (DIN) at 5700 rpm
*Maximum torque:* 167 lbs/ft at 3500 rpm
*Carburettors:* Twin Stromberg

**Transmission:**
Borg Warner Type 35 3-speed automatic

**Chassis/Body:**
*Wheelbase:* 8ft 4in (2.54 metres)
*Overall length:* 14ft 5.75in (4.42 metres)
*Overall width:* 5ft 3.5in (1.61 metres)
*Height:* 4ft 1.5in (1.26 metres)
*Weight:* 2981 lbs (1355 kg)
*Suspension:* Coil spring independent all round; MacPherson strut and anti-roll bar front, semi-trailing arms rear
*Wheel dimensions:* 5.5J wide x 14 inches diameter five-spoke alloy

**Performance:**
*Top Speed:* 115 mph
*0-60 mph:* 10.7 seconds
*Standing quarter mile:* 18.3 secs at 77 mph

**General:**
*Production total 1976:* 2,466 (UK models)